AN INTRODUCTION TO POLLEN ANALYSIS

BY

G. ERDTMAN, Ph. D.
Västerås, Sweden

Foreword by ROGER P. WODEHOUSE
Author of "Pollen Grains"

CONTENTS

Chapter 16

Chapter 17

Chapter 18

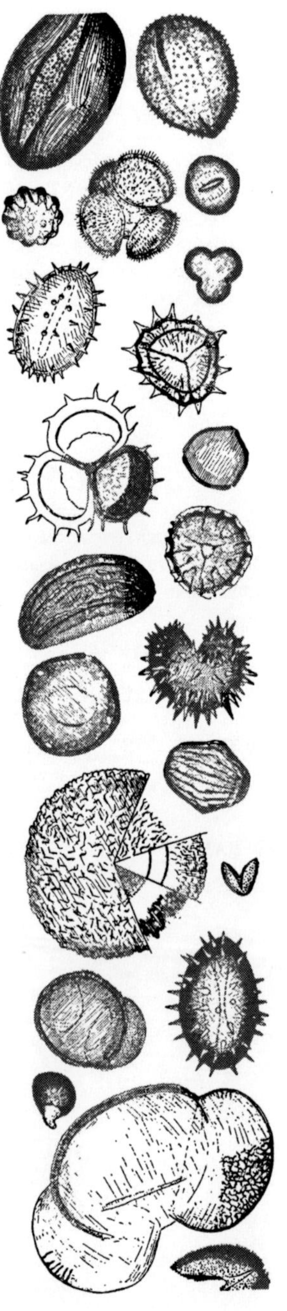

Chapter I

INTRODUCTION

It has been claimed that skilful hunters of old could bring their prey down even if they caught only a fleeting glimpse of its shadow. But still more remarkable appear the performances accomplished to-day by the pollen analyst. Out of pollen from crumbled clay or minute pieces of peat, taken from bits of earthenware or a stone axe, may be constructed a picture of the primeval forests which flourished in the region at the time when the pot or axe dropped into the bog. This is, however, not the whole story. In a carefully investigated region it is also possible to determine the relative, in certain places even the absolute, age of a pollen-bearing sample and to ascertain its place in a system of curves, illustrating changes in vegetation and climate, during ages long past.

Pollen analysis is a young science. At the sixteenth Meeting of Scandinavian Naturalists in Oslo in 1916 it received its birth certificate by LENNART VON POST* who, at that occasion, read a paper on the pollen of forest trees in bogs of southern Sweden. The science of pollen analysis became of age only a few years ago, having already made a brilliant record in its childhood and youth. This is amply attested by approximately 1500 publications which have been published on the subject to date.

Pollen analysis is essentially founded on the following facts. At the time of flowering many trees shed great quantities of pollen. A single birch catkin may produce in excess of ten million pollen grains. In general, pollen grains are very small, as a rule averaging between one hundredth and one tenth of a millimeter in diameter. Easily carried by the wind, some of them are transferred into higher regions by vertical air currents and remain there for days, weeks, or even months, before they settle back to earth. In the meantime this " plankton of the air " may have moved over great distances. Greenland peats contain pollen grains of pine and spruce, that must have been carried at least 100 kilometers (the distance of the nearest coniferous forest in Labrador). CHARLES LINDBERGH trapped pollen grains and spores by means of " sky-hooks " during a flight over Greenland. In 1937, the author collected pollen grains and spores by means of vacuum cleaners practically the whole way across the Atlantic between Gothenburg and New York. Among the pollen grains thus obtained American specimens were found at least as far as 700 kilometers east of Newfoundland. For practical purposes of pollen analysis, however, such far-travelling grains are of minor importance. After their aerial journey the pollen grains may settle on the surface of a bog or lake or in a bay, where sediments are deposited, and be

* Born in Västerås 1884, geologist at the Geological Survey of Sweden 1910–1930, Professor of Geology at the University College of Stockholm since 1930; portrait page 231.

embedded in the accumulating peat or ooze and preserved. In this way the peat bogs and the sediment banks on lake-bottoms become archives of vegetational history imprisoning pollen grains, season after season, millennium after millennium.

A trained pollen analyst, studying these forest archives, may not only be able to tell the family or genus to which the pollen grains belong but also the species, and, in some cases, even the subspecies. The identification of the pollen grains is, however, only one side of the matter. Another, and no less important, is the interpretation of the fossil pollen flora. This is sometimes a very difficult problem. Thus a peat sample may provide hazel nuts but no hazel pollen; and bogs within an area with aspen and poplar predominating, may contain much the same pollen flora as bogs of a nearby area with pine and spruce predominating. The latter case may be explained by the fact that the exines of aspen and poplar pollen grains are so thin that they are preserved only in exceptional cases, whilst those of pine and spruce pollen are very resistant to decay. Furthermore, pine and spruce pollen is easily carried by the wind and may be scattered over considerable areas outside of the coniferous region. Therefore, it goes without saying that the evidence of pollen grains sometimes must be taken with a grain of salt.

The development of pollen analysis might, in a certain sense, be compared to the development of a river system. The Oslo paper was the source and stimulus. Successively, and in ever increasing numbers, tributaries — collaborators — joined the main stream, the course of which was directed by von Post and his assistants of the Geological Survey of Sweden. At times uncritical overproduction increased the stream more in breadth than in depth. At present its volume has increased to such an extent, that a cleavage of the stream into different channels must be expected. From these channels an increasing number of fields of scientific activity — from quaternary geology to climatology and archaeology — may be fertilized as a result of a more and more efficient system of irrigation.

Improved research methods have contributed to the present situation. For example, by certain chemical methods it is now possible to dissolve away more of the matrix in which the fossil grains occur, thus greatly facilitating their analysis. Different soils, previously considered more or less devoid of pollen (clay, raw humus, etc.) have become amenable to pollen analysis. And by concentrating the pollen grains of a sample, which in itself is rich in pollen, rare kinds of pollen may be discovered.

To the pollen geologist pollen grains are merely index fossils, by means of which a particular horizon may be traced from one peat deposit to another, from province to province, or even from one country to another. To the pollen botanist, on the other hand, identification is an important consideration. The more pollen species identified, the more complete will be the picture of the former vegetation.

Pollen and spore research is concerned not only with the late Quaternary history of the vegetation, but also with the history of the plant kingdom during more distant epochs, with the elucidation of the enigma of the origin of the angiosperms, etc. A successful attack on these great problems demands, above all, an extensive knowledge of the pollen and spore types among contemporary plants. Surprisingly little, however, has so far been accomplished in this field.

This outline of the potentialities of pollen analysis may easily be augmented inasmuch as pollen research is not confined exclusively to palaeobotany, plant systematics, and plant geography. It is also linked up with many other fields of scientific activity, such as forestry, soil science, limnology, and different geological disciplines. We may also mention archaeology, climatology and aerobiology — with its ramifications in medicine, hygiene, phytopathology, etc. — and even pharmacognosy (the detection of honey and drug adulteration through pollen analysis).

Historical: — Modern pollen analysis, as already mentioned, goes back to 1916, when L. von Post wrote his paper on fossil pollen of Swedish bogs. This does not imply, however, that pollen analysis, in contradistinction to most developments of scientific thought, is a product of spontaneous creation. On the contrary, it had a long line of forebears, of parents and grandparents, a period of gestation, when it was unknown except among a few hopeful workers. Its birth was attended by doubts and hesitation. But now pollen analysis is firmly established and universally accepted, and the skeptics have had to confess that their dire predictions have come to naught.

The Swiss geologist J. Früh was one of the pioneers of pollen analysis. In his paper "Kritische Beiträge zur Kenntnis des Torfes" (1885) not only were enumerated most of the common tree pollen types but also a considerable number of spores and herb pollen grains. This indicates a keenness seldom met even among modern workers. In connection with a study of bottom samples from Swedish lakes, F. Trybom (1888), a Swedish zoologist, encountered pollen grains of pine and spruce in profusion. As they were very resistant against decay he considered them serviceable as index fossils in palaeontology.

Ftom 1895 onward, Germany's leading peat botanist and peat stratigrapher, C. A. Weber (1856–1931), made a considerable number of pollen analyses. He proceeded along the same lines as Früh, but in some respects his work was even more detailed. Thus Weber (1918, p. 259) did not always confine himself to qualitative analyses; a paper published in 1896 contains calculations of the ratio of pine to spruce pollen and in an investigation undertaken fourteen years later, the pollen frequencies were expressed as percentages of the total forest tree pollen.

In 1897 the Danish archaeologist G. Sarauw gave a cursory description of a Postglacial submarine peat dredged near Copenhagen. In this peat *Pinus* pollen grains were found together with a number of grains resembling those of *Betula* and *Corylus*. The grains were counted, but no percentages were established. A few years later G. Lagerheim* (1860–1926), an eminent microscopist and one of the most versatile of micropalaeontologists since the days of Ehrenberg, published a series of papers and notes which make him, without doubt, the "father

* Professor of Botany at the University College of Stockholm 1895–1925; portrait page 233.

of modern pollen analysis." While examining ooze samples for algae and protozoa he frequently encountered fern and moss spores (LAGERHEIM 1901, 1902*a*) which were determined by means of a reference set of lactic acid preparations (LAGERHEIM 1902*c*) made by his assistant T. VESTERGREN. During the course of these investigations LAGERHEIM gradually devoted more attention to fossil pollen. His first analyses were entirely qualitative. Later quantitative analyses were made. The first results of these were published in a paper by WITTE (1905). Further results were published by Dr. N. O. HOLST (1909) in "Postglaciala tidsbestämningar," a work containing some remarks of fundamental importance concerning the aims and potentialities of pollen analysis which deserves therefore to be mentioned here in some detail.

N. O. HOLST (1846–1918)* frequently approached specialists in order to obtain as exhaustive information as possible on certain critical points. One of these specialists was LAGERHEIM, who identified rhizopods and plant microfossils. In the Kallsjö bog in the extreme south of Sweden, HOLST encountered a number of lacustrine beds which he considered to be particularly interesting. The lower part of these contained remains of such algae as *Codonella cratera*, *Anabaena Lemmermannii*, *Botryococcus*, and *Pediastrum* spp., while the upper part was rich in *Scenedesmus*, chiefly *S. quadricauda*, and also yielded fruits of *Trapa*. The pronounced difference between these layers made it particularly desirable to calculate the relative amounts of tree pollen. The percentages, given by LAGERHEIM, were as follows: —

	Lower part:	Upper part:
Decreasing upwards:		
Pinus	21.25 per cent	4.23 per cent
Betula	17.50 —	11.11 —
Alnus	44.37 —	40.17 —
Ulmus	14.37 —	13.67 —
Corylus (calculated separately)	61.75 —	25.78 —
Increasing upwards:		
Fraxinus	0.62 —	2.56 —
Quercus	1.25 —	16.25 —
Tilia	0.62 —	11.96 —

In his comments on this table HOLST writes (*l.c.* p. 291; translation from the original Swedish with but slight modifications), "It appears from the investigations by LAGERHEIM that the hazel pollen frequency has declined strongly, *viz.* from 61.75 per cent in the lower layers, to 25.78 per cent in the upper (the total of all tree pollen grains making 100 per cent). This can hardly designate anything but the suppression of the hazel by the shading of other trees, particularly the oak, in equal pace with their expansion and growth. It also seems that the following trees have decreased in number upwards, *viz.* pine (strongly), birch (less), alder, and elm (slightly). On the other hand the ash, oak and linden, show an increase.

LAGERHEIM's investigations gave more than the result aimed at, *viz.* — as LAGERHEIM himself pointed out — "a reliable method with which to follow, step by step, from one layer to another the immigration of

* Geologist of the Geological Survey of Sweden 1877–1905, portrait page 235.

all plants whose pollen or spores are preserved as fossils as well as the relative frequency of these species. When the rate of formation of all the different layers can be determined, it will also be possible to calculate the speed with which the plants in question immigrated ".

The pine, according to HOLST (*l.c.* p. 31), is " very suppressed in the upper, *Trapa*-bearing layers, but does not disappear either there or in the superimposed peat, continuing in the latter up to the top bed. Thus it seems likely that the pine disappeared from southernmost Sweden in historical times, exterminated probably by the activity of man ". It may be added that HOLST expounded several other ideas entirely new to Swedish phytogeography, such as the early immigration of spruce from the south. Furthermore he stressed the necessity of investigating microfossils as well as megafossils in order to get a complete picture of the fossil flora of peat bogs. "An entirely megascopical investigation is naturally limited to remains of plants which have grown on the bogs or at least in their immediate vicinity. An investigation of this kind will therefore give better results near the boundaries than in the central parts of the bogs. Microscopical investigation does not suffer this limitation. It also embraces things carried by the wind even from distant surroundings " (*l.c.*, p. 21).

Analyses by LAGERHEIM were also published in other reports of the Geological Survey of Sweden and in papers by VON POST, SAMUELSSON (1910), SERNANDER (1911), and others. It is obvious now, however, that but few persons realized the importance of these investigations at that time. Among the few who did was WESENBERG–LUND, the Danish limnologist. " It may be reasonably anticipated ", he writes, " that pollen investigations will play an ever-increasing rôle in Quaternary geology." But at the same time he warns against drawing hasty conclusions from the pollen flora of lacustrine deposits without considering the part that might be played by the bottom fauna in carrying pollen grains from one layer to another (WESENBERG–LUND 1909, p. 468).

By LENNART VON POST, primitive pollen analysis was transformed into a refined method in Quaternary geology. As regards microfossils VON POST was a pupil of LAGERHEIM, and he spent much time in the botanical laboratory of the University College of Stockholm following, among other things, the experiments which LAGERHEIM undertook in order to make recent pollen grains assume the same appearance as fossil grains. As LAGERHEIM did not publish anything on pollen analysis itself, it is often difficult to distinguish between the work done by him and that done by VON POST. I may well mention that he did not object to the heading of a chapter, "Die pollenanalytische Arbeitsmethode nach L. VON POST", in a paper by ERDTMAN (1921, p. 15). But this heading, as the author now — eighteen years later — sees it, does not render LAGERHEIM due justice. We may perhaps most impartially express the situation as follows: LAGERHEIM ("the spiritual father of modern pollen analysis") was truly a pioneer in micropalaeontological and chemical-technical questions, while VON POST, a trained and keen-eyed peat stratigrapher, indicated how pollen analysis should be applied in order to give information on problems related to Quaternary geology and palaeontology.

Pollen diagrams, giving a visual representation of the composition of the fossil pollen flora at different levels of a deposit, were first drawn by VON POST. In the beginning he did not use any diagrams at all, but he soon realized, as did H. LINDBERG in Finland, that the horizon where the pollen grains of spruce, one of the last immigrants to Sweden, appeared, "the spruce horizon" (VON POST 1909*a*, p. 283) or "the spruce pollen limit" (VON POST 1909*b*), could be used as a datum line in determining the relative ages of different bog strata, and thus serve as a means of tracing synchronous horizons in different bogs within a minor area (*cf.* also VON POST 1913). More extensive investigations were begun in 1915 and their results, including a number of pollen diagrams, offered to the meeting of Scandinavian naturalists at Oslo in 1916.

In this historical review, it seems only fair to devote some attention to a thesis by Dr. ULRICH STEUSLOFF, now almost forgotten and difficult to obtain outside Germany ("Torf- und Wiesenkalk-Ablagerungen im Rederang- und Moorsee-Becken. Ein Beitrag zur Geschichte der Müritz," reprinted from "Archiv der Ver. der Freunden der Naturgeschichte in Mecklenburg", vol. 59, Güstrow 1905). The theme was suggested to STEUSLOFF by E. GEINITZ, geologist and professor at Rostock University, who hoped that STEUSLOFF, a trained botanist and conchologist, would be able to elucidate the age of certain terraces, peats, and lake deposits in northwestern Mecklenburg. STEUSLOFF took a particular interest in certain beds of lake lime which he searched for diatoms, bluegreen algae, etc. When trying different stains, he noticed that Magdala red, which at that time was frequently used for staining algae, caused the fossil pollen grains to stand out in a brilliant red colour. STEUSLOFF then decided to make a special study of the pollen grains in order to obtain micropalaeontological evidence as a complement to the results already obtained by peat macropalaeontologists, such as G. ANDERSSON, NATHORST, and others. For fossil pollen identification, recent pollen was treated with nitric acid, Schulze maceration mixture, or with some other chemicals. This material was stored in alcohol and used for comparison with fossil pollen whenever required.

In a study of banded lake marl, samples were collected from seven strata, light and dark alternating. In each sample between five and seven hundred pollen grains were counted (*Alnus*, *Betula*, *Pinus*, *Quercus*, *Salix*, *Tilia*, *Cyperaceae*, *Ericaceae*, *Gramineae*, *Nymphaea*, *Polygonum*, *Typha*). There were also counted a number of algae, certain fungus spores, and animal remains, *e.g.* cladocera antennae and shells of *Bosmina*, *Lynceus*, and *Sida*. STEUSLOFF believed that the banding was of seasonal origin with spring layers, poorer in lime, alternating with autumn layers richer in lime. The fossil pollen flora, however, was nearly the same in the supposed "spring" as in the supposed "autumn" layers in spite of the fact that the pollen grains shed during spring and early summer are entirely different from those shed later. Thus, there must be a special reason for their conformity, and STEUSLOFF suggests in this connection that pollen grains may float in the water some time before sinking.

The comparatively low pine pollen frequency of a sedge-peat was considered to be due to the fact that the pine sheds its pollen at a time when the sedges have attained full growth, thus overshadowing the water between the tussocks. On the other hand the surface of the water is more freely exposed to the pollen grains of alder, birch, and other trees, which shed their pollen at an earlier date.

The history of the forests is outlined in accordance with the composition of the fossil pollen flora. In the oldest layers only pine, alder, birch, and willow pollen were found. The microfossils of somewhat younger layers testify that conditions for plant life gradually grew more genial. Dense pine forests with a sparse growth of oak and hazel covered the surrounding sandy country while swamp forests, with alder and birch, fringed the Rederang lake. Later *Tilia* appeared, and the frequency of *Quercus* and *Corylus* increased. A detailed table, showing the results of the pollen analyses, is attached.

Steusloff subjected different materials, such as lake lime, phragmites peat, carex peat, and even sand to pollen analysis, and he touched several questions bearing on the theory of pollen statistics — such as the buoyancy of pollen and the surface receptivity of different bog types — more than a decade before the extensive treatment of the same problems by Swedish scientists.

It may finally be mentioned that a paper by Stark (1912) contains notes on fossil pollen, including a quantitative analysis, and that Rudolph (1917) in the first part of his description of Bohemian peatlands devoted a chapter to fossil pollen.

References: —

Cain, S. A., 1939: Pollen Analysis as a paleo-ecological research method (Bot. Rev. 5:627–654).

Erdtman, G., 1921: Pollenanalytische Untersuchungen von Torfmooren und marinen Sedimenten in Südwest-Schweden (Ark. f. Bot., vol. 17).

Florschütz, F., 1940: Resultaten der Pollenanalyse (Vakbl. v. Biol. 21, 14).

Früh, J., 1885: Kritische Beiträge zur Kenntnis des Torfes (Jahrb. K. K. Geol. Reichsanstalt, vol. XXXV).

Godwin, H., 1934: Pollen Analysis, an outline of the problems and potentialities of the method (New Phytologist 33).

Holst, N. O., 1909: Postglaciala tidsbestämningar (Sveriges Geol. Unders., ser. C, no. 216; with analyses by Lagerheim).

Lagerheim, G., 1901: Om lämningar av rhizopoder, heliozoer och tintinnider i Sveriges och Finlands lakustrina kvartäraflagringar (Geol. Fören. Förhandl., vol. 23).

—— —— 1902*a*: Bidrag till kännedomen om kärlkryptogamernas forna utbredning i Sverige och Finland (*Ibid.*, vol. 24).

—— —— 1902*b*: Torftekniska notiser (*Ibid.*, vol. 24).

—— —— 1902*c*: Metoder för pollenundersökning (Botan. Notiser).

von Post, L., 1906: Norrländska torfmossestudier, I (Geol. Fören. Förhandl., vol. 28; with analyses by Lagerheim).

—— —— 1909*a*: Skarbysjö-komplexet och dess dräneringsområdes postglaciala utveckling (*Ibid.*, vol. 31).

—— —— 1909*b*: Stratigraphische Studien über einige Torfmoore in Närke (*Ibid.*, vol. 31).

—— —— 1913: Über stratigraphische Zweigliederung schwedischer Hochmoore (Sveriges Geol. Under., ser. C, no. 248).

—— —— 1918: Skogsträdspollen i sydsvenska torfmosselagerföljder (Forhandl. ved 16. Skand. Naturforskermøte 1916).

Rudolph, K., 1917: Untersuchungen über den Aufbau Böhmischer Moore, I (Abhandl. Zool.-Bot. Ges. Wien, vol. IX).

Samuelsson, G., 1910: Scottish peat mosses. A contribution to the knowledge of the late-quaternary vegetation and climate of North Western Europe (Bull. Geol. Inst. Univ. Upsala, vol. X; with analyses by Lagerheim).

Sarauw, G., 1897: Cromer-skovlaget i Frihavnen of Traelevningerne i de ravførende Sandlag ved København (Medd. Dansk Geol. Foren., no. 4).

Sernander, R., 1911: Om tidsbestämningar i de scano-daniska torfmossarna (Geol. Fören. Förhandl., vol. 33; with analyses by Lagerheim).

Stark, P., 1912: Beiträge zur Kenntnis der eiszeitlichen Flora und Fauna Badens (Ber. Naturf. Ges. Freiburg i. Br., vol. XIX).

STEUSLOFF, U., 1905: Torf- und Wiesenkalk-Ablagerungen im Rederang- und Moorsee-Becken. Ein Beitrag zur Geschichte der Müritz (Arch. d. Freund. d. Naturgesch. in Mecklenburg, vol. 59).

TRYBOM, F., 1888: Bottenprof från svenska insjöar (Geol. Fören. Förhandl., vol. 10).

WEBER, C. A., 1896: Die fossile Flora von Honerdingen und das nordwestdeutsche Diluvium (Abh. Nat. Ver. Bremen, vol. XIII).

WEBER, C. A., 1918: Über spät- und postglaziale lakustrine und fluviatile Ablagerungen in der Wyhraniederung bei Lobstädt und Borna und die Chronologie der Postglazialzeit Mitteleuropas (*Ibid.*, vol. XXIV).

WESENBERG-LUND, C., 1909: Om limnologiens betydning for kvartaergeologien saerlig med hensyn til postglaciale tidsbestemmelser og temperaturangivelser (Geol. Fören. Förhandl., vol. 31).

WETZEL, O., 1937: Geschichtliche Umschau über die Mikropaläontologie (Zeitschr. f. Geschiebeforsch. u. Flachlandsgeol., vol. 13).

WITTE, H., 1905: *Stratiotes aloides* L. funnen i Sveriges postglaciala aflagringar (Geol. Fören. Förhandl., vol. 27).

Chapter II

CHEMISTRY OF PEAT*

In order to facilitate the understanding of the chemistry of peat and similar products, its mode of formation, and its peculiar properties — which are of great importance in the preservation of pollen grains and spores in bogs — it is necessary to review briefly the chemistry of the chief constituents of living plants.

Herbaceous plants are composed of a great variety of chemical substances some of which are quantitatively predominant and, therefore, must be considered as potential precursors of peat. Many of the constituents of living herbaceous plants are subjected to microbiological decomposition almost immediately after death and, consequently, do not directly contribute to the formation of peat, — at least not to any considerable extent. Examples of these are sugars, starch, and proteins. The mechanical tissues are more resistant and hence of greater interest for our present study. The chemistry of herbaceous plants is very similar to that of woody plants. Because of their technical importance, woody plants have been investigated in more detail than most herbaceous plants and this is the reason why in the following discussion the chemistry of wood, especially of coniferous wood, is almost exclusively taken into consideration.

Carbohydrates of Wood: — The most important constituent of wood (40–45 per cent) is cellulose, a polysaccharide, which is built up entirely from glucose by a process involving continuous dehydration. Conversely, cellulose is hydrolysed to glucose by the action of acids or specific enzymes (cellulase):

$$\underset{\text{cellulose}}{(C_6H_{10}O_5)_n} + nH_2O \rightarrow \underset{\text{glucose}}{nC_6H_{12}O_6}$$

" n " is a very high number, the exact magnitude of which has not yet been ascertained. It is probable that " n " may vary not only from plant to plant but also within one and the same cellulose-fibre. Cellulose, therefore, is not a completely homogeneous compound.

Chemical as well as physical methods have been employed for the determination of the molecular weight of cellulose. The values of " n " thus obtained vary from about 200 to 1000. This number of glucose units, therefore, takes part in the formation of a single cellulose molecule.

Except as regards the uncertainty concerning the exact molecular size, the problem of the chemical structure and configuration of cellulose has recently been solved. Cellulose forms straight chains possessing the structure demonstrated in the following scheme in which but four glucose units are pictured.

* Kindly contributed by Dr. H. Erdtman of Stockholm, a brother of the author of this book. — Dr. S. A. Waksman has rendered valuable aid in making this chapter ready for the press.

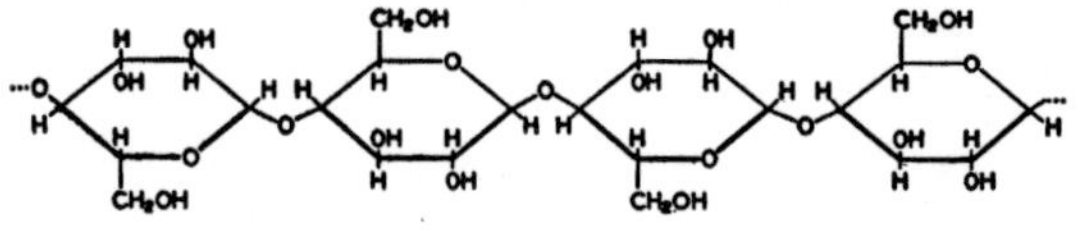

(I)

It is possible that in the fibre some cellulose chains may be connected by ether-linkages. The fibre-structure and mechanical properties of natural cellulose are due to the chain nature of the molecules.

Another 20–25 per cent of the wood is very similar to cellulose and this fraction is generally referred to as "hemicelluloses". These are complex polysaccharides, some of which are derivatives of glucose but are nevertheless different from cellulose. Others are derivatives of mannose or galactose, both of which possess the same constitution as glucose but differ stereochemically. They are called glucans, mannans, and galactans. Starch is built up from glucose units which are connected in chains, but it differs from cellulose in its stereochemical configuration, and the chains are probably branched. Due to this peculiar structure, starch possesses less mechanical strength than cellulose and is non-fibrous.

One of the most important types of hemicellulose is xylan. This polysaccharide is hydrolysed by acids to a simple sugar, xylose. Unlike those already mentioned, xylose contains only five instead of six carbon atoms.

Other important plant constituents are the pectins, acidic carbohydrates frequently occurring in leaves, stalks, and fruits of herbaceous plants and to a lesser extent in wood. They are of a very complex nature and are hydrolysed to different sugars and acid sugar derivatives, or uronic acids, hence they are known as polyuronides.

Many lower and higher fungi, as well as insects contain chitin, a nitrogenous carbohydrate which is closely related to cellulose. Instead of one of the hydroxyl groups of each glucose unit, chitin contains an acetylated amino-group ($CH_3CO — NH —$). Chitin is of considerable interest in the chemistry of peats and soils since it is very resistant to chemical and biological decomposition, and may be considered as one of the sources of the nitrogen content of humus.

Lignin: — About 30 per cent of wood consists of an amorphous substance or mixture of closely related substances which is chemically less understood than cellulose and hemicelluloses. This fraction is called lignin. Unlike the polysaccharides, lignin does not undergo hydrolytic cleavage with acids. In fact, lignin is generally defined as the insoluble product obtained by hydrolysis of wood with strong mineral acids. Common reagents for its preparation are 72 per cent sulphuric, phosphoric, superconcentrated hydrochloric or liquid hydrofluoric acid. However, the lignin obtained in this way is not the unchanged " native lignin " of the wood. It is a more or less condensed, polymerised or otherwise modified reaction product. The amorphous nature, high molecular weight and insolubility in ordinary solvents of these lignin preparations, as well as the difficulties encountered in degradation experiments aiming at the isolation of defined cleavage products make the elucidation of the structure of lignin extraordinarily difficult.

Lignin contains aromatic nuclei. Apart from physical evidence, this is demonstrated by the fact that it contains about 15 per cent of methoxyl groups which, undoubtedly, are attached to aromatic nuclei. Aromatic products are obtained from lignin by the application of several degradation methods. Fusion with caustic potash yields about 10 per cent of protocatechuic acid. Catechol, guajachol, pyrogallol, pyrogallol dimethyl-ether and homologues of these phenols are more or less regular constituents of wood tar. Since they are not formed by destructive distillation of the carbohydrates of wood, it must be concluded that they are derived from the lignin fraction.

Analytical considerations show that lignin is probably a derivative of phenyl propane (II):

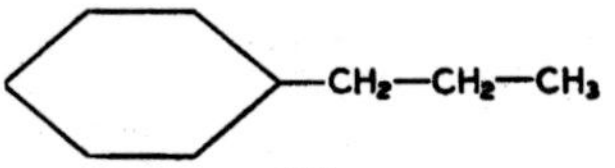

$-CH_2-CH_2-CH_3$

(II)

This conclusion is in harmony with early suggestions of P. T. Klason, who believed lignin to be related to coniferyl alcohol (III):

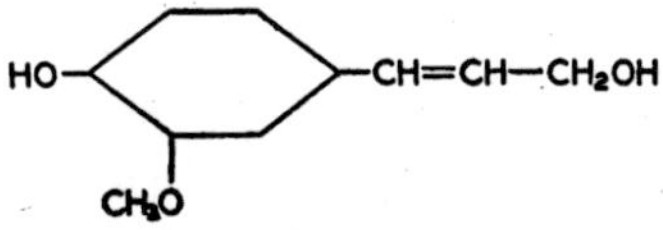

HO— $-CH=CH-CH_2OH$

CH_3O

(III)

The glucoside of coniferyl alcohol, coniferin, has been isolated from the cambial sap of many trees.

Special interest is attached to a compound recently isolated by Hibbert (Cramer, etc., 1939) from the lignin fraction of wood. It possesses the following structure:

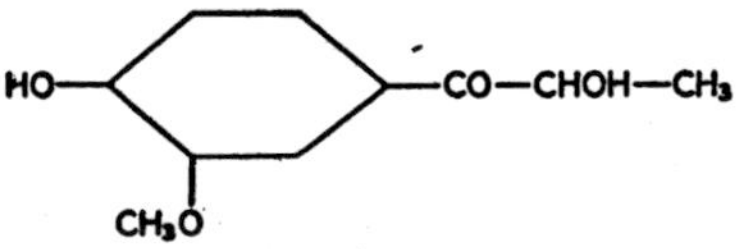

HO— $-CO-CHOH-CH_3$

CH_3O

(IV)

Hibbert considers this substance to be the building-stone of lignin and agrees with Freudenberg, who, previous to the discovery of this simple substance, arrived at the conclusion that lignin is a polymerisation product of such substances. Freudenberg (1933, 1939) has advanced a number of suggestions as to the probable structure of lignin; all the lignin units are said to be derivatives of phenyl propane containing, in addition to alcoholic hydroxyl groups, methoxyl and phenolic hydroxyl groups.

Continuous connection of such units by means of ether-linkages would result in the formation of chain molecules part of which may have the following structure (V):

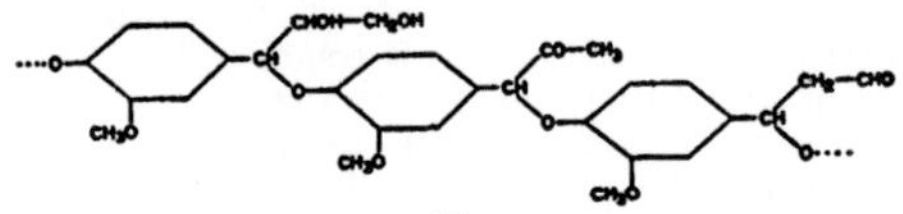

(V)

In order to account for the stability of lignin to hydrolysis, it is suggested that the side chain of one phenyl propane unit is condensed with the aromatic nucleus of the following unit with the formation of carbon-carbon linkages stable towards hydrolytic agents (VI):

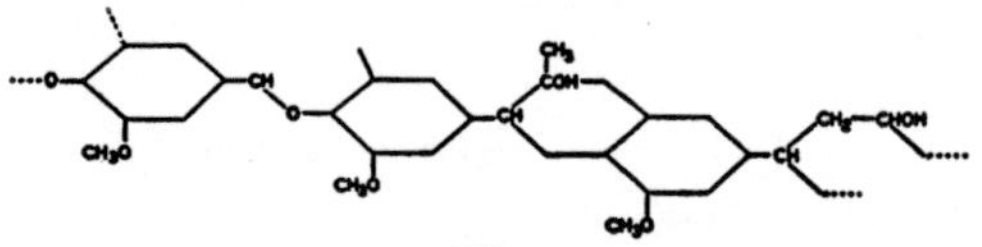

(VI)

Such condensation reactions may occur even in the living plant or during the isolation of the lignin. Intermolecular condensation will lead to the formation of three-dimensional structures of the same type as those occurring in certain artificial resins (bakelite). This would account for the mechanical properties of lignified cellulose (wood).

Scheme VI is certainly superior to V in explaining the properties and reactions of native and isolated lignin. The formula is, however, highly speculative. For example, the condensation reactions are said to proceed in a direction which is not easy to reconcile with the known reactivity of substances related to the hypothetical units. One would rather expect closed rings within one and the same unit, not from one unit to the neighbouring one.

Erdtman (1933) suggested, as a result of his studies of the dehydrogenation of isoeugenol (VII) to dehydro-di-isoeugenol (VIII), that heterocyclic structures similar to that in his scheme B might arise from coniferyl-alcohol or related substances by way of enzymatic dehydrogenation.

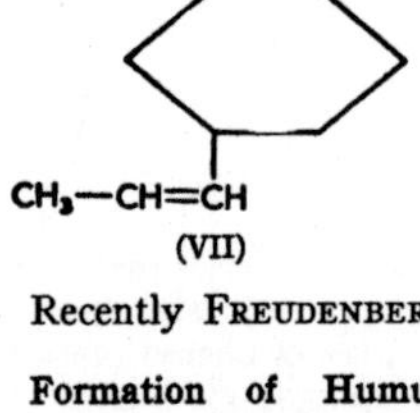

(VII) 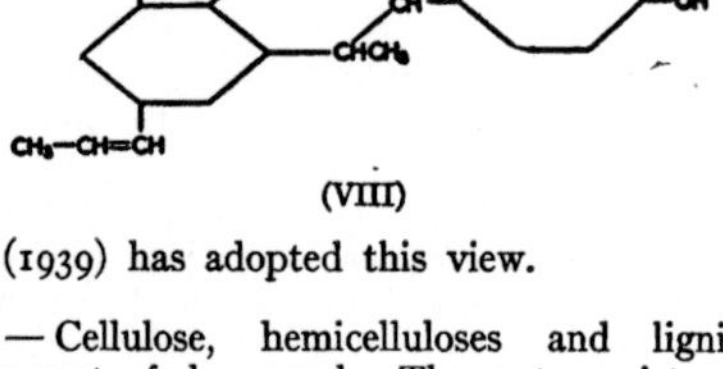 (VIII)

Recently Freudenberg (1939) has adopted this view.

Formation of Humus: — Cellulose, hemicelluloses and lignin account for about 90–97 per cent of dry wood. The rest consists of proteins, resins, fats, waxes, and ash. Most of these substances have but little importance, *e.g.* as precursors of the bitumen fraction of peat and coal.

The degradation of wood and of other plant material in nature is a process in which microorganisms play an important rôle. This should be emphasized since most chemists have a tendency to overlook it. On the other hand, the biologist is apt to overestimate the function of these microorganisms. Humus formation is neither a purely chemical nor a purely biological, but a biochemical process.

Under aerobic conditions, dead plants are quickly decomposed by the action of microorganisms with the formation, mainly, of carbon dioxide and water. When the supply of air (oxygen) is suppressed by the presence of water, the decomposition process proceeds much more slowly and in certain cases does not advance with the same speed as the formation of new plant material. In this way the peat bog grows, the process being influenced by climatic and geological factors. In the surface layer of a bog, the conditions are mainly aerobic for varying periods of time. During this time the activity of microorganisms is considerable. In the deeper layers, the conditions become increasingly anaerobic, followed by decreasing occurrence of microorganisms. Not even the bottom layer of deep raised-bogs, however, is perfectly sterile (WAKSMAN and PURVIS 1932).

The organic material produced from decaying plants under a restricted supply of oxygen generally possesses a characteristic brown or brownish-black colour, and the term humus should be restricted to this product. The composition of humus varies widely, and the term has no chemical significance. Humus has only a biological, agricultural, and geological meaning. Due to the practical importance of humus, it has been subjected to careful chemical study for more than a century. Humus of different origin and age has been investigated with the intention of isolating definite chemical compounds and elucidating their structures. It has been possible to isolate chemically pure substances from humus and to identify them with known compounds. They are, however, quantitatively of very little importance. The main fraction is amorphous and of high molecular weight (ZEILE 1935), and consequently possesses properties which are very unfavourable for exact chemical studies.

Humic Acids: — The most striking feature of humus is its colour; this fact has not received enough attention from those who have tried to speculate concerning its chemical nature. If humus is extracted with dilute hydrochloric acid to remove cations (especially calcium ions) and then with dilute alkali the material is more or less completely dissolved. Acidification of the brown or brownish-black filtrate causes the precipitation of abundant flocculent material. The substances obtained in this way are termed "humic acids." Under certain conditions, humic acids disperse in water to colloidal solutions of brown colour. The alkaline salts are much more coloured than are the acids themselves. They are almost black. The change of colour on addition of alkali is a characteristic and important property of humic acids.

Salts of humic acids with other bases are usually insoluble in water.

Calcium salts are sometimes deposited in nature as dopplerite (DOPPLER 1849, LEWIS 1881, TIDESWELL and WHEELER 1922). Iron salts are of importance in connection with the formation of certain iron ores (limonite; *cf.* SENFT 1862, ASCHAN 1932, NAUMANN 1921).

When treated with 95 per cent alcohol only a part of the humic acid preparations isolated according to the above method is brought into solution. This fraction is referred to in the literature as "hymatomelanic acid." The alcohol insoluble fraction is termed "humic acid." Some authors believe that these fractions are chemically homogeneous. However, this is certainly not the case and unpublished observations of the author of this chapter show that the fractionation of humic acids with alcohol is only of imaginary value. In these experiments, freshly precipitated humic acids from various samples of soil or peat were digested with cold 95 per cent alcohol. Part of the acid went into solution ("hymatomelanic acid"), part remained on the filter as an alcohol insoluble residue ("humic acid"). It is clear that the alcohol with which the humic acid was digested was not 95 per cent but much more dilute due to the great amounts of water present in the humic acid jelly. When a sufficient amount of 95 per cent alcohol was added to the "hymatomelanic acid" solution, a precipitation of "humus acid" occurred, and when the "humus acid" was washed with dilute alcohol it went completely into solution — as "hymatomelanic acid." It must, therefore, be concluded that no separation had been accomplished and that the "hymatomelanic" and "humus acid"-fractions are identical.

That humic acids are not homogeneous compounds is evident. This can be demonstrated by treatment with acetylbromide (KARRER and WIDMER 1921, KARRER and BODDING-WIGER 1923) in which humic acid is only partly soluble. In these experiments, acetylbromide acts not simply as a solvent, but reacts with the acids and forms acetylated and brominated products.

When humus or brown coal is extracted with dilute alkali, as described for the isolation of humic acids, it is frequently observed that not all of the colouring matter is soluble. The insoluble matter is generally called humin. The relation of this humin fraction to the humic acids — if any — is not yet understood. The humins contain more carbon than humic acids. When heated with strong alkali, they pass into solution with the formation of products showing properties similar to those of ordinary humic acids. Although insoluble in dilute alkali, the humins possess acidic properties and are capable of salt formation. This is of importance in the base exchange capacity of soil.

Chemical Nature of Humic Acids: — It has already been pointed out that one of the most characteristic properties of humic acids is their colour. This colouring matter must play some structural rôle in the chemical picture. The author (ERDTMAN 1926) observed that humic acids easily undergo reduction when treated with sodium amalgam in the absence of air. In this way, colourless or only faintly brownish solutions of the sodium salts of leucohumic acids are obtained. Exposed to air, these solutions absorb oxygen and the colour reappears. These findings were later confirmed by ZETZSCHE (ZETZSCHE 1932,

ZETZSCHE and REINHART 1939). The reaction strongly indicates a quinoid nature of humic acids. Additional evidence in favour of this view is that humic acids, when treated with acetic anhydride and zinc dust yields stable leucoacetates possessing only feeble colouration (ERDTMAN 1933, 1934). These leucoacetates are, as such, insoluble in dilute alkali, but undergo hydrolysis with the formation of leucohumic acids. If oxygen be present, they are oxidized to humic acid. Humic acids are methylated with dimethyl sulphate and alkali or diazomethan (FUCHS 1928, 1929, FUCHS and STENGEL 1929, PLUNGUIAN and HIBBERT 1935). This indicates the presence of hydroxyl groups.

Treatment with alcoholic hydrogen chloride also yields alkyl derivatives generally believed to be esters, indicating the presence of carboxyl groups. Since there is strong evidence in favour of the hydroxyquinone nature of humic acids, the "esterification" with alcoholic hydrogen chloride does not prove that humic acids are carboxylic acids. Many hydroxyquinones are about as strong as humic acids and most carboxylic acids, and are converted into ethers when treated with alcoholic hydrogen chloride.

Very little insight into the structure of the components of the humic acids has hitherto been gained by degradation experiments. A small yield of protocatechuic acid may be obtained by fusion with alkali. In the same way, this acid has been obtained from lignin. Since natural humic acids undoubtedly possess quinoid groups, it follows that they contain aromatic systems. The formation of aromatic polycarboxylic acids has been reported as the result of high pressure oxidation of humic acids and of coal (FISCHER, etc., 1920). In these experiments high temperatures are employed, and the polycarbonic acids isolated undoubtedly arise from the thermal decomposition products of the humic acids and not from unchanged humic acids.

Origin of Humic Acids: — A great number of more or less speculative hypotheses regarding the origin of the humic acids in humus have been advanced, and at times advocated with an emphasis which the weak supporting observations hardly justify. According to the oldest theory, which is still favoured by some authors, the humic substances arise from cellulose and similar carbohydrates. Recently BERGIUS (1927, 1928) has emphasized the fact that cellulose, like most organic substances, is thermodynamically unstable. Such substances have a tendency to decompose spontaneously with the formation of products possessing less free energy than the parent substance. This decomposition is accelerated when the temperature is increased. BERGIUS studied the thermal decomposition of cellulose at high temperatures and believed that products obtained on alkaline oxidation of the resulting charcoal are identical with humic acids. However, geological as well as chemical evidence shows that the formation of coal in nature is not a process in which high temperatures are involved. This is conclusively demonstrated by the occurrence of spores and of thermolabile substances in coal, such as porphyrines derived from chlorophyll and certain resins.

The spontaneous breakdown of cellulose is a process which requires millions of years. Humic acids are formed in a few years. BERGIUS' views, therefore, do not explain the formation of humus in nature but the spontaneous decomposition of organic material may perhaps be of importance in the later stages of coal formation.

It has long been known that cellulose as well as simple sugars, when treated with strong mineral acids, yield products showing similarities to natural humic acids. This is due to dehydration with the formation of furan derivatives which undergo polymerisation to brown or black acidic substances. This reaction has led several chemists to believe that carbohydrates are the parent substances of natural humic acids and to postulate a furanoid nature for humic acids. The comparison, however, of a natural product formed under mild conditions with decomposition products formed in such a drastic reaction is unacceptable to the biologist or biochemist. It can safely be said that there is as yet no valid experimental proof in favour of the view that natural humic acids are derivatives of furans. Careful studies of the properties of natural humic acids and humic acids derived from sugars reveal distinct differences. Therefore, it is doubtful whether carbohydrates take any direct part in the formation of humic acids in nature. It is not impossible, nevertheless, that natural humic acids and sugar humic acids may be chemically — not biologically — related. This will be discussed further in connection with the lignin theory of the origin of humic acids.

According to the theory which is most widely accepted, lignin is the precursor of humic acids. This theory is old and goes back to the early work of LESQUEREUX, FRÉMY, HOPPE-SEYLER, and KLASON. It has recently been advocated again by FISCHER and SCHRADER (1922; *cf.* also FUCHS 1928; WAKSMAN 1936). During the decomposition of dead plants in nature, cellulose and hemicelluloses are attacked first and are converted into gaseous or water-soluble products. The lignin is much more resistant and is accumulated during the process. Therefore, the surface layer of a peat bog contains a larger percentage of carbohydrates and less lignin than the lower layers. The methoxyl content of peat increases with increasing depth, but reaches a maximum and then drops. It has already been mentioned that lignin is considered to be a polymer of phenolic units containing methoxyl groups. Hydrolysis of these methoxyl groups during the course of humus formation causes the liberation of phenolic hydroxyl groups followed by alkali solubility of the resulting products. The process is accompanied by development of the characteristic colour of humus, and this phenomenon, undoubtedly, is due to oxidation (dehydrogenation). It is a very familiar fact that fresh peat when exposed to air usually darkens considerably due to the absorption of oxygen. The oxygen pressure of intact peat bogs is extremely low, and it is probable that part of the humic acids of the peat occurs in the reduced form as leucohumic acids.

HOLMBERG (1935) investigated the composition of some samples of elm wood. His results are shown in TAB. 3 (p. 17).

Since humic acids probably are of a hydroxyquinone nature, it is of interest to note that several hydroxyquinones of known structure possess properties which closely resemble those of natural humic acids (PUMMERER and HUPPMANN, 1927).

Artificial humic acids of hydroxyquinone nature were obtained by

Table 1: CHEMICAL COMPOSITION OF A HIGHMOOR PEAT PROFILE (PER CENT OF DRY MATERIAL). (WAKSMAN, 1936, p. 271):—

DEPTH (CM)	HEMICELLULOSES	CELLULOSE	LIGNIN AND HUMIC ACIDS	ASH
1 – 8	26.5	16.9	27.2	2.0
8 – 20	25.2	14.7	29.2	1.1
20 – 30	24.6	16.0	28.9	1.0
30 – 46	22.3	13.7	32.2	0.9
46 – 61	18.5	14.7	33.2	1.1
183 – 214	15.9	15.6	37.4	1.0
460 – 480	12.7	11.9	44.8	1.1
550 – 580	6.0	5.1	54.1	2.8

Table 2: CHEMICAL COMPOSITION OF PEAT AND COAL (PER CENT OF DRY MATERIAL). (WAKSMAN, 1936, p. 284):—

	PEAT	LIGNITE	BITUMINOUS COAL	SAPROPEL COAL	ANTHRACITE COAL
Carbon	50–60	60–65	75–90	76–82	92–96
Hydrogen	5–7	5–6	4–6	8–11	2–3
Oxygen	30–40	20–30	4–15	8–12	2–3
Nitrogen	1–4	0.4–2.5	0.4–2.0	0.6–1.2	0.5–0.7
Sulphur	0.2–2	0.1–3	0.3–2.0	0.6–3.0	0.4–0.8
Methoxyl	1.3–3.5	0.5–5.0	0–0.8	—	—
Hydroxyl	2–5	2–8	trace to 0.1	—	—
Carbonyl	1–3	2–3	—	—	1
"Carboxyl"	4–6	1–10	0–0.2	—	—

Table 3: CHEMICAL COMPOSITION OF ELM WOOD (HOLMBERG, 1935):—

	RECENT ELM WOOD	FOSSIL ELM WOOD (ABOUT 4000 YEARS OLD)	
		INNER FRESH PORTION	OUTER DECOMPOSED PORTION
Water extract	5.3	1.4	3.6
Acetone extract	1.2	1.4	2.0
Methoxyl	4.8	5.6	9.85
Acetyl	5.5	0.2	—
Pentosan	20.9	15.3	7.2
Lignin*	22.5	27.1	—
"Hemicellulose"	32	26	—
Cellulose	49	54	17
Lignin**	21	27	64

* Lignin estimation with 72 p.c. sulphuric acid.
** " " " thioglycolic acid.

ELLER and his coworkers by the oxidation of phenols or quinones in weakly alkaline or neutral solutions (ELLER and KOCH 1920, ELLER 1921–1925, ELLER and SCHÖPPACH 1926). These "phenol humic acids" serve as valuable models for the investigation of natural humic acids. The result of these studies indicates a close relationship between these two groups. The preparations agree not only as regards colour, acidity, and colloidal nature but in addition the similarity of their derivatives is also remarkable. However, they show marked differences in chemical behaviour from the humic acids obtained by the action of strong acids on sugars (PLUNGUIAN and HIBBERT 1935). ELLER believes natural humic acids obtained by oxidative polymerisation of phenols to be structurally related.

The question now arises: do phenols occur in plants in amounts corresponding to the quantity of humic acids isolated from the peat formed from them? Many plants turn brown or black on drying: *Monotropa*, *Orobanche*, several orchids, and many others. This colouration is due to the oxidation of phenols, and the colouring matter is undoubtedly similar to the artificial phenol humic acids. A number of plants are rich in tannins, which are complex phenols, or contain glucosides of phenols, all of which are easily converted by oxidation, either directly or with enzymes, to phenol humic acids.

However, it is evident that these sources of humic acids are not sufficient to account for the humic acids of soil, mud, and peat. Therefore, it is suggested that the phenolic substances obtained by the decomposition of lignin are transformed into humic acids by oxidation. To some extent, this conversion may proceed in another way. Microorganisms, especially lower fungi, first assimilate the lignin products transforming them into phenols which undergo oxidative polymerisation to humic acids after death. Some bacteria and fungi are known to contain considerable amounts of phenolic constituents.

Assuming the lignin theory of the origin of humic acids to be essentially correct, the question of the chemical structure of phenol humic acids becomes increasingly important. The elucidation of the structure of lignin and phenol humic acids may provide the key to the understanding of natural humic acids.

The mechanism of the oxidative polymerisation of simple phenols, which are in all probability related to lignin, has been subjected to a systematic study by the author (ERDTMAN 1933). The phenols to which special interest is attached are catechol, hydroxyquinone, and pyrogallol. The first phase in the conversion of these phenols into phenol humic acid appears to be their oxidation to the corresponding quinones. The quinones thus obtained couple with derivatives of diphenyl. Pyrogallol (IX) *e.g.* is converted into a hexahydroxydiphenyl (X):

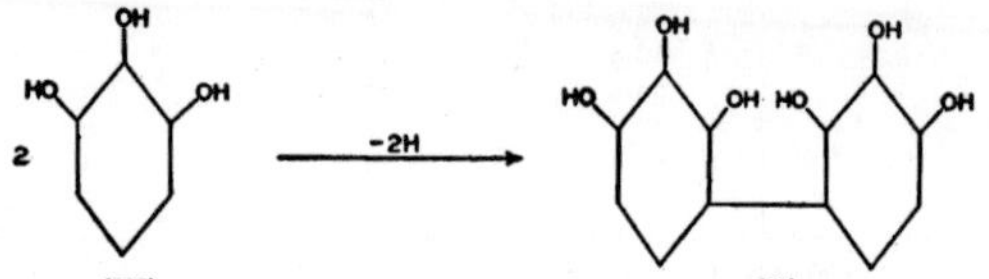

(IX) (X)

Hydroxyquinol yields an isomer of this, dipyrogallol (XI).

Oxidation of these diphenyl-derivatives to quinones followed by renewed coupling affords products of higher, still unknown molecular weight. The probable structure of the humic acid obtained from hydroxyquinol may be represented as shown below (XII) (H. ERDTMAN 1934).

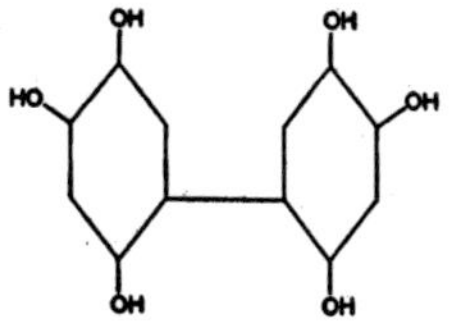

(XI)

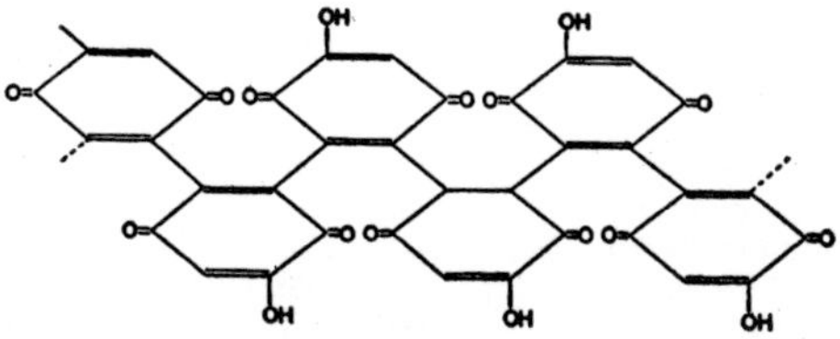

(XII)

This formula accounts for the dark colour of the humic acid, its colloidal nature, its reduction with zinc and acetic anhydride to form stable leucoacetates. It also explains the acidic nature of the substance.

According to most workers in this field, humic acids are true carboxylic acids. There are no conclusive proofs in favour of this view, however. The fact that, at higher temperatures, humic acids split off carbon dioxide does not give proof of the presence of carboxyl groups. The acidity of hydroxyquinones equals most carboxylic acids, in fact the grouping $HO-\underset{|}{C}=CH-\underset{|}{C}=O$ occurring in hydroxyquinones may be looked upon as being composed of the elements of the carboxyl group HO and CO separated by a – CR = CR – group.

It has been suggested that the esterification of humic acids with alcohols and hydrochloric acid would prove their carboxylic nature. However, this is not the case. It is a well known fact that many hydroxyquinones are capable of the same facile "esterification" with formation of esters readily hydrolysed by the action of weak alkali (Scholl and Dahll 1924, Anslow and Raistrick 1939). Further, quinonecarboxylic acids are extremely unstable and, usually, decompose spontaneously, yielding quinones and carbon dioxide.

Humic acids are complex mixtures of substances which are most probably of closely related structure. This accounts for their amorphous nature. To this may also be due the restricted rotation of the individual nuclei around their connecting link. This offers possibilities for the coexistence of innumerable stereoisomers (atropisomerism).

The importance of nitrogen in humus has recently been pointed out (Waksman and Iyer 1932, Waksman 1936). The nitrogen content of humic acid preparations has never been overlooked but has at times been considered a result of impurities (Detmer 1871, Eggertz 1888, Odén 1919). Usually the organic matter in the soil contains more nitrogen than does peat. This is very easy to understand since soil has undergone much greater decomposition because of the prevailing aerolic conditions. Bacteria and fungi are rich in proteins and nitrogenous carbohydrates (chitin). Since soil is being continuously decomposed, it may attain a state of biological equilibrium, resulting in a constant ratio between nitrogen and carbon in the soil.

Humic acid and the amphoteric proteins are able to form salt-like complexes, which are due to the amorphous state of both components. These are not easily separable. Such complexes actually do exist in nature, especially in well-manured soils. Only seldom, however, has the presence of proteins been clearly demonstrated. This has been done by their hydrolysis to amino-acids. Usually, the amount of protein has been calculated from the nitrogen content of the soil. This is a simple method, which is being extensively used with respect to nitrogen content in the investigation of foodstuffs and similar products. Its application is limited, however, to such material already known to be free from non-protein nitrogen. This is not the case with soil, peat, or humic acids. Since many investigators (Detmer 1871, Eggertz 1888, Odén 1919) have been able to isolate humic acids containing only traces of nitrogen, the opinion has been prevalent that the nitrogen content is due to nitrogenous impurities. An important contribution to the elucidation of the origin of the nitrogen in humus or humic acids was given by Eller (*l.c.*), who demonstrated that humic acids not only form ammonium salts when treated with ammonia but also products in which the nitrogen has entered the anion. Part of this nitrogen is so firmly linked that it cannot be split off by means of boiling alkali.

These reactions have close analogies among the hydroxyquinones. Carbonyl oxygen or hydroxyl-groups may be replaced by imino or amino groups when hydroxyquinones are treated with ammonia or amines, and alkyl groups may be split off and replaced with nitrogen containing radicals. Interaction between amino-groups or between amino- and hydroxyl-groups in different nuclei also results in the formation of alkali-stable heterocyclic rings (carbazol-rings).

Melanins: — In this connection, the so-called melanins, dark-coloured substances occurring in both animals and plants, should be briefly mentioned. Certain amino-acids contain phenolic hydroxyls.

HO— [ring] —CH_2—CH(—COOH)—NH_2

(XIII)

HO—, HO— [ring] —CH_2—CH(—COOH)—NH_2

(XIV)

An example is tyrosine (XIII), an important constituent of many proteins. By the action of the enzyme tyrosinase, it is hydroxylated to dihydroxy-phenyl-alanine (XIV) which is not uncommon in plants. Further oxidation proceeds by way of several intermediates to the indole derivative (XV).

This yields melanins by oxidative polymerisation. The melanins

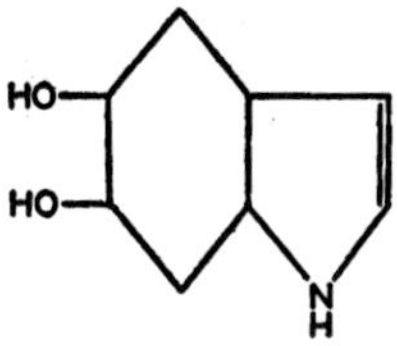

(XV)

are closely related to phenol humic acids and contain a high percentage of nitrogen (Raper 1927). A mixture of such melanins and humic acids would be difficult indeed to separate by present methods. Thus, the nitrogen content of most humic acid preparations appears to be easily accounted for without assuming the presence of proteins.

Preservation of Organized Material in Peat: — Peat and brown coal, as well as many other types of coal, are complex mixtures of more or less decomposed residues of plants and, to a lesser degree, of lower animals. Lack of oxygen, due to the abundance of water, is essential to the process of their formation. Apart from mineral particles and undecomposed remains, the most thoroughly investigated and most characteristic components of these materials are the brown or brownish-black humic substances which possess a more or less pronounced acidic nature. These are aromatic substances, probably polymers of hydroxyquinones derived from phenolic constituents of plants among which lignin is predominant. Some peats appear to contain relatively large amounts of proteins.

About 80–95 per cent of an undrained peat bog is water. The oxygen pressure in the bog is extremely low, because of the presence of the easily oxidisable humic acids or their leucoderivatives.* The amorphous nature of most components of peat makes it very efficient as a filter for solids. All these properties contribute to the preservation of material which is not rapidly decomposed on the surface of the bog and which is resistant to weak acids. Cutinised tissues, such as the epidermis of many plants, guard-cells of stomata, and hairs (*e.g.* of *Hippophaë*) are, therefore, preserved in peat, as well as leaf spines of *Ceratophyllum* and idioblasts of *Nymphaea*. The lime shells of molluscs are more easily attacked by the acids of the peat. The silica shells of diatoms, as well as the more or less chitinous parts of ostracodes (*e.g.* spermatophores), ephippiae of *Daphnia*, statoblasts of *Bryozoa*, eggs of many worms, cysts of tardigrades and chitinous skeletons of insects are resistant. Keratin exhibits high resistance and hair of higher animals (woolen clothes), nails, skin, and leather may be long preserved in peat. The tanning properties of humic acids are of importance in this respect.

Preservation of Pollen and Spores in Peat: — Special attention must

* The consumption of oxygen by microorganisms active in the decomposition processes and the slow diffusion of oxygen through the bogs are partly if not largely responsible for the low oxygen content.

be devoted to pollen and spores which, as a rule, are very well preserved in bogs. It has already been mentioned that the physical conditions in the bog are such as to restrict the drift of pollen and spores. The success of the method of pollen analysis depends upon these two fundamental facts.

The chemistry of pollen and spore membranes is, unfortunately, still very poorly understood. Most pollen or spores are not easily obtained in quantity.* This fact, as well as their insolubility in all solvents and their high resistance to most chemical reagents, makes the study of their chemical nature very difficult. The spores of *Lycopodium clavatum*, however, can be obtained in great quantities and have been more closely investigated than most pollen of more direct interest in connection with pollen analysis. Early work of JOHN (1814), BRACONNOT (1829), and BERZELIUS did not result in the isolation or characterisation of definite components. BERZELIUS pointed out the great resistance of pollen to alkali. The classical method of preparing peat for pollen analysis is based upon this discovery. That pollen is usually very rich in fat has long been known.

Recently, F. ZETZSCHE and his collaborators have subjected pollen and spores to closer examination and, during these studies, some facts of importance related to our present problem have been learned. These investigators isolated the characteristic substance of the *Lycopodium* spores (sporonine) by extraction of the spores with boiling alkali (ZETZSCHE and HUGGLER 1928, ZETZSCHE and VICARI 1931). In this way, fat (about 50 per cent), acidic substances and proteins (about 8 per cent) are removed. The insoluble residue consists mainly of cellulose and a specific substance termed sporonine. The spores contain only 2 per cent cellulose. The sporonine (25 per cent of the spores) was isolated by hydrolysis of the cellulose with mineral acids. This material is not related to lignin. It contains only very little methoxyl (1.25 per cent), splits off water, decomposes on heating, and burns with a sooty flame. It contains 65 per cent carbon and 8.5–8.7 per cent hydrogen. The rest is oxygen, and the amount of this element is surprisingly high. The reaction with boiling acetic anhydride proceeds with the formation of acetylated products containing 28–29 per cent of acetoxyl. This indicates that about half the number of oxygen atoms is present in the form of hydroxyl groups. The function of the other oxygen atoms is not known. They probably belong to ether linkages.

Sporonine is obviously highly unsaturated. It reacts readily with halogenes. Bromination affords products containing 50–60 per cent bromine, part of which is easily removed by alkali. The brominated products are far more resistant to oxidation than is sporonine. ZETZSCHE has developed a method for the isolation of pollen and spores from coal which involves successive bromination and oxidation of the material with fuming nitric acid (ZETZSCHE and KÄLIN 1932). According to ZETZSCHE, the sporonine is related to certain terpenes or similar compounds, *e.g.* rubber. Although some observations seem to

* The pollens of the ragweeds (*Ambrosia elatior* and *A. trifida*) may be obtained from many collectors and drug companies in the United States for from six to twenty cents a gram. That from corn (*Zea mays*), hickory and some of the oaks may be collected in any quantity without much effort. (WODEHOUSE). — *Cf.* also KELLEY, J. W., 1928: Methods of Collecting and Preserving Pollen for Use in the Treatment of Hay Fever (Circul. 46, U.S.D.A.).

support this view, the matter is by no means settled. Not even the obvious determination of methoxyl groups attached to carbon — so characteristic of terpenes — appears to have been carried out.

ZETZSCHE and VICARI (1931) examined pollen of some trees of more special interest to pollen analysis. The pollen grains of *Pinus silvestris* contained 21.9 per cent sporonine (since pollen and microspores are homologous, the introduction of the term pollenine appears to be unnecessary) and 2.0 per cent cellulose. The pollen grains of *Picea orientalis* contained 20.0 and 2.2, those of *Corylus avellana* 7.3 and 1.1 per cent, respectively. The sporonines are very similar to that of *Lycopodium* spores but contain a little less oxygen. It is interesting to note the difference between the sporonine content of *Corylus* and that of the conifers, *Pinus* and *Picea*. Such differences might be of considerable interest in connection with the evaluation of pollenstatistical data, because the preservation of pollen in peat appears to be due largely to their sporonine-bearing membranes. Unfortunately, practically nothing is known regarding the chemical composition of other pollen, and an investigation of those pollens, which one would suspect to be occurring in peat (*e.g. Juncaceae, Populus*, etc.), but which have never been encountered therein, is urgently needed. Pollen grains of such plants decompose easily, on account of their thin exine and perhaps also as a result of the absence of sporonine.

An observation of ZETZSCHE and KÄLIN (1931) is especially interesting for pollen analysts. These workers found that sporonine of *Picea orientalis*, when subjected to prolonged action of air, takes up oxygen with an increase of the oxygen content from 25 to 65–70 per cent. This autooxidation appears to be a photochemical process involving the addition of oxygen molecules to double linkages in the sporonine, with the formation of peroxides. This conclusion is supported by the fact that the autooxidized sporonine liberates iodine from hydriodic acid. Sporonine of *Lycopodium* did not undergo autooxidation although it was expected to do so. The sporonine did not suffer any morphological change during this process, but deep-seated chemical changes must have occurred since the autooxidized sporonine was easily soluble in dilute alkali, in striking contrast to the nonautooxidized membrane which resists the action of boiling concentrated alkali solutions. It should be noted, however, that these observations were made on isolated sporonine and not on fresh pollen. A close investigation of the autooxidizability of pollen is needed and would be of general interest in connection with the theory of pollen analysis.

In the meantime, this phenomenon ought to be borne in mind especially when soil or similar products to which air (and light) has free access are subjected to pollen analysis. Due to the lack of oxygen in the undrained peat bog and the abundance of easily oxidizable substances present, pollen grains which have "survived" the first critical period on the surface and eventually became embedded in the peat, are well preserved and suffer very little change in thousands or millions of years.

References: —

ANSLOW, W. K. and RAISTRICK, H., 1939: The action of alcoholic monomethyl amine on benzo- and toluquinone derivatives (Journ. Chem. Soc.).

ASCHAN, O., 1932: Water humus and its rôle in the formation of marine iron ore (Ark. kemi, min., geol., vol. 10 *a*).

BERGIUS, F., 1927: Die chemischen Grundlagen der Kohleverflüssigung (Svensk kem. tidskr., vol. 69).

—— —— 1938: Beiträge zur Theorie der Kohleentstehung (Naturwissensch., vol. 16).

CRAMER, A. B., HUNTER, M. J., and HIBBERT, H., 1939: Studies on lignin and related compounds. XXXV. The ethanolysis of spruce wood (Journ. Amer. Chem. Soc., vol. 61).

DETMER, W., 1871: Die natürlichen Humuskörper des Bodens und ihre landwirtschaftliche Bedeutung (Landwirtsch. Vers. Stat., vol. 14).

DOPPLER, C., 1849: Über eine merkwürdige in Österreich aufgefundene gelatinöse Substanz (Sitz.-Ber. Akad. Wiss. Wien, vol. 3).

EGGERTZ, M. A., 1888: Studien und Untersuchungen über die Humuskörper der Acker- und Moorerde (Meddel. Landtbruksakad., no. 3; Stockholm).

ELLER, W., 1921/22: Künstliche und natürliche Huminsäuren (Brennstoffchemie, vols. 2, 1921, pp. 129–133, and 3, 1922, pp. 49–52, 55, 56; *cf.* also Liebig's Annalen d. Chemie, vol. 431, 1923, and 442, 1925).

—— —— and KOCH, K., 1920: Synthetische Darstellung von Huminsäuren (Ber. Deutsch. Chem. Ges., vol. 53).

—— —— and SCHÖPPACH, A., 1926: pp. 17–20 in Brennstoffchemie, vol. 7.

ERDTMAN, H., 1926: Nyare undersökningar över naturliga och konstgjorda huminsyror (Svensk kem. tidskr., vol. 38).

—— —— 1933*a*: Dehydrierungen in der Coniferylreihe, II. Dehydri-isoeugenol (Liebig's Ann. Chem., vol. 503).

—— —— 1933*b*: Studies on the formation of complex oxidation and condensation products of phenols. A contribution to the investigation of the origin and nature of humic acid (Proc. Roy. Soc., London, Ser. A, vol. 143).

—— —— 1934: Phenoldehydrierungen, IV (Liebig's Ann. d. Chem., vol. 513).

FISCHER, F. and SCHRADER, F., 1920: Allgemeine Bemerkungen über den chemischen Abbau von Stein- und Braunkohlen, Lignin und Zellulose durch Druckoxydation (Ges. Abh. z. Kenntn. d. Kohle, vol. 5).

—— —— —— —— 1922: Entstehung und chemische Struktur der Kohle (Essen).

—— —— —— —— and TREIBS, W., 1920: Über den chemischen Abbau von Braunkohlen durch Druckoxydation (Ges. Abh. z. Kenntn. d. Kohle, vol. 5).

FREUDENBERG, K., 1933: Tannin, Cellulose, Lignin (Berlin).

—— —— 1939: Lignin (Fortschr. d. Chem. org. Naturstoffe, Wien).

FUCHS, W., 1928: Über Fortschritte in der Chemie der Huminsäuren und der Kohle (Ztschr. angew. Chemie, vol. 41).

—— —— 1928/29: Unsere derzeitige Kenntnis von der chemischen Natur der Huminsäure, des Hauptbestandteiles der Braunkohle (Ges. Abh. z. Kenntn. d. Kohle, vol. 9).

—— —— and STENGEL, W., 1929: Zur Kenntnis der Hydroxyl- und Karboxylgruppen der Huminsäuren (Brennstoffchemie, vol. 10).

HOLMBERG, B., 1935: Über subfossiles Ulmenholz. Ligninuntersuchungen IX (Norske Vidensk. Selsk. Forhandl., vol. VII, no. 37).

KARRER, P. and BODDING-WIGER, B., 1923: Zur Kenntnis des Lignins (Helv. Chim. Acta, vol. 6).

—— —— and WIDMER, F., 1921: Zur Kenntnis der Zellulose und des Lignins (*Ibid.*, vol. 4).

LEWIS, H. C., 1881: On a new substance ressembling dopplerite from a peat bog at Scranton (Proc. Amer. Phil. Soc., vol. 20).

NAUMANN, E., 1921: Die Bodenablagerungen des Süsswassers (Arch. Hydrobiol., vol. 13).

—— —— 1929: Die Bodenablagerungen der Seen (Verhandl. Int. Ver. Limnologie, vol. IV).

ODÉN, S., 1919: Die Huminsäuren (Kolloidchem. Beih., vol. 11).

PLUNGUIAN, M. and HIBBERT, H., 1935: Studies on lignin and related compounds, XI. The nature of lignite humic acid and the so-called "humic acid" from sucrose (Journ. Amer. Chem. Soc., vol. 57).

PUMMERER, R. and HUPPMANN, G., 1927: Die Kondensation von Chinonen mit Phenolen (Ber. Deutsch. Chem. Ges., vol. 60).

RAPER, H. S., 1927*a*: The tyrosinase-tyrosine reaction, VI. Formation of 5:6-dihydroxyindole-2-carboxylic acid from tyrosine (Biochem. Journ., vol. 21).

—— —— 1927*b*: Die Einwirkung von Tyrosinase auf Tyrosin (Fermentforsch., vol. 9).

SCHOLL, R. and DAHLL, P., 1924: Über die Einwirkung von Ferricyankalium auf Purpurin in alkalischer Lösung (Ber. Deutsch. Chem. Ges., vol. 57).

SENFT, F., 1862: Die Humus-, Marsch-, Torf- und Limonitbildungen als Erzeugungsmittel neuer Erdrindelagen (Leipzig).

TIDESWELL, F. F. and WHEELER, R. V., 1922: On dopplerite (Journ. Amer. Chem. Soc., vol. 121).

WAKSMAN, S. A., 1936: Humus. Origin, chemical composition and importance in Nature (Baltimore).

—— —— and IYER, K. R. N., 1932: Synthesis of a humus nucleus, an important constituent of humus in soil, peat, and composts (Soil Sci., vol. 34).

——— and PURVIS, E. R., 1932: The microbiological population of peat (Soil Sci., vol. 34).

ZEILE, K., 1935: Zur Kenntnis der Huminsäure (Kolloidzeitschr., vol. 72).

ZETZSCHE, F., 1932: Fossile Pflanzenstoffe (*in* KLEIN, G.: Handbuch der Pflanzenanalyse, Wien).

——— and HUGGLER, K., 1928: Untersuchungen über die Membran der Sporen und Pollen, I. *Lycopodium clavatum* L. (Liebig's Ann. Chem., vol. 461).

——— and KÄLIN, O., 1931: Zur Autoxydation der Sporopollenine (Helv. Chim. Acta, vol. XIV).

——— ——— 1932: Eine Methode zur Isolierung des Polymerbitumens (Sporenmembranen, Kutikulen usw.) aus Kohlen (Braunkohle).

——— and REINHART, H., 1939: Beitrag zur Reduktion der Huminsäuren (Brennstoffchemie, vol. 20).

——— and VICARI, H., 1931a: Untersuchungen über die Membran der Sporen und Pollen, II. *Lycopodium clavatum* (Helv. Chim. Acta, vol. XIV).

——— ——— 1931b: Untersuchungen über die Membran der Sporen und Pollen III. *Picea orientalis, Pinus silvestris, Corylus avellana* (*Ibid.*, vol. XIV).

Chapter III

POLLEN PREPARATIONS

Laboratory Outfit: — The items enumerated below are listed in alphabetical order. A few items which are essential when treating only fossil material are also included and are marked with asterisks.

Centrifuges: — An ordinary handworked centrifuge is adequate. A well-equipped laboratory should, however, be provided with an electric centrifuge, *e.g.* of the Corda pattern, model C, with removable lid, four cases and adjustable resistance. Its maximum speed amounts to about 3000 revolutions per minute. Half of this speed is normally sufficient. Minimum speed should be used when working with ordinary glass tubes and copper cases (see below).

Centrifuge Tubes: — made from Jena or Pyrex glass; capacity about 14 c.c. The glasses should be provided with lines, marking a volume of 5 and 10 c.c. respectively. The glasses should be washed in running water and cleaned with a good brush. If necessary, some cotton wool should be pushed down into the tapering end of the glasses and turned round a few times by means of a forceps. The glasses are then rinsed with alcohol, eventually followed by ether, or simply left to dry on the pins of a drying rack.

Centrifuge Cases: — The ordinary aluminium cases should have some cotton wool at the bottom to avoid breaking the centrifuge glasses. The cotton wool has to be changed at intervals to prevent it from absorbing water and different chemicals as a result of which it may gradually be transformed into a hard body, corroding the aluminium and becoming difficult to remove. For centrifuging fluids containing hydrofluoric acid, special copper cases should be available.

Chemicals: —

Acetic acid: concentration not less than 95 per cent.

Acetic anhydride: technical, or — better — chemically pure; not the expensive kind used for analytical work.

Acetone.

Alcohol: 95–96 per cent.

Anisol: used as an immersion fluid instead of cedar oil. Wiping off is unnecessary as the anisol gradually evaporates. Refractive index about the same as that of cedar oil (1.515).

Caustic potash or soda: 10 per cent and ½ per cent solutions.

Glycerine.

Glycerine jelly, according to Kisser (Zeitschr. f. wiss. Mikroskopie, vol. 51, 1935). Ingredients: gelatine 7 grms., distilled water 19 c.c., glycerine (82 per cent) 33 grms., phenol (crystals) 1 grm. The gelatine is dissolved in the water, phenol and glycerine added, the whole thoroughly mixed by stirring with a glass rod. The fluid is then filtered through spun glass pressed into a heated funnel.

Hydrochloric acid: specific gravity 1.19.

* Hydrofluoric acid: concentration 50 to 60 per cent.

* Nitric acid: weak (concentration about 5 to 10 per cent), concentrated, and fuming.

Sodium chlorate: solution prepared by dissolving 10 grms. sodium chlorate in 20 c.c. distilled water.

Sulphuric acid: concentrated.

* *Crucibles:* — china and copper (or platinum); capacity about 30 to 35 c.c.

Forceps: — about 15 cm long, nickeled; shanks with very thin and smooth ends.

Funnels: — diameter about 6 cm; spout about 2 cm.

Metal screening: — brass screening (as used for straining milk, etc.) with about 300 meshes to the sq. cm, cut to fit the funnels.

Pipettes: — capacity 5 c.c. and 1 c.c. They are perfectly cleaned if kept for some time in thick-walled glass tubes with ground stoppers and spun glass soaked with fuming nitric acid at the bottom.

Rack for Storage Glasses or Vials: — When collecting pollen or spores of living plants a square plywood box (outside dimensions 130 by 130 by 47 mm; wall thickness 5 mm) containing 100 cells made from ten leaves of cardboard (120 by 35 by 1 mm) intersecting another set of ten leaves of the same material may be used as a rack for small storage vials (30 by 5 mm). The vials are corked with rubber stoppers or ordinary corks and given 1 c.c. glacial acetic acid each.

Stains: — *Cf.* p. 30.

Storage Glasses: — See Rack for Storage Glasses or Vials.

Water Bath: — For several years the author has used a bath made from copper plate [height 115 mm, diameter 83 mm; removable lid with five apertures (diameter 2 cm)]. The bath has three horizontal handles riveted to its middle, by means of which it is able to rest on a ring in a stand. On top of the lid is a brass net with five apertures corresponding to those of the lid. A thermometer is hung through the central aperture, whilst the four marginal apertures are made to fit and, at the same time, hold the centrifuge glasses securely. During heating, the glasses should be kept with their opening about 1 cm above the net. The bath is filled to about 3 cm from its rim. Boiling over is prevented by floating cotton wool in the water.

Acetolysis Method: —

Acetolysis of Fresh Material: — Pollen and spore-bearing parts of fresh plants are put in vials with glacial acetic acid and may be kept so for days or even months awaiting final preparation. A part of a catkin of *Corylus*, a dozen flowers of *Sagina procumbens*, one flower of *Pyrola rotundifolia*, a few stamens of *Nelumbo*, half a dozen sori of *Polypodium vulgare*, a few fertile thalli of *Riccia sorocarpa*, etc., are sufficient.

When the preparations are to be made, the vials are slipped into the aluminium cases of the centrifuge and the centrifuge is slowly rotated. The acetic acid is then decanted and a mixture of acetic anhydride and concentrated sulphuric acid is added, only enough to make it possible to empty the glass and transfer the pollen-bearing material to an ordinary centrifuge tube. The mixture of acetic anhydride and concentrated sulphuric acid is prepared by adding the acid drop by drop or in small jets to nine times as many cubic centimeters of the anhydride. Only as much of the mixture as is expected to be used during the day should be made each time. After transferring the pollen-bearing material to the centrifuge tubes, the acetolysis mixture is brought up to the 5 c.c. line. A glass rod is inserted into each tube and the tubes transferred to the water bath.

Heating begins with the water at room temperature or, in order to perform the heating as quickly as possible, at a temperature between 70° and 80° C. The temperature should not exceed the last figure lest the centrifuge tubes crack and a reaction between the hot water and the acetolysis mixture occurs. The bath should, if possible, be placed in a fume cupboard behind a sliding window. When the boiling point has been reached, heating is immediately stopped, and the fluid in the centrifuge tubes stirred. The tubes are then transferred to the centrifuge. After centrifuging, which should take about half a minute, the reaction mixture is decanted into a reserve receptacle. Distilled water is added to the sediment until the centrifuge tube is filled to the 10 c.c. line. The tube is then throughly shaken until the mixture foams. The foam is removed by adding a few drops of acetone or alcohol kept in a drop flask. After centrifuging and decanting, the sediment may be washed with water again or directly suspended in a mixture of glycerine (1 part) and distilled water (1 part). It should stay there at least ten minutes. Meanwhile a new set of centrifuge tubes may be transferred to the water bath and the procedure just mentioned repeated. Or the material may be left for several hours or even to the following day. After centrifuging and decanting, the centrifuge tubes are placed upside down on filter paper.

Preparation of microscopic slides is then carried out in the following way. The glycerine jelly, which is kept in a glass tube covered by a cork with a glass rod going through it, is heated in a water bath to about 50° C. With a diamond, the number of the centrifuge tube or the name of the plant, etc., is written on a slide. A drop of the melted glycerine jelly is transferred to the slide after it has been heated gently over an alcohol lamp and placed on a white paper. With the forceps, some of the pollen-bearing sediment is taken from the centrifuge tube and deposited in the jelly. Occasional large particles are removed with the forceps. Large particles may also be removed at an earlier stage, *viz.* when the pollen-bearing material is still in the glycerine-water mixture. To this end, the centrifuge tube is held loosely between the thumb and the index finger and repeatedly knocked with the other hand at the tapering end. This causes the fluid to be thrown up on the inner walls of the tube. The pollen grains slide down with the fluid, while larger particles stick to the wall and may easily be removed with cotton wool which is pushed into the tube with the forceps. A round cover glass is heated over a lamp and placed on the jelly. The slide is then quickly turned upside down (Willrath in Potonié 1934) to allow the pollen grains to settle on the cover glass.

Sealing does not seem to be necessary and should in any case not be performed during the first days. At the palaeobotanical department of the State Museum of Natural History at Stockholm, pollen slides prepared in the way described above have been sealed with Rützou's "Praeparatlak" from the chemical laboratory of H. STRUER, Skindergade 38, Copenhagen.

Some pollen types, particularly large and conspicuous ones (such as *Cirsium*, *Geranium*, *Ipomoea*, *Malva*, and *Scabiosa*) turn deep reddish brown or, in extreme cases, almost black after acetolysis. They must be chlorinated to make it possible for the elements of the exine to appear with due clearness. Chlorination is quickly effected by adding about 5 c.c. of glacial acetic acid to the glycerine-soaked sediment. Three to five drops of sodium chlorate solution are added and the whole stirred with a glass rod. After adding about ½ c.c. of concentrated hydrochloric acid, the fluid is again stirred. Chlorine immediately appears. Its action is particularly effective in its nascent state and an effective bleaching is, as a rule, obtained in less than half a minute. After centrifuging, the reaction mixture is decanted, the sediment washed with distilled water, the tube thoroughly shaken, and the foam removed with acetone or alcohol. After another centrifuging, the water is decanted, dilute glycerine added, etc., etc.

A small fraction of the acetolysed material will usually suffice for a microscopic preparation. The rest of the material may be stored in concentrated glycerine for future use. It seems advisable to decant the glycerine later, eventually after centrifuging, and to replace it with a few drops of glycerine jelly. New microscopical slides may then be made by simply dipping a glass rod into a heated storage vial and transferring a drop of the pollen-bearing material to a slide.

Acetolysis of Herbarium Specimens: — Pollen- and spore-bearing parts of herbarium specimens are removed by means of forceps and put into small paper envelopes. Previous to chemical treatment, the material has to be powdered. For this purpose, the dry flowers, stamens, etc., are spread out on a brass screen on a funnel and ground against the screen. The plant powder on the inside of the funnel is then washed down into a centrifuge tube with a mixture of acetic anhydride and concentrated sulphuric acid supplied drop by drop from a 5 c.c. pipette. The centrifuge tube is then filled with acetolysis mixture to the 5 c.c. line. The subsequent treatment is the same as that previously described in connection with fresh material.

The results obtained by means of the acetolysis method are usually very satisfactory. Only the delicate pollen grains of the *Cannaceae*, *Juncaceae*, *Lauraceae*, *Musaceae*, and *Thurniaceae* come out in a more or less shrivelled and wrinkled condition.

Other Methods: — More than a century ago, different acids, such as hydrochloric, nitric, and sulphuric acid, were already in use by SPRENGEL, BRONGNIART, RASPAIL, FRITZSCHE, and others in the study of pollen grains. A more recent development of the use of sulphuric acid is described by WILLRATH (POTONIÉ 1934):

„Eine kleine Probe des recenten Pollens wird auf dem Objektträger mit einem Tropfen Wasser benetzt, dem mit Hilfe eines Glasstabes etwas ± verdünnte H_2SO_4 hinzugefügt wird. Der Objektträger wird dann vorsichtig über der Flamme erwärmt, bis der Tropfen fast gänzlich verdunstet ist. Die Pollen erhalten hierbei einen dunkleren Farbton. Dieser entspricht nicht nur bis zu einem gewissen Grade den Farbtönen, die manchmal den aus der Kohle gewonnenen Exinen zukommen, sondern lässt auch die Musterung der Exine deutlicher hervortreten. Zu beachten ist aber, dass die Pollen nicht zu dunkel werden dürfen, da sonst eine Vergleichsmöglichkeit nicht mehr vorhanden ist. Das mit einem Tropfen Glyzerin-Gelatine betupfte Deckglas wird auf den Objektträger gebracht, und zwar mit grösster Vorsicht, da die Pollen leicht nach der Seite wegrutschen. Nach dem Erkalten werden die Präparate mittels Pinsel und heissem Wasser gesäubert und das Deckglas nach einigen Tagen mit Asphaltlack oder Deckglaskitt abgedichtet. Asphaltlack ist besonders dann vorzuziehen, wenn unter Ölimmersion gearbeitet wird und die Gefahr besteht, dass das Objektiv das Präparat berührt und in diesem Falle der spröde Deckglaskitt abspringt".

Concentrated sulphuric acid has been used by FISCHER (1890) and lactic acid by LAGERHEIM (1902).

More than 25 years ago L. VON POST and his assistants had a reference set of pollen preparations made by boiling flowers from herbarium specimens with ten per cent caustic potash solution and embedding the pollen grains in glycerine jelly. Under this treatment the pollen grains, however, usually do not become entirely transparent. In spite of this, the method has been used considerably among pollen analysts (*e.g.* by L. R. WILSON, Rhodora, vol. 36, 1934, and LEWIS and COCKE, Journ. Elisha Mitchell Sci. Soc., vol. 45, 1929, who performed "artificial fossilization" by extracting fresh pollen with ether followed by boiling in caustic potash. Such slides are claimed to be of " utmost value in identifying important American species ").

In a study of grass pollen, FIRBAS (1937) availed himself of a modification of the alkali method:

„Bei der Aufbereitung wurde so vorgegangen, dass reife trockene Antheren zerkleinert und dann 10 Min. in 10 proz. KOH gekocht wurden. Die Flüssigkeit wurde zentrifugiert, der Rückstand mit $\frac{1}{10}$ n-HCl schwach angesäuert, 2 mal mit H_2O gewaschen, mit ein wenig Glyzerin versetzt, durch 24stündiges Verdunsten auf offenen Objektträger in stark konzentriertes Glyzerin übergeführt und schliesslich in Glyzeringelatine eingebettet. Beim grössten Teil der Pollenkörner waren nach dieser Behandlung Zellinhalt und Intine weitgehend zerstört und ein Zustand erreicht, der dem des fossilen, mit KOH aufbereiteten Materials gut entsprach."

In conclusion it seems appropriate to consider the methods applied by two very experienced pollen morphologists, FISCHER (1890, 1912) and WODEHOUSE (1935). Their pollen preparations are among the most beautiful preparations of their kind. From the point of view of a pollen analyst, however, the fact that the pollen grains in these preparations are as a rule not quite transparent must be regarded as a disadvantage.

FISCHER left the pollen grains to dry on a slide, until they adhered to it. After the pollen had been washed with benzene or xylol on the slide, a drop of weak fuchsin solution (fuchsin 1 part, alcohol about 10,000 parts) was added. Differentiation was then accomplished by using a chloral-hydrate solution or weak hydrochloric or acetic acid. The preparations were completed by heating or gently boiling the pollen grains in glycerine jelly. Fuchsin was the favourite stain, but safranin, methylene blue, iodine green, malachite green, gentian violet, and Bismarck brown were also used. The inner construction of the exines was made more visible by embedding the pollen grains in a medium with a high refractive index such as concentrated chloral hydrate solution, or, preferably, a solution of potassium-mercury-iodide.

WODEHOUSE's method is as follows (*l.c.*, pp. 106–108):

"A small amount of pollen . . . is placed on the center of a microscope slide and a drop of alcohol added and allowed partly to evaporate. A second and third and even fourth drop may be added if necessary. The alcohol spreads out as it evaporates and leaves the oily and resinous substance of the pollen deposited in a ring around the specimen. The oily ring is wiped off with cotton moistened with alcohol, and, before the specimen has had time to dry completely, a drop of hot, melted methyl-green glycerin jelly is added, and the pollen stirred in with a needle and evenly distributed. During the process the jelly is kept hot by passing the slide over a small flame. . . . If naturally shed pollen is not available, satisfactory material can generally be obtained from herbarium specimens, provided they were quickly and completely dried. Often it is only necessary to tap the dry flowers over the slide or crush a few anthers on it. If pollen cannot be removed in this way, a few anthers or, with the Compositae, a few florets may be removed from the specimen and placed on the slide. These are then moistened with alcohol, followed by a drop of water, and heated to boiling. The pollen may then be teased out, and the anthers and other debris removed, leaving the pollen in the water. The water is then drawn off with filter paper, and the jelly added as before. . . . The slides retain their brilliancy for several months unimpaired but unfortunately are not permanent, for the dye fades slowly out after a period varying from about 9 months to 2 years after which they are entirely bleached. Such preparations may always be restored, however. . . . Aqueous fuchsin has the same selective properties as methyl green, is much more vigorous in its action, and is permanent. The only objection to its use is that the red color is theoretically not quite so satisfactory for observation with high-power lenses as the blue color, though in actual practice there is little difference. It may be used in place of methyl green as above".

References: —

ALFORD, R. I., 1941: Pollen Studies. A contrast color method of examining unstained pollen grains (J. Allergy 12: 572).

ERDTMAN, G., 1934: Über die Verwendung von Essigsäureanhydrid bei Pollenuntersuchungen (Svensk Bot. Tidskr., vol. 28).

FIRBAS, F., 1937: Der pollenanalytische Nachweis des Getreidebaus (Zeitschr. f. Bot., vol. 31).

FISCHER, H., 1890: Beiträge zur vergleichenden Morphologie der Pollenkörner (Breslau).

—— 1912: Botanisch-mikrotechnische Mitteilungen, I. Über Untersuchung von Pollenkörnern (Zeitschr. f. wiss. Mikroskopie, vol. 29).

LAGERHEIM, G., 1902: Metoder för pollenundersökning (Bot. Notiser).

POTONIÉ, R., 1934: Zur Mikrobotanik der Kohlen und ihrer Verwandten, II. Zur Mikrobotanik des eocänen Humodils des Geiseltals (Arb. Inst. Paläobot. Petrogr. Brennst., vol. IV).

WODEHOUSE, R., 1935: Pollen grains (New York).

Chapter IV

PREPARATION OF FOSSIL POLLEN-BEARING MATERIAL

Collecting of Peat Samples: — The directions and methods discussed below are essentially those of L. VON POST, published in the instructions to the members of the qualitative and quantitative statistical peatland survey conducted by the Geological Survey of Sweden during the years 1917–1924. These instructions have been reprinted in an extensive work on the peat resources of southern Sweden by VON POST and GRANLUND (1926).

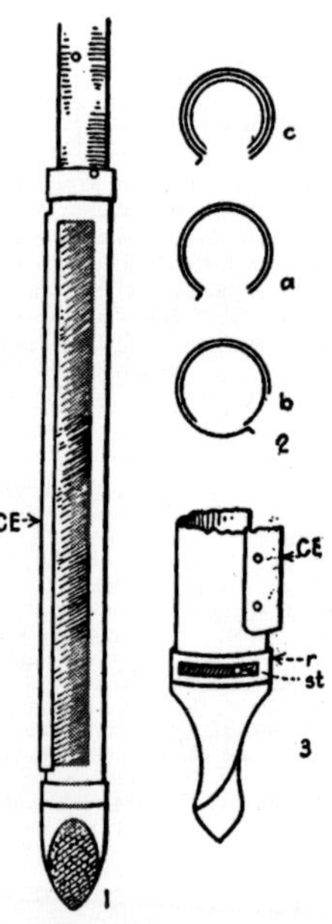

TEXTFIGURE 1. — PEAT AUGER, HILLER MODEL. — (1) Lower end of auger with chamber open. The cutting edge, CE, should be fairly sharp. The opposite edge may be marked at intervals of 5 cm. (2) Cross sections of the chamber: *a*, open, *b*, closed, *c*, to show the position of the zinc lining which is represented in the sketch by the inner arc. (3) Lower part of the chamber and screw point. The outer sleeve of the chamber is riveted to the ring, *r*, but the inner is free to revolve within. *st* is a pin fastened to the inner sleeve; it can traverse the length of the slot in the ring, thus checking at the proper places the opening and closing movements of the sleeve (from WODEHOUSE 1935, adapted from KRÄUSEL).

For collecting peat samples the following equipment is needed: spade, peat-knife, forceps, glass tubes corked at both ends (*e.g.* 75 mm long, 13 mm inside diameter), waxed paper, oakum, and a peat auger. There are probably as yet no better augers than those designed by HILLER, a Swedish peat engineer. They are manufactured by the Beus and Mattson Company, Mora, Sweden. There are two models, a smaller one with a relatively short container (32 cm) and extension rods measuring 100 cm each, and a larger one with a long container (40 cm) and longer extension rods (150 cm). The construction of the lower end of the auger is diagrammatically shown in TEXTFIG. 1.

Much time may be wasted and useless material obtained if peat samples are not taken in the most suitable spot (or spots) in the bog under investigation. Thus, at the outset, the main stratigraphical features of the bog must be ascertained by a number of trial borings. Peat-cuttings, if available, may be used with much profit. Peat is directly picked with forceps from the cleaned walls and wrapped in waxed paper. A few cubic centimeters or even less should be enough, particularly if the peat is highly disintegrated. The peat may also be stored in glass tubes of the model mentioned. These are easily filled by pushing them horizontally into the peat.

When a boring is to be made a big sod is removed from the surface of the bog with the spade. From the walls of the sod the first samples are taken, say from 2, 5, 10, 15, and 20 cm below the surface. Then the auger is put into the hole left by the sod and forced down into the peat. Meanwhile the handle of the auger should be kept turning slightly to the right (clockwise) to prevent the container from opening. When the desired depth is reached the container is opened and a good compact core obtained by turning the handle swiftly four to six revolutions to the left. It is then closed again by turning the handle twice to the right. The auger is finally pulled up out of the peat with a slight continued revolution to the right to be sure that the chamber will stay closed. The chamber is then opened, and the outer layer of the enclosed peat, which might have been contaminated with material from higher levels, is removed with the knife. Samples are then taken with the forceps from the interior of the core. The greatest caution is necessary to eliminate any risks of contamination. The spade should stand with its blade thrust into the ground near the boring-place. The lower end of the auger with the chamber is put through the handle of the spade while the samples are being taken.

It may sometimes be found profitable, when working with an auger of the larger model, to place a thin removable zinc lining inside the chamber. When the container is opened during the boring, the lining is filled. The lining with the peat core enclosed is then removed from the chamber and may be taken intact into the laboratory. By the proper use of zinc linings, a complete series of such cores from the surface to the bottom of the bog can be obtained.

Chemical Treatment of Peat Samples: —

Alkali Method: — The process of digesting peat by means of boiling with dilute alkalies was used by LAGERHEIM and adopted by L. VON POST as a standard method in pollen analysis. About half a cubic centimeter of the material to be analysed is placed on a slide and boiled in 10 per cent KOH or NaOH until most of the water has evaporated. (If the peat is enclosed in a glass tube a small amount should be taken from each end of the sample.) A few drops of glycerine are added to prevent drying. After repeated stirring a few drops of the material are transferred to another slide and a cover glass applied.

This method is simple and effective and still widely used. In case the peat is poor in microfossils coarse plant débris are removed by straining and finer particles, including the pollen grains, are concentrated by centrifuging (*cf. e.g.* RUDOLPH and FIRBAS 1924, p. 15). During recent years, the alkali method has been modified in the following way by VON POST (1933, pp. 524, 525):

In slowly decomposed kinds of peat, the smaller pollen grains may be hidden by leaves, rootlets, etc., and considerable trouble can result thereof. In such cases, however, there can be used, as done in the Geological Department of the University of Stockholm, a metal strainer (meshes about 0.25 mm) for removing the coarser components of the peat. The sample is boiled rapidly in a test-tube with a fairly large quantity of weak KOH solution and shaken repeatedly with the hot liquid until entirely dispersed in the latter. The emulsion thus obtained is poured on the strainer and the fluid collected into a glass container. Then the residue on the strainer is squeezed and stirred with a spatula, the strainer removed to another glass, and the liquid poured through it once more. This procedure is repeated a number of times until the residue on the strainer has been washed free satisfactorily from the finer substances. Finally, the suspension has to be condensed in a centrifuge, and analysis preparations to be made in the ordinary way. By this procedure, there may result such a concentration of the pollen content that analyses are easily made, even though the actual pollen-frequency of the soil is an extremely slight one. Of course, this method has less effect the more the peat is decomposed. As long as decomposition has not progressed too much it works well.

Practically the same technique has been adopted by B. POLAK (1933, p. 23).

If the peat is very poor in pollen, if it contains silica grains which are likely to obscure the pollen grains, and finally if the pollen grains are not well preserved, FAEGRI (1936) recommends staining with a fuchsin-methylene blue mixture (about ½ to ¼ per cent solutions):

„Die Pollenkörner werden intensiv lila gefärbt (*Salix*-Körner farben sich im allgemeinen bedeutend schwächer als diejenigen von *Betula, Alnus* usw., worauf man beim Analysieren aufmerksam sein muss, diejenigen von *Pinus* und *Cyperaceen* blaulila oder schmutzig-violett), *Sphagnum*-Blätter graugrün bis grau, Grundmasse der limnischen Feindetritussedimente braunviolett bis schmutziglila (aber weniger intensiv als die *Cyperaceen*-Pollenkörner), Radizellen bläulich-grünlich, Scheidenepidermen des *Vaginatum*-Torfes schwachlila, Sklerenchym-Fasern grau usw. *Sphagnum*-Sporen werden nicht oder nur schwach lila gefärbt, während Farnsporen eine lilablaue bis bräunliche Farbe annehmen".

Chlorination-Acetolysis Method: * — Peats and other kinds of fossil-bearing material are sometimes so poor in pollen that the modified alkali method, in spite of staining, centrifuging, etc., is very time wasting and sometimes completely useless. Apparently, the only helpful method in such cases is to dissolve as much as possible of the peat without damage to the pollen grains. Experiments along these lines were made in collaboration with Dr. H. ERDTMAN.

In these experiments, there were used certain " standard substances," such as ordinary peats, or other fossiliferous substances, which had been dried, ground in a mortar, and divided into very small parts by rubbing through metal screens. The homogeneity of the powder thus obtained allows direct comparison, both quantitatively and qualitatively, between the results obtained by analysis of any part of a standard substance. Consequently, the action of different chemicals can be objectively established.

Among the constituents of peat are cellulose and hemicellulose which on acid-hydrolysis are transformed into water-soluble products such as glucose. Other constituents of peat are lignin and humic acids (*cf.* chapter II), neither of which can be hydrolised by acids, but which may be easily destroyed by oxidation. The more resistant elements, such as pollen grains and spores, are thus isolated if all these elements are removed by means of combined hydrolysis and oxidation. For the oxidative destruction of lignin and humic acids, treatment with chlorine dioxide (SCHMIDT and GRAUMANN, Ber. Deutsch. Chem. Ges., vol. 54:2, 1921) is useful. The remainder of the peat is then treated with sulphuric acid (80 per cent) in order to hydrolyse the polysaccharides (*cf.* G. and H. ERDTMAN 1933). These procedures, however, have

* *Cf.* also BARGHOORN, E. S., and BAILEY, I. W., 1940: A useful method for the study of pollen in peat (Ecology 21: 513).

a serious drawback, *viz.* the length of time (15 hours and more) required in the chemical treatment of the peat. This led to continued experiments carried out by the senior author. During these experiments, it was found that quick and satisfactory oxidation may be obtained by using chlorine *in statu nascendi* and that the polysaccharides were quickly and easily broken down by means of acetolysis (ERDTMAN 1936*a* and *b*).

The chlorination-acetolysis method includes the following steps.

Powdering of the Peat: — Powdering is effected by rubbing dry peat against a brass or nickel screen with approximately 300 meshes per square cm (area of meshes about one third of a square mm). Wet peat is dried in vacuo over concentrated sulphuric acid. Heating, particularly to high temperatures, should be avoided because of the risk of forming insoluble humines. However, if wet peat has to be investigated before drying for one reason or another, it is placed on the sieve and kneaded through the meshes with the addition of water. The peat suspension is collected in a centrifuge glass, centrifuged, and the fluid decanted. The sediment is washed once or twice with glacial acetic acid and then treated as described below under "Chemical treatment." Peat wrapped in waxed paper is usually dry enough for powdering after a few days. Particularly good results are apparently obtained if the peat is powdered when it just has reached the exact stage of dryness which will permit it to be powdered easily. This may be done in the field, and the powder, ready for chemical treatment, stored in small test tubes. Much time may also be saved if the chemical treatment of the samples is done in the field, *e.g.*, during rainy days when outdoor work is impossible.

The advantages gained by powdering the peat are several. Because of the uniform size of the particles and the uniform content of water, powdered material is reacted upon more readily by the chemicals than are wet and unpowdered samples. Furthermore it is known that the frequencies of pollen grains and other microfossils may fluctuate considerably even within samples of a few cubic centimetres or less. Thus if part of a sample be picked at random for analysis it may be that, whether from the presence of whole stamens or sporangia or because of other reasons, the frequencies of the microfossils differ more or less from the frequencies correctly representing the horizon of the sample. This disadvantage is much reduced if the sample be powdered and thus made practically homogeneous in its pollen and spore content.

Chemical Treatment and Final Preparation: — The powdered peat (about 0.1 grm. would be enough) is chlorinated as described on p. 28. Bleaching is usually accomplished in half a minute or slightly more, depending on the amount and the physico-chemical properties of the peat. The chlorination mixture with the suspended peat is then centrifuged, the fluid decanted, and the sediment washed with glacial acetic acid (at least 5 c.c.). After centrifuging and decanting, the sediment is suspended in the acetolysis mixture (p. 28) which should fill the centrifuge tubes to the 10 c.c. (or, eventually, to the 5 c.c.) line. The procedure is then identical with that previously described (pp. 27–29). After the last centrifuging and decanting, a part of the sediment is taken with forceps and stirred on a slide with a mixture of 1 part glycerine and 1 part water. A cover-glass is then applied without pressure. Pressure may cause currents within the liquid, carrying the pollen grains — particularly the smaller ones — towards the edges or even outside the preparation, and thus cause a more or less serious source of error as shown by the table below (G. and H. Erdtman 1933, p. 357).

Table 4: Effect of pressure: —

Species and diameter (μ) of pollen grains	A (per cent)	B (per cent)
1. Mixture of *Corylus* (29.2) and *Pinus*	50 Co, 50 P	34 Co, 66 P
2. Mixture of *Fagus* (50.0) and *Myrica* (27.9)	67 F, 33 My	76 F, 24 My
3. Mixture of *Fraxinus* (27.5) and *Picea*	46 Fx, 54 Pc	28 Fx, 72 Pc
4. Mixture of *Fraxinus* (27.5) and *Myrica* (27.7)	30 Fx, 70 My	30 Fx, 70 My

In the four preparations the pollen percentages of which are shown in column A of the table, no material was squeezed out from below the cover glasses. In the four preparations in column B, the cover glasses were pressed against the slides and some of the pollen-bearing fluid pressed out at the edges of the cover glasses. Mixtures of fresh pollen, boiled with 10 per cent NaOH, and washed with water, were used. The dimensions quoted in the table are averages of ten measurements for each pollen species only.

Many kinds of peat, raw humus, swamp and lake mud (" dy ") as well as certain types of limnic ooze ("gyttja") may profitably be treated by means of the chlorination-acetolysis method. The pollen frequency is sometimes considerably increased if the chemical treatment is terminated by washing the acetolysed residue with cold ½ per cent KOH or NaOH for a few seconds. Prolonged action by alkali may seriously attack the pollen grains.

Sandy and Clayey Samples: — Sandy and clayey samples are usually subjected to special treatment. Their content of silica dioxide is removed either mechanically by separation with carbon tetrachloride or calcium chloride solution (Zetzsche 1932, pp. 311, 340), or chemically with hydrofluoric acid (*cf.* Assarsson and Granlund 1924). The last method seems to be the better one. Before treatment with hydrofluoric acid, the largest mineral particles should be disposed of by pouring alcohol on the powdered sample, followed by shaking and decanting. After centrifuging the alcohol-soaked sediment is suspended in hydrofluoric acid which fills a copper crucible to approximately one third. Here, the silica dioxide is gradually dissolved. Heating of the reaction mixture should be applied cautiously and stopped as soon as boiling begins. After cooling and dilution with water, the fluid is poured into copper tubes and centrifuged. The sediment is washed with hydrochloric acid (2 vols. acid, sp. gravity 1.19, 1 vol. water) after decanting, and the suspension transferred to an ordinary centrifuge glass, where it is centrifuged again and the fluid decanted. The sediment is then washed with distilled water, and finally with glacial acetic acid. After centrifuging and decanting, the material is bleached and acetolysed in the usual way.

Samples containing much calcium carbonate are soaked with alcohol and treated with dilute nitric acid. The residue is washed with distilled water or, in order to obtain a higher concentration of the pollen grains, with glacial acetic acid (followed by chlorination and acetolysis).

Vivianite with much silica dioxide is treated with hydrofluoric acid as described above. If there is only a slight admixture of silica, the powdered samples may simply be shaken with dilute hydrochloric acid, washed first with water, then with glacial acetic acid, and finally bleached and acetolysed.

Material that cannot be powdered by rubbing against a sieve — *e.g.* mud strongly compacted by drying and certain very compressed and dry substances, such as submersed peat (" Moor-log ") and certain inter- and preglacial peats etc. — must be ground in a mortar and the ensuing fine material worked through the sieve. If it cannot be ground, it may be softened by means of nitric acid (10 per cent or stronger, according to the character of the material), pressed against the net and kneaded through the meshes with the addition of water and then collected in a centrifuge glass and washed with glacial acetic acid before being bleached and acetolysed.

In investigations of diatomite, the shells of the diatoms may be rendered invisible by treating the samples with a fluid of the same refractive index as that of the shells, *e.g.* highly concentrated chloral hydrate solution (Gistl 1928). In alkaline preparations, any diatoms present are dissolved.

The Mechanical Dispersion Method: — A recently proposed method is that of McCulloch (1939) which does not depend upon chemicals for deflocculation. He recommends the following treatment:

1. Four-tenths of a gram of peat is placed in the cup of a mixing machine with 40 c.c. of distilled water and 3 drops of 1 per cent water-soluble safranin. The electric mixing machine is of the type commonly employed at soda fountains. The cup is fitted with baffles to facilitate dispersion. This equipment, which is used in the hydrometer method of mechanical analysis of soil, is described by Bouyoucos (1927): The hydrometer as a new and rapid method for determining the colloidal content of soils (Soil Sci., vol. 23, pp. 319–331).

2. The mixing cup is blocked up until the agitator is within 1 mm. clearance of the bottom of the cup. Ten minutes of agitation is usually ample to effect separation of the pollen. In strongly fibrous peat, however, 20 to 30 minutes may be required.

3. The suspension is then divided between two centrifuge tubes and centrifuged for ½ hour, at 3600 rotations per minute. With 20 g. of suspension in each tube, this gives a sedimentation force of approximately 57,000 times gravity at the tip of the tube.

4. Most of the supernatant liquid is decanted, and the mount is made from the surface of the sediment.

For sandy materials, McCulloch recommends a modified technique. The cup of the mixing machine is lowered until the agitator is ¼ inch from the bottom. The suspension is strained through a cheesecloth filter which retains the sand but through which the pollen grains can be washed with distilled water. The entire filtrate is then centrifuged again. In this way mounts, comparable to those from pure peat, can be obtained. McCulloch states that the pollen grains were sufficiently numerous on a microscope slide, and at the same time sufficiently free of adherent material, to permit ready detection.

Lake Investigations (*cf.* Lundqvist 1927): — The stratigraphy of lake sediments is ascertained by means of profile borings. The direction of the profiles is marked by lines provided with cork floats *e.g.* at every third meter. The lines must be steadily anchored to prevent their being shifted from their position by wind or currents. With a Hiller auger, borings are then made along the profile line at adequate intervals (from less than 1 m to 100 m or more, if the stratification obviously is very uniform). The difficulties met with in handling the heavy boring tool from a skiff or punt are considerable. In comparatively shallow waters the difficulties may be overcome, but in case of deeper water the boring tool must be handled from a pontoon. In countries where the lakes are frozen during winter the field work may be conducted to advantage during that season in order to obtain as great accuracy as possible. From the youngest sediments, including those in formation, samples may be obtained by means of different types of sound-leads (pipe-leads), etc.

Ice Investigations: — Pollen analysis in its original sense is founded on the fact that pollen grains may be incorporated and preserved in the peat of living bogs or in growing sediment accumulations of lakes, lagoons, or bays, etc. However, pollen grains may also be embedded in the snow in growing parts of glaciers and finally incorporated in the ice. Ehrenberg (Passatstaub und Blutregen, Bericht d. Akad. d. Wiss., Berlin 1849) encountered pollen grains in ice, but it was not until a few years (1932) ago that Vareschi found well preserved pollen grains in samples from a Swiss glacier and started a systematic investigation of glacier ice. Quantitative pollen analyses were carried out in order to obtain new facts for the elucidation of old problems, such as the stratification and movements of glacier ice, wind carriage of pollen grains, etc.

Samples were collected from six glaciers and in most cases pollen grains were so numerous as to make pollen analyses possible. Samples with no pollen at all were seldom encountered and, in some cases, samples were found with more than 5000 pollen grains per cubic decimetre. The mean absolute number of pollen grains per cubic decimetre varied from 350 in the Aletsch glacier to 1200 in the Gepatsch glacier in the Oetztal Alps. The pollen grains were well preserved, the largest (*Abies*, *Picea*), however, frequently being flattened and torn. All small size pollen grains were perfectly preserved, even such forms which are not generally considered to occur in a fossil state in peat or clay. The last statement is somewhat at odds with the detailed pollen records given by Vareschi, since pollen grains of *Juniperus* and *Populus* apparently have not been found in ice.

Collecting ice samples is effected as follows:

„Die Erfahrungen der Pollenanalyse dürfen natürlich nicht ohne weiteres auf die Eisanalysen übertragen werden. Schon zur Probeentnahme muss die dort übliche Arbeitsweise aufgegeben werden. Man benützt keinen Bohrer, sondern die zahlreichen Aufschlüsse wie Gletscheroberfläche, Spalten, Staffelbrüche, Randklüfte und zugängliche Teile der Gletschersohle. Das oberflächliche Eis ist dabei stets sorgfältig soweit abzutragen, als eine Verunreinigung von der Oberfläche her zu befürchten ist. Erfahrungsgemäss hat sich ergeben, dass die Pollen kaum durch das Haarspaltennetz ins Innere eindringen, und dass schon nach Abtragung von 1–2 dm Eis brauchbares Material gewonnen werden kann. Jede entnommene Probe soll nur Eis von durch und durch gleicher Beschaffenheit enthalten. Es muss deshalb die Kornstruktur innerhalb einer Probe ungefähr konstant sein, denn nur dann besteht die Wahrscheinlichkeit, dass alle Teile der Probe aus derselben Witterungsperiode stammen, das heisst, ungefähr gleich alt sind. Ebenso müssen die Farbe des Eises, die gleichmässige Verteilung der Verunreinigungen beachtet und blaue und weisse Bänder und Schlieren gesondert untersucht werden. Aber mit all diesen Vorsichtsmassregeln können wir nicht mehr erreichen, als die Wahrscheinlichkeit, dass die Probe wirklich in allen ihren Teilen gleichzeitig entstanden und gleichmässig verändert worden ist. Die Sicherheit über die Homogenität gibt erst die Pollenanalyse. Da stellt sich dann manche scheinbar reine Probe als Mischeis heraus: statt der Pollen einer bestimmten Jahreszeit findet sich ein Zufallsgemisch.

Bei jeder Probenentnahme wird notiert: Ort, Korngrösse, Schichthöhe, Beimischungen, Farbe, Charakter des umgebenden Eises und evt. Achsenrichtung der meisten Kristalle (bestimmt nach den Tyndallschen Schmelzfiguren). Die Eisstücke werden in reiner Wachsleinwand gesammelt und in Weithalsflaschen gefüllt. Als besonders zweckmässig hat sich die 2-Liter-Bülacher-Flasche mit Patentverschluss erwiesen. Ein bis zwei solcher Flaschen pro Probe genügen.

Nun kann die Probe zur weiteren Verarbeitung mitgenommen werden. Im nächsten Standquartier wird das Eis geschmolzen (Primuskocher!) und in einer grossen Handzentrifuge von ca. ½ Liter Fassungsraum und mindestens 2500 Umdrehungen pro Minute zentrifugiert. Der Bodensatz wird dann in kleine Fläschchen oder Stoffbehälter gefüllt, und so zubereitet kann man den Staubinhalt von mehreren hundert Litern Eiswasser bequem im Rucksack zu Tal tragen. Falls im Laboratorium nicht bald an die weitere Verarbeitung der Proben geschritten werden kann, muss man sie durch ein paar Tropfen einer Konservierungsflüssigkeit vor Zersetzung und Verpilzung schützen" (Vareschi 1935).

Microscopical preparations are made after treatment with hydrofluoric acid etc. The perfect state of pollen grains in ice makes it desirable to collect samples from ancient ice in Alaska, Siberia, the New Siberian Isles, etc. From studies of such pollen much information concerning the former distribution of many plants may be obtained.

Investigations of Brown-Coal: — According to Bode (1931) brown coal may be prepared for pollen analysis in the following way:

„Eine kleine, nicht zu stark zerkleinerte Kohlenprobe, etwa 5 g., wird am besten in einer weithalsigen Flasche ungefähr einen Tag lang mit Salpetersäure von der Konzentration 1:6 bis 1:10 behandelt. Danach wird gut ausgewaschen und in verschlossener Flasche etwa 12 bis 24 h mit Ammoniak (oder besser verdünnter Kalilauge) ausgezogen. Nach sorgfältigem Auswaschen der Humusbestandteile (durch Dekantieren und längeres Absitzenlassen) bleiben nur die bituminösen Bestandteile zurück, darunter die Pollen. Durch Ausschleudern in der Zentrifuge kann man die Präparate leicht vom Wasser befreien und in Glycerin übertragen".

The following method is recommended by Potonié (with additions by Willrath; cited from Willrath in Potonié 1934, pp. 108, 109; *cf.* also Potonié 1931, p. 6):

„Auf eine bestimmte Gewichtsmenge (ca. 1 g.) der zerkleinerten lufttrockenen Braunkohle giesst man zunächst HNO_3 konz. und beobachtet die Reaktion. Erfolgt ein sofortiges kräftiges Aufbrausen und Entweichen roter Dämpfe, so wird die HNO_3 mit Wasser solange verdünnt, bis ein Überschäumen der Probe nicht mehr möglich ist. Tritt keine starke

Reaktion ein, kann mit HNO_3 konz. weiter gearbeitet werden. Nach ca. 1½–2 Std. prüft man, ob die Kohle genügend oxydiert ist. Ein kleiner Kohlenbrocken wird auf den Objektträger gebracht und 1 Tropfen $\frac{n}{1}$ KOH hinzugefügt. Unter dem Mikroskop wird beobachtet, ob Pollen oder Cuticulen zu sehen sind. Löst sich der Kohlenbrocken nicht, so ist die Mazeration noch nicht genügend fortgeschritten und man lässt die Probe weiterhin stehen. Andernfalls wird dekantiert. Die Mazerationsdauer ist je nach dem Inkohlungsgrad grossen Schwankungen unterworfen, z.B. bei Weichbraunkohlen einige Stunden, bei Flammkohlen aus dem Ruhrgebiet ca 2 Tage, bei höher inkohlten oft bis zu 14 Tagen. Ist sie beendet, dann gibt man $\frac{n}{1}$ KOH hinzu und lässt ca ¼–1 Std. stehen, bis die oxydierte Substanz genügend gelöst ist. Um ein schnelleres Absetzen der feinen Teilchen herbeizuführen, wird zentrifugiert. Im Zentrifugenglas wird das jetzt ziemlich fest auf dem Boden sitzende Gut solange ausgewaschen, bis keine alkalische Reaktion mehr auftritt. Der Rest wird mit Glyzerin-Gelatine gut ausgeschüttelt und in ein Glasröhrchen gegeben, welches bei den Rohproben in der Sammlung verbleibt. Mit Ammoniak lassen sich die aus der Kohle durch die Einwirkung der HNO_3 gebildeten Huminsäuren ebensogut herauslösen wie mit KOH oder NaOH. Zur Verhütung einer Entfaltung bakteriellen Lebens muss aber dem Ammoniak etwas Formalin beigefügt werden."

„Zur Herstellung eines Dauerpräparates entnimmt man (um möglichst vergleichbare Fossilienzahlen zu erhalten) mit Hilfe eines Glasstabes bestimmter Dicke einen Tropfen des in Glyzeringelatine aufgeschwemmten Mazerationsproduktes und bringt ihn auf ein Deckglas, welches dann vorsichtig auf den Objektträger gelegt wird. Weist das Präparat Luftblasen auf, so wird über kleiner Flamme erwärmt. Nach Möglichkeit ist ein Erwärmen des Präparates jedoch zu vermeiden, da die gleichmässige Verteilung der Pollen im Präparat darunter leiden könnte, wodurch diese eventuell für pollenanalytische Zwecke unbrauchbar werden. Das Präparat wird dann mit dem Deckglas nach unten zum Erkalten hingelegt, damit die Pollen ans Deckglas gelangen; so wird eine Untersuchung mit starkem Objektiv ermöglicht. Gefärbte Präparate werden hergestellt, indem man dem Mazerationsprodukt gleich nach dem Zentrifugieren einige Tropfen z.B. alkoholischer Fuchsinlösung hinzufügt".

However, the character of pre-quaternary pollen- and spore-bearing material is so varied that other methods must frequently be employed. Certain coals may be treated with pyridine in order to soften the matrix and facilitate subsequent treatment. The pyridine is removed from the coal by washing with dilute hydrochloric acid followed by water (Raistrick 1934). On certain occasions, it may be advisable to powder the coal in order to assure rapid and uniform action of the chemicals. Bleaching may be effected by gently heating coal powder in a test tube with concentrated nitric acid to which have been added a few crystals of sodium chlorate. Further methods have been described by Zetzsche and Kälin (1932) and others.

Single-Grain Preparations (*cf.* Faegri 1940): — In pollen analysis pollen grains are sometimes found, which should be kept permanently for later reference. This may be done by sealing the preparation and marking off the position of the grain desired. This procedure has two disadvantages: the sealing is never absolutely effective and even if the preparation does not deteriorate the pollen grains soon move away from their original position and are lost. In other cases, a drawing or a description of the pollen grain in question can be made, but unless very skilfully done, neither of them is of much value except as a mental aid to the person who made them.

However, it is not difficult to isolate single pollen grains from an ordinary preparation. Usually, the pollen grain type in question is very scarce; there may be only one specimen or a few available. In such cases the position of the grains must be marked so that they can be found at the end of the analysis. During the operations outlined below it is essential to keep the pollen grain under continual microscopic control.

The first process consists of gently pushing the cover glass away. A hooked needle is useful for this purpose. If the preparation contains too much liquid, especially if the liquid is not sufficiently viscous, it is difficult to achieve good results without draining off part of the liquid by means of blotting-paper. If the pollen grain adheres to the cover glass and is pushed outside the slide, it can be removed from the under side of the cover glass. However, it is easier to mark off the position of the grain on the upper surface of the glass with ink, remove the glass completely and then place it upside down on another slide.

When the pollen grain has been uncovered, it must be isolated on the slide or cover glass. The surrounding substance is removed by means of a preparation needle. It is necessary to remove also the remaining liquid, as a floating grain can be " fished out " only with difficulty. The appearance of a sharp marginal shadow indicates suitable conditions.

The next step is the " fishing-out." A very small piece of glycerin jelly is taken on the tip of the preparation needle and placed on the pollen grain which usually adheres immediately. It can then be removed to another slide and more glycerin jelly added; the preparation is then ready for the cover glass. The best preparation needle for the purpose is an ordinary sewing needle the end of which has been ground to the shape of a lancet.

The only difficulty in the whole process arises at the moment the grain passes the edge of the cover glass or slide. Owing to the heavy marginal shadow the grain cannot be observed and, if the quantity of liquid be too great, the grain is likely to float away along the edge.

When the pollen grain in question is rather frequent in the sample, it is desirable to make a special preparation without a cover glass for the purpose of isolating single grains. The whole procedure then consists of clearing away the surrounding substance and " fishing out " the grain. It is also recommended to take more than one grain, inasmuch as the grains cannot be turned around in permanent preparations and, consequently, only one view is available with a single grain.

The method of single-grain preparations is of use in the following cases: *1.* For documentation. Sometimes the discovery of certain pollen grains is of great phytogeographical significance, even if they occur singly. Examples of such grains are those of *Ilex* or *Hedera* in the marginal areas of or outside the regions of their present distribution. Such grains should be isolated and their preparations kept in a botanical collection for the purpose of future reference.

2. For identification. Sometimes, unknown pollen grains occur in such quantities and under such circumstances that it is desirable to identify them. A single-grain preparation, but preferably one with a number of grains, can be kept for future revision or can be sent to colleagues or specialists for determination.

3. For comparison. It is often obvious that fossil grains give a better impression of the constitution of the pollen grains than artificially "fossilised" grains. Even though a reference collection can hardly be made up exclusively of such preparations, it is nevertheless desirable that the more important types should be represented in this condition. The beginner's work would be much facilitated in this way.

4. Finally, abnormal pollen grains may be retained for possible reference.

References: —

Assarsson, G. och Granlund, E., 1924: En metod för pollenanalys av minerogena jordarter (Geol. Fören. Förhandl., vol. 46).

Bode, H., 1931: Die Pollenanalyse in der Braunkohle (Internat. Bergwirtsch. u. Bergtechn., 24. Jahrg.).

Erdtman, G., 1935: Pollen statistics, a botanical and geological research method (Pp. 110–125 *in* Wodehouse, R.: Pollen grains, New York).

— — 1936*a*: New methods in pollen analysis (Svensk Bot. Tidskr., vol. 30).

— — 1936*b*: Neue pollenanalytische Untersuchungsmethoden (Ber. geobot. Inst. Rübel f. 1935).

Erdtman, G. and H., 1933: The improvement of pollen-analysis technique (Svensk Bot. Tidskr., vol. 55).

Faegri, K., 1936: Einige Worte über die Färbung der für die Pollenanalyse hergestellten Präparate (Geol. Fören. Förhandl., vol. 58).

— — 1940: Single-grain pollen preparations. A practical suggestion (*Ibid.*, vol. 61, 1939).

Gistl, R., 1928: Die letzte Interglazialzeit der Lüneburger Heide pollenanalytisch betrachtet (Bot. Arch., vol. 21).

Lundqvist, G., 1927: Bodenablagerungen und Entwicklungstypen der Seen (Die Binnengewässer, vol. II).

McCulloch, W. F., 1939: A postglacial forest in central New York (Ecology, vol. 20).

Polak, B., 1933: Über Torf und Moor in Niederl. Indien (Verh. Akad. Wet. Amsterd., vol. XXX:3).

von Post, L., 1933: On improvements of the pollen-analysis technique (Geol. Fören. Förhandl., vol. 55).

von Post, L. och Granlund, E., 1926: Södra Sveriges torvtillgångar, I (Sveriges Geol. Unders., ser. C. no. 335).

Potonié, R., 1931: Zur Mikroskopie der Braunkohlen. Tertiäre Blütenstaubformen (Braunkohle, H. 16).

— — 1934: Zur Mikrobotanik der Kohlen und ihrer Verwandten, II. Zur Mikrobotanik des eocänen Humodils des Geiseltals (Arb. Paläobot. u. Petrogr. d. Brennsteine, vol. 4).

Raistrick, A., 1934: The correlation of coal-seams by microspore-content (Trans. Inst. Min. Engin., London, vol. LXXXVIII).

Rudolph, K. und Firbas, F., 1924: Die Hochmoore des Erzgebirges (Beih. Bot. Centralbl., vol. II:XLI).

Vareschi, V., 1935: Blütenpollen im Gletschereis (Ztschr. f. Gletscherkunde, vol. XXIII).

— — 1942: Die pollenanalytische Untersuchung der Gletscherbewegung. (Veröffentlichungen des Geobotanischen Instituts Rübel in Zürich 19).*

Zetzsche, F., 1932: Fossile Pflanzenstoffe (Pp. 293–344 *in* Klein, G.: Handbuch der Pflanzenanalyse, vol. 3, 1).

— — und Kälin, O., 1932: Eine Methode zur Isolierung des Polymerbitumens (Sporenmembranen, Kutikulen usw.) aus Kohlen (Braunkohle, H. 21, 22).

* Of this important recent publication of which no copies are yet available in the New World Rieche reports as follows in a number of *d. Biologe* which reached us a few days before this book went to press: "Diese Arbeit stellt eine große Leistung und einen ebenso großen Fortschritt der durch Beobachtung gewonnenen Erkenntnisse auf dem Gebiete der Gletscherkunde dar und interessiert in biologischer Hinsicht insofern sehr, als in ihr die paläobotanische Arbeitsmethode der Pollenanalyse mit überraschendem Erfolg auf ein Spezialgebiet der Geologie angewandt wird. Die bei Torfuntersuchungen entwickelte Methode der Pollenanalyse wurde von Vareschi (von den Botanischen Staatsanstalten in München-Nymphenburg) sinngemäß für die Eisanalyse abgewandelt, so daß an Hand der gewonnenen Pollendiagramme ein einwandfreier Einblick in den Aufbau der Eismassen aus Eisteilen mit verschiedenen Pollenaspekten gewährleistet wird. Die Untersuchung der aspektstratigraphischen Bilder verschiedener Gletscherteile führte zum Nachweis der Jahresschichtung im Gletschereis bzw. zur Feststellung von Störungen der ursprünglichen Schichtung in anderen Gletscherteilen und gipfelt darin, daß durch praktische Beobachtung nachgewiesen wird, in welchem Maße und in welchem Bereiche die beiden großen, einander widersprechenden Gletscherbewegungstheorien von S. Finsterwalder und H. Philipp zu Recht bestehen. Damit ist auf diesem strittigen Gebiet endlich ein entscheidender Schritt von der theoretischen Spekulation, die bislang von noch nicht genügender Beobachtung ausging, zur Empirie getan und es ist eine neue, äußerst fruchtbare Diskussionsgrundlage geschaffen, da es jetzt möglich ist, für ein bestimmtes Eisstück im Gletscher eine klare Aussage zu machen, ob es sich in seiner ursprünglichen Lage befindet oder nicht. Die von Vareschi eingeführte Methode brachte mit ihrer empirischen Klärung der Frage, wie sich ein Gletscher bewegt, einen sehr zu beachtenden Fortschritt und verspricht für künftige glaziologische Untersuchungen, zu denen damit herausgefordert ist, wesentliche Erfolge." (F. V.)

Chapter V

POLLEN AND SPORE MORPHOLOGY

Terminology: — In the field of pollen analysis, two tendencies in connection with pollen morphology can be traced: *a.* some workers describe pollen grains singly as they see them and frequently use but a few, if any, conventional terms; *b.* others avail themselves of a more or less complicated nomenclature in an effort to build up an orderly system of pollen description and morphology.

R. POTONIÉ is the chief representative of the last group, after his important "Zur Morphologie der fossilen Pollen und Sporen" (1934). It appears that a nomenclature, of the type proposed by POTONIÉ, can be used to advantage only when the actual construction of the pollen grains is known in some detail (RUDOLPH in Beih. Bot. Centralbl., vol. LIV, 1935). We are still, however, far from attaining that goal. Microtome technique has not been systematically applied to the study of fossil pollen and spores or to the study of recent material artificially brought into a more or less "fossil" condition. Until sufficient work has been done along these lines, it is more reasonable to allow ourselves a less elaborate terminology for salient features of pollen and spore morphology, particularly those features which are readily interpreted without the aid of microtome sections.

As to the terms "pollen" and "spore", it is necessary to keep in mind that they are not homologous — the microspore is the immediate product of division of the mother cell, the pollen grain contains within its wall the microgametophyte developed from the microspore. In view of the fact that pollen analysis commonly deals also with spores, there would seem to be no need for a separate term, "spore analysis".

The majority of pollen grains have two coats — an outer, the exine, and an inner, the intine (FRITZSCHE 1837, p. 28). These terms may also be used in the description of the spores of mosses and ferns instead of the longer terms "exosporium" and "endosporium". The intine and its eventual subdivisions will not be considered here for they are not found in the fossil state, at least not in peat, clay, browncoal, etc. The exine, as pointed out by FRITZSCHE (*l.c.*), frequently consists of two layers, but the subdivision of the exine (*sensu lat.*) into exine (*sensu str.*) and intexine, as proposed by him, may lead to confusion. Therefore, some authors have used the term exoexine for exine *sensu stricto*. To avoid this unwieldy term as well as to have a convenient term in case the exine should exhibit three layers instead of two, we may speak of ektexine (for exine *sensu stricto*, or exoexine), mesexine (or a layer between the outermost and the innermost exine layers), and endexine (for intexine). The outer surface of the exine may sometimes be provided with some kind of sculpturing or ornamentation. The ornamentations of sculptured pollen grains are exceedingly varied: the exines may be echinate, reticulate, etc. (see glossary, p. 49). Unsculptured grains, lacking spines or projections of any kind, are known as psilate (WODEHOUSE 1935).

Pollen grains may broadly be grouped in three classes according to their general shape: tricolpate radiosymmetrical grains with three furrows; monocolpate bilateral grains with one furrow; and acolpate grains without furrows. Tricolpate grains are found mainly in dicotyledons, monocolpate grains in monocotyledons and gymnosperms. Acolpate are rarer than monocolpate grains; they occur in gymnosperms and angiosperms, both monocotyledons and dicotyledons.

The spores of bryophytes and pteridophytes may also be grouped into three classes: trilete radiosymmetrical spores with a triradiate tetrad scar; monolete bilateral spores with a single unbranched scar; and alete spores without scars. Spores belonging to different classes may occur in the same family (*e.g. Polypodiaceae*) or even in the same genus or species (*e.g. Isoëtes*, where the megaspores are trilete and the microspores monolete).

Pollen grains of the dicotyledonous type — tricolpate grains — are produced in fours by pollen mother-cells. The two necessary nuclear divisions take place in rapid succession, almost simultaneously, at right angles to each other. The daughter nuclei usually tend to take up positions as far from each other as possible within the confines of the pollen mother-cell. This results in their being arranged tetrahedrally. Subsequently, ridges grow inwardly from the wall of the pollen mother-cell, dividing it into four spaces, each of which corresponds to a pollen grain cell. The polar axes of pollen grains of this type are defined as lines extending through the centres of the grains and directed towards the centre of the tetrad, where all four would meet if extended. Thus, each pollen grain has an inner, or proximal, and an outer, or distal, pole and a proximal and a distal half, meeting at the equator of the grain. In the tetrad with the daughter-cells still in close contact with one another, the proximal part of each grain has three contact areas with its neighbours. Each contact area with its extension to the equator and further on to the distal pole is provided with a central furrow, or colpa. These furrows cross the equator at right angles and are, therefore, called meridional furrows. Tricolpate grains have four planes of symmetry, one transverse, coincident with the equatorial plane, and three vertical (longitudinal), extending from the furrows past the polar axes to the wall opposite the furrows.

The shape of these grains is that of an ellipsoid of revolution with the polar axis as axis of rotation. In polar view, the outline of the grain is circular, in some cases somewhat triangular, with furrows at the angles. In equatorial view, the outline is elliptical. With the polar axis comprising the major axis of the ellipse, the pollen grains with decreasing eccentricity may be termed perprolate, prolate, subprolate, prolate spheroidal, and spherical, while those — in case the polar axis comprises the minor axis of the ellipse — with increasing eccentricity may be termed spherical, oblate spheroidal, suboblate, oblate, and peroblate (compare TEXTFIG. 2 and TAB. 5 where also the suggested relations between polar axis and equatorial diameter have been given). The details of this terminology have been discussed with Dr. PAUL RICHARDS (Trinity College, Cambridge) and some outstanding English mathematicians to whom the writer is indebted for valuable suggestions.

Table 5: SHAPE CLASSES AND SUGGESTED RELATIONS BETWEEN POLAR AXIS (P) AND EQUATORIAL DIAMETER (E): —

SHAPE CLASSES	P : E
Perprolate	> 2 (> 8:4)
Prolate	2–1.33 (8:4–8:6)
Subprolate	1.33–1.14 (8:6–8:7)
Spheroidal	1.14–0.88 (8:7–7:8)
prolate spheroidal	1.14–1.00 (8:7–8:8)
oblate spheroidal	1.00–0.88 (8:8–7:8)
Suboblate	0.88–0.75 (7:8–6:8)
Oblate	0.75–0.50 (6:8–4:8)
Peroblate	< 0.50 (< 4:8)

In single tricolpate grains, it is almost always impossible to make any distinction between the two poles. However, FIG. 223, PL. XIII, shows a young pollen grain of *Trapa natans*, where it was possible to make a distinction because of a distinct triradiate scar denoting the last place of contact between the four daughter-cells of a tetrad.

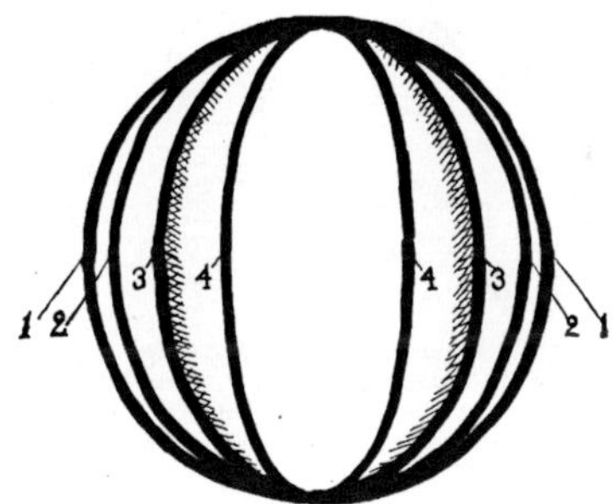

TEXTFIGURE 2. — SHAPES OF RADIOSYMMETRICAL POLLEN GRAINS. — With the polar axis of the grains (vertical in the diagram) comprising the *major* axis of the ellipse: 1 by 1, spherical; between 1 and 2, prolate spheroidal; between 2 and 3, subprolate; between 3 and 4, prolate; 4 by 4, perprolate. — With the polar axis of the grains (horizontal in the diagram) comprising the *minor* axis of the ellipse: 1 by 1, spherical; between 1 and 2, oblate spheroidal; between 2 and 3, suboblate; between 3 and 4, oblate; 4 by 4, peroblate.

Diagrammatic outlines of tricolpate grains in four different positions are presented in TEXTFIG. 3: V.

Pollen grains of the monocotyledonous type — monocolpate grains — are also produced in fours by pollen mother-cells. The two nuclear divisions take place successively and the resulting grains are usually arranged in a single plane and not in tetrahedral tetrads. The grains are typically bilateral, boat-shaped, provided with two planes of symmetry, one of which extends from " prow to stern ", the other from " port to starboard." The keel corresponds to the proximal part of the grain. The opposite distal part is provided with a single longitudinal furrow, usually dividing the deck in equal portions. The waterline represents the equator of the grain and a mast shipped on deck at the intersection of the two symmetry planes would form an elongation of the polar axis of the grain. (However, the terms equator and polar axis are but seldom used in connection with monocolpate grains.) In descriptions of monocolpate grains, it is customary to speak of their distal and proximal part, their outline in lateral (apical or transversal) view, etc. (compare TEXTFIG. 4: VIII).

In addition to tri- and monocolpate grains — already described — and acolpate grains (which need no special description), many other pollen types have been distinguished: di-, tetra-, hexa-, octo-, nona-, dodeca-,

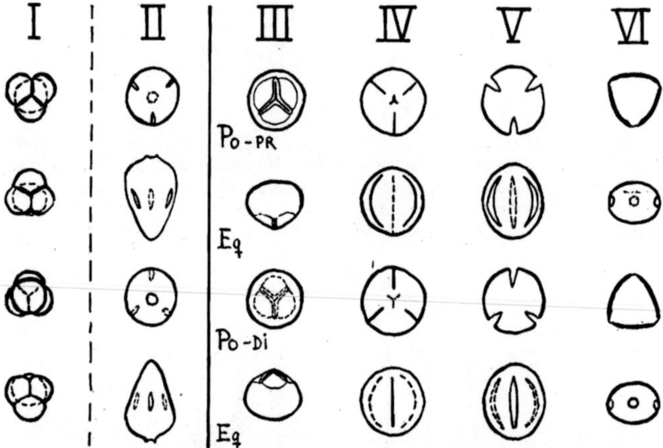

TEXTFIGURE 3. — POLLEN AND SPORE TYPES. — The pollen grains and spores are drawn in four different positions. The drawings are arranged in six columns, *viz.* I, tetrahedral tetrad; II, *Carex* pollen; III, trilete spore; IV, tricolpate pollen (*Trapa* pollen with triradiate scar); V, tricolpate pollen; VI, three-pored pollen, *e.g. Engelhardtia.* — Po, polar view; Eq, equatorial view; Pr, proximal part; Di, distal part.

pentadeca-, and triaconta-colpate grains. In hexacolpate grains, the colpae are equivalent in number and orientation to the six lines of contact between the four triangles corresponding to a tetrahedron; those of a dodecacolpate grain to the lines of contact between the six squares corresponding to a cube; and those of a triacontacolpate grain equivalent to the lines of contact between the twelve pentagons corresponding to a pentagonal dodecahedron; etc. (WODEHOUSE *l.c.*).

The pores in cribellate grains (*i.e.* grains with a varying number of pores more or less uniformly scattered over the surface) are generally considered to be short germinal furrows, rounded in form and coinciding in size with their enclosed germ pores (exits). Further information regarding these and other types is given by FISCHER (1890) and WODEHOUSE (*l.c.*).

The pollen grains of many *Cyperaceae* are quite aberrant. Here such terms as proximal, distal, etc. are not applicable, at least not in the same sense as used above, since these pollen grains are not homologous with ordinary pollen grains but with pollen mother-cells instead. After the nuclear divisions within the pollen mother-cell, three of four daughter nuclei are pressed down into the thick intine of the apex of the more or less tetrahedral grains and finally degenerate, a feature which was observed earlier by ELFVING (1878). Thus the wall of the mature grain is nothing but the wall of the pollen mother-cell. TEXTFIG. 3: II shows the probable correlation between this type of sedge pollen grain and a tetrahedral pollen tetrad.

The trilete spores of mosses and ferns are produced in fours by spore mother-cells and, as the nuclear divisions are similar to those in

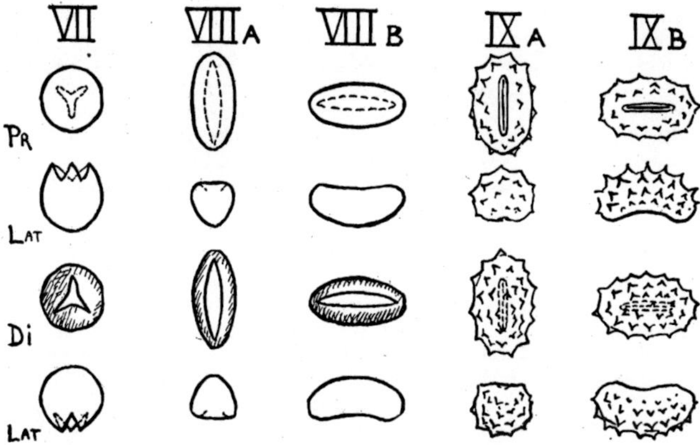

TEXTFIGURE 4. — POLLEN AND SPORE TYPES (*contd.*). — The pollen grains and spores are figured in four different positions. The drawings are arranged in five columns, *viz.* VII, pollen with three-slit opening (*e.g. Johnsonia*); VIII A and VIII B, monocolpate pollen grains (longitudinal and transverse positions); IX A and IX B, monolete spores (longitudinal and transverse positions). — Pr, proximal part; Di, distal part; Lat, lateral view.

the pollen mother-cells of plants with tricolpate pollen, the daughter-cells are arranged in tetrahedral tetrads. The proximal part of every spore has three contact areas (*areae contagionis*, WICHER 1934). These areas touch each other laterally along three lines, which meet at the proximal pole, forming a triradiate streak which is usually known as the tetrad scar. Each arm of the scar is severed longitudinally by a dehiscence fissure. Trilete spores have three planes of symmetry, all meridional, one for each dehiscence fissure (TEXTFIG. 3: III).

Monolete spores are similar to monocolpate pollen grains in several respects. However, monolete spores resemble closely boats turned upside down; the keel corresponds to the distal part, while the proximal part is provided with a straight longitudinal streak which is severed by a dehiscence fissure. When describing monolete spores, as well as trilete, it is customary to speak, just as in monocolpate pollen grains, of their distal and their proximal part, of their shape in lateral (apical or transverse) view, etc. (compare TEXTFIG. 4: IX).

Not only spores but also pollen grains may be trilete. Some pollen grains (*e.g.* in *Trapa natans*) may, at a certain stage of development, present a triradiate scar, which later disappears. But, on the other hand, some pollen grains (especially among the pteridosperms and possibly also among other classes of extinct spermatophytes) were provided with a permanent triradiate scar and did not develop any colpae at all. For this reason, it is impossible in many cases — at least at the present stage of our knowledge — to decide whether a spore *sensu lat.* is a pollen grain or a spore *sensu str.*

It should also be emphasized that it sometimes, particularly when dealing with old and poorly preserved material, may be difficult to make a distinction between monocolpate pollen grains and monolete spores, or to decide whether a certain grain be a pollen of *Nuphar*-type (FIG. 257, PL. XV) or a spore of the *Dryopteris thelypteris*-type (FIG. 482, PL. XXVIII), or again if it be a palm or a lily pollen grain with a three-slit opening or a fern spore with a triradiate scar (compare TEXTFIGS. 4: VII and 3: III).

When quoting the dimensions of pollen grains and spores, all possibility that may lead to a misinterpretation must be avoided. In radiosymmetrical grains, the size is expressed simply by quoting the length of the polar axis and the equatorial diameter. In monocolpate grains, on the other hand, the length may be expressed as the distance between the extreme points of a central longitudinal section running in the same direction as the *colpa* (furrow). The maximum breadth is usually equal to the distance between the extreme points of a central transversal section through the grain or spore. When speaking of winged conifer pollen grains, a special terminology should be used. The width of a fully expanded grain (a figure which, incidentally, usually does not seem to be of much diagnostic value) may be defined as the distance between the extreme parts of the two opposite wings. The width of the body (*i.e.* the distance between the two points where the proximal root of the bladders meet the body) is more reliable as a diagnostic character. The breadth of the body and wings can only be measured in grains in polar view. Their height is measured in grains in end view with both bladders fully exposed. The height of the body is identical with the length of the polar axis, while the height of the bladder is identical with the length of a perpendicular line stretching from the convex extremity of the bladder to the endexinous floor constituted by the body. Both figures are of minor importance. Several measurements concerning winged conifer pollens hitherto published are of no value since there are no precise descriptions regarding the way the actual measurements were made.

The process of the measurement of pollen grains under the microscope will not be dealt with here. However, attention may be drawn to a method of measuring without the aid of a microscope (KÖHLER 1933, pp. 15–22; *cf.* also MECKE 1920). Its value in calculating the size of pollen grains and spores is not as yet thoroughly tested. It seems to be particularly suitable in dealing with very small, isodiametric bodies. If a number of rays of parallel light are allowed to fall upon a slide with the bodies to be measured evenly distributed, more or less clearly defined interferencial rings will appear to an observer closely examining the slide from behind. The room should be dark and the slide held or fastened onto a screen provided with a hole of about the same size as an ordinary round cover glass. By measuring the interferencial rings, the average radius of the pollen grains or spores is calculated by means of the following formulae:

a) $r_1 = \frac{1.220 \cdot \lambda}{2 \cdot \sin \Theta_1}$ b) $r_2 = \frac{2.233 \cdot \lambda}{2 \cdot \sin \Theta_2}$ c) $r_3 = \frac{3.249 \cdot \lambda}{2 \cdot \sin \Theta_3}$

Θ_1, Θ_2, and Θ_3 denote the angular distance from the luminary to the limit between red and blue. The indices denote different interferencial rings. λ is the wave length. For white light, λ is considered equal to 0.571 μ. Some of the results obtained by KÖHLER are as follows:

Spores of *Lycopodium* (radius in μ; species not stated): —

Average of interferencial measurements: 15.231 ± 0.181; number of spores 486.
Averages of measurements under the microscope: 15.229 ± 1.151; 109 spores.
14.915 ± 0.865; 100 spores.

Spores of *Ustilago maydis* (radius in μ): —

Averages of interferencial measurements:

a) ring 1: 3.667 ± 0.043; number of spores 141.
b) ring 2: 3.646 ± 0.038; " " " 130.
c) ring 3: 3.650 ± 0.047; " " " 59.

Averages of microscopical measurements:

3.717 ± 0.286; number of spores 100.
3.674 ± 0.278; " " " 100.
3.627 ± 0.250; " " " 90.

In spite of its extreme simplicity, this method has not received as much attention as it deserves. It ensures results which, at the same time, are rapidly obtained and, what is more, entirely reliable. Tedious and time-wasting measurements under the microscope can be replaced by a simple observation.

The chief terms in pollen and spore morphology which are used in this book are arranged alphabetically according to the Latin names and briefly explained in the following glossary which is chiefly an excerpt from POTONIÉ (1934) and WODEHOUSE (1935).

Glossary: —

Aequator: equator; the great circle midway between the two poles of radiosymmetrical pollen grains, dividing the pollen grains into polar hemispheres.

Aletus: alete; spores without a tetrad scar.

Annulus: exine area immediately surrounding a pore and differing from the general surface of the exine.

Arci (sing. *arcus*): band-like parts of the exine, extending in sweeping curves from pore to pore (*cf.* TEXTFIG. 7, p. 69).

Area contagionis (plur. *areae contagionis*): contact area, "Kontakthof," "Pyramidenfläche"; following the simultaneous divisions of the nucleus of the spore mother-cell, the spores are in contact with each other along the contact areas. Each spore has three such areas. The contact areas meet at the proximal pole of the spore. They are laterally separated by the tetrad scar and, as a rule, distally limited by more or less curved lines or fringes (*curvaturae*).

Aspidatus: aspidate; bearing aspides.

Aspis (plur. *aspides*): a shield-shaped area surrounding a germ pore. Aspidate germ pores protrude as rounded domes. The protrusions are due to a thickening of the intine underlying the region of the pore and sometimes also to a lesser annular thickening of the exine.

Colpa (plur. *colpae*): germinal furrow; a longitudinal groove or opening in the exine of a pollen grain either enclosing a germ pore or serving directly as the place of emission of the pollen tube.

Colpa transversalis: transverse furrow; an elliptical or elongated opening in the intexine, underlying the true furrow and with its long axis crossing that of the latter at right angles. There may be any number of transitions between transverse furrows and germ pores underlying ektexinous furrows.

Colpatus: colpate; possessing germinal furrows. Generally used with numerical prefixes as mono-, di-, and tri-, signifying the number of furrows.

Cribellatus: cribellate; possessing a number of rounded pores more or less equally spaced.

Cristae: crests; different kinds of sculptural elements may join laterally to form rather intricate ridges or crests.

Curvaturae: see *Area contagionis.*

Distalis: distal; that part of a pollen grain or spore which is turned outward in its tetrad. In monopored or monocolpate grains, it is the side upon which the pore or furrow is borne. In other grains, the distal and proximal sides are generally not distinguishable from each other after the tetrad has been broken up into individual grains. In spores, the distal side is opposite the tetrad scar (= ventral, WODEHOUSE).

Dyas: dyad; pollen grains united in pairs (*cf.* PLATE III, FIG. 29).

Echinatus: echinate; see *Ornamentatio.*

Ektexinium: ektexine; the outer of the two main layers of the exine.

Endosporium: intine.

Exinium: exine; the outer, very resistant layer of the pollen (or spore) wall.

Exitus: exit; germinal aperture, a hole in the furrow membrane through which the germ pore — the place of emergence of the pollen tube — protrudes.

Exolamella: see *Ornamentatio.*

Exosporium: exine (fern spores excepted).

Fissura dehiscentis: fissure of dehiscence; a central longitudinal fissure in the scar of monolete spores; also the three-armed fissure in the tetrad scar of trilete spores.

Intercolparis: intercolpar; situated between the furrows.

Interlacunaris: see *Lacunae.*

Intinium: intine; the inner, slightly resistant layer of a pollen or spore wall. At the germination of a pollen grain the intine protrudes, forming the membrane of the pollen tube.

Lacunae: if the meshes (*lumina*) of a reticulum are large and arranged regularly, they may be termed *lacunae*. The *lacunae* are separated by interlacunar ridges (*muri*) or crests (*cristae*).

Lumen (plur. *lumina*): the space between the ridges (*muri*) of a *reticulum*.

Mesexinium: mesexine; denotes, topographically, a layer between the ektexine and endexine though it may be formed by (and possibly should be referred to as) these layers or one of them.

Monocolpatus: monocolpate; having a single germinal furrow. If the grain is encircled by a single furrow it is regarded as dicolpate or zonate.

Monoletus: monolete; a spore with a single straight tetrad scar.

Muri: low ridges separating the *lumina* of an ordinary *reticulum*.

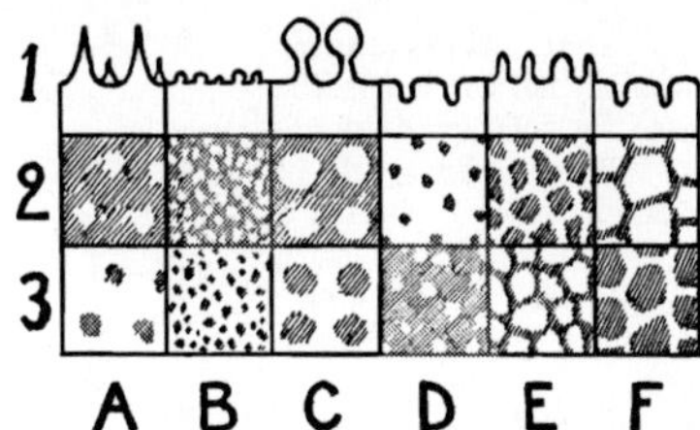

TEXTFIGURE 5. — TYPES OF EXINE ORNAMENTATION. — 1, Ornamentation in optical section (lateral view); 2, in surface view at high adjustment; 3, in surface view at low adjustment; A, echinate exine (with spines); B, granular exine (with grana; there is a gradual transition from grana to the warts of verrucate exines); C, piliferous exine (with pila); D, pitted, or scrobiculate, exine; E, reticulate exine; F, exine with "negative reticulum" (exine isles separated from each other by a network of grooves).

Operculum: a thickening — clearly defined and of measurable bulk — of the pore membrane. On rare occasions the operculum may be represented by a number of more or less separate thickenings in the pore membrane.

Ornamentatio: ornamentation, sculpturing; form elements appearing as a relief on the surface of the exine. The exine may be provided with spines, spinules, warts, granules, *pila* (small rods with rounded, swollen top-end), pits, streaks, reticulations, etc., and the pollen grains are accordingly described as echinate, subechinate, verrucate, granulate, piliferous, scrobiculate, striate, reticulate, etc. TEXTFIG. 5 is a diagram illustrating some of these types of ornamentation. In the upper row (1), they are seen from the side; in the middle row (2), from above at a high adjustment; and in the lower row (3), from above, but at a lower adjustment. The distribution of dark and bright areas in the lower rows is due to the character of the ornamentation, but similar patterns may also be produced by the texture of the exine. A network, or *reticulum*, is formed by anastomosing ridges (*muri*) on the surface of the exine, enclosing small, frequently more or less irregular, spaces (*lumina*). Anastomosing grooves in the surface of the exine, enclosing small elevated exine surfaces, would constitute a negative, or inverse, *reticulum* (TEXTFIG. 5: F).

Perinium: perine; the outermost layer, outside of the exine, in certain spores.

Perisporium: see *Perine.*

Porus: pore; the rounded apertures which frequently occur in the general surface of the exine in the absence of germinal furrows (*colpae*). There is a gradual transition from a *porus simplex,* which does not lead into a *vestibulum,* to a *porus vestibuli,* which forms the entrance to such a vestibule (TEXTFIG. 7). Sometimes the aperture of a pore leads into a more or less elongated exine collar rising above the general surface of the grain (*porus collaris*).

Porus collaris, p. simplex, p. vestibuli: see *Porus.*

Proximalis: proximal; that part of the grain or spore which is turned inward in the tetrad, opposite the furrow in monocolpate grains. The proximal side of the spores is usually provided with a tetrad scar, in trilete spores also with contact areas (= dorsal, WODEHOUSE).

Psilate: smooth (pollen grains without ridges, spines, etc.).

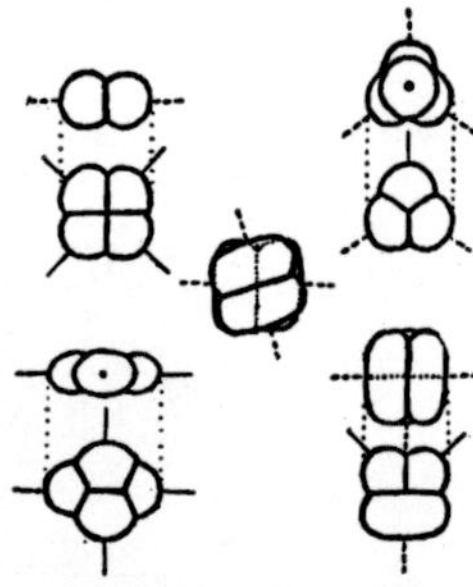

TEXTFIGURE 6. — TETRADS (SCHEME). — Upper left corner: tetragonal tetrad in lateral and surface view. Lower left corner: rhomboidal tetrad in lateral and surface view. Upper right corner: tetrahedral tetrad in surface and lateral view. Lower right corner: hexahedral tetrad in two different positions. Center: transitional form between tetragonal and hexahedral tetrad. — If situated in the plane of the figure the polar axes of the single grains are indicated by full lines, otherwise by broken lines or a dot.

Reticulum: see *Ornamentatio.*

Reticulum cristatum: crests arranged in networks.

Reticulum simplex: a network of the ordinary type, consisting of low and smooth ridges (*muri*).

Scrobiculatus: scrobiculate; see *Ornamentatio.*

Spinae: see *Ornamentatio.*

Structura: structure, texture; different patterns, in surface view usually more or less " granular ", not produced by eventual sculpturing of the exine but by formative elements within the exine.

Tetras: tetrad; a union of four pollen grains or spores formed by one mother-cell. If the grains are arranged in one plane, the tetrads may be tetragonal (TEXTFIG. 6, upper left corner), or rhomboidal (*ibid.*, lower left corner). In the first case all four cells meet at the center of the tetrad, in the second case only two of them meet. If the grains are arranged in two planes, the tetrads are tetrahedral (*ibid.*, upper right corner) or, in exceptional cases, hexahedral. In the last case, the grains are about twice as long as broad and arranged crosswise in two stories. This case and transitions to tetragonal tetrads (*ibid.*, centre) may be found *e.g.* in *Picea.*

Tricolpatus: tricolpate; possessing three meridionally arranged germinal furrows.

Triletus: trilete; possessing a triradiate tetrad scar.

Verrucae, verrucatus: see *Ornamentatio.*

Vestibulum: a small chamber under an aspidate pore.

Pollen diagnoses: — The pollen morphology of many genera and even of some families is still almost entirely unknown. In many cases, pollen grains of our most common plants are but imperfectly known. Several exhaustive plant monographs dismiss the morphology of the male gametophyte in a few lines if they consider the pollen at all. Therefore, it seems logical to insist that a pollen or spore diagnosis be given in descriptions of every species. A short pollen diagnosis would be somewhat as follows:

Diagnosis pollinaria (imaginata): —

Pollina subprolata, tricolpata; exinium reticulatum. Axis polaris circiter 38–42 μ; diameter aequatorialis c. 31–35 μ; colpae longae (c. quattuor quintas continentes distantiae interpolaris), angustae, extremitatibus acutis, nihil seu parum immersae, singulae cum pori germinali aequatoriali circulari instructae diametri c. 3 μ; lumina reticuli hexagonalia, in medio areae intercolparis maxima (ad 1.8 μ), deinde paulatim decrescentia adversus polos et margines colparum; altitudo maxima murorum c. 0.9 μ; crassitudo exinii (muribus exceptis) c. 2.5 μ, cuius tertiam partem endexinium explet.

As Wodehouse (Bull. Torr. Bot. Club, vol. 60, 1933, p. 480) has suggested, introduction of pollen diagnoses warrants the building up of reference collections of permanently mounted pollen slides which will prove to be valuable additions to our herbaria. Pollen diagnoses will no doubt prove a valuable achievement in plant systematics and facilitate the identification of pollen grains for pollen analytical research.

Notes concerning the Descriptions and Illustrations in Chapters VI–X: — Most of the pollen grains and spores described subsequently are such as have been encountered in European late-quaternary deposits (*cf.* Erdtman, Geol. Fören. Förhandl., vol. 59, pp. 158, 159). Added to these are a number of others, the pollen or spores of which have been found in older (interglacial, Tertiary, etc.) deposits, in extra-European late-quaternary deposits, or which for some other reason may be interesting from certain points of view.

The illustrations have all been drawn to the same scale (× 1600; Leitz 2 mm. apochromatic objective, numerical aperture 1.32, compensation eye-piece no. 6) and later reduced about two and a half times for reproduction. In the figures, plates I–XXVIII, the scale is 645:1 throughout.

The drawings simply are meant to convey suggestions as to the identification of pollen grains and spores and are not meant to take the place of pollen and spore preparations. Hatching has been freely used to accentuate the shape and may not necessarily be a feature of the grains. Many illustrations are provided with small accessory sketches showing certain details of the sculpturing, texture, etc., much enlarged. Numbered sketches exhibit the changing appearance of these details according to the adjustment of the microscope from high to low.

In the descriptions size-figures refer, if not otherwise stated, to the size of that pollen grain or spore pictured. These figures may not be representative of the actual mean size, as the pollen grains or spores may have been influenced by the chemical treatment, the mounting medium, etc. The majority of the illustrations are drawn from microscopical preparations made from acetolysed herbarium specimens. The description of these illustrations gives the locality or country of origin of the plants. Descriptions referring to illustrations of fresh grains (soaked with acetic acid and acetolysed) give both locality and year of collection.

Pollen descriptions from " Pollen grains, their structure, identification and significance in science and medicine " by R. P. WODEHOUSE (1935) have — with kind permission — been freely used and frequently literally cited. If simply " WODEHOUSE " is cited, and no year added, the quotation is always from this book. Descriptions and illustrations are arranged as follows with the families in accordance with ENGLER-DIELS: Syllabus der Pflanzenfamilien (11. Auflage, 1936):

Monocotyledons: families in alphabetical order: *Alismataceae — Typhaceae* (chapter VI, pp. 55–65; plates I–III, figs. 1–33);

Dicotyledons: families in alphabetical order: *Aceraceae — Ericaceae* (chapter VII, pp. 66–97, plates III–X, figs. 34–177); *Fagaceae — Violaceae* (chapter VIII, pp. 98–128, plates XI–XX, figs. 178–386);

Gymnosperms: genera in alphabetical order: *Abies — Welwitschia* (chapter IX, pp. 129–145, plates XXI–XXVI, figs. 390–449);

Pteridophytes (families in alphabetical order) (chapter X, pp. 146–151; plates XXVII–XXVIII, figs. 450–488).

References: —

ELFVING, F., 1878: Studien über die Pollenkörner der Angiospermen (Jenaische Zeitschr. f. Naturwissenschaften, vol. XIII).

FISCHER, H., 1890: Beiträge zur vergleichenden Morphologie der Pollenkörner (Breslau).

FRITZSCHE, C. J., 1837: Über den Pollen (Mém. Sav. Étrang. Acad. St. Petersb., vol. 3).

KÖHLER, H., 1933: Über die Chlorverteilung und die Tropfengruppen im Nebel und über Farbenberechnung der Kränze im weissen Lichte nebst einigen kritischen Bemerkungen der Koagulationstheorien der Nebeltropfen (Ark. f. mat., astron. och fysik, Svenska Vetenskapsakademien, vol. 24 A, no. 9).

MECKE, R., 1920: Experimentelle und theoretische Untersuchungen über Kranzerscheinungen im homogenen Nebel (Ann. Phys., vols. 61, 62).

POTONIÉ, R., 1934: Zur Mikrobotanik der Kohlen und ihrer Verwandten, I. Zur Morphologie der fossilen Pollen und Sporen (Arb. Inst. f. Paläobot. u. Petrogr. d. Brennsteine, vol. IV).

WICHER, C. A., 1934: Sporenformen der Flammkohle des Ruhrgebietes (*Ibid.*, vol. IV).

WODEHOUSE, R. P., 1935: Pollen grains. Their structure, identification and significance in science and medicine (New York, McGraw-Hill).

Chapter VI

POLLEN MORPHOLOGY — MONOCOTYLEDONS

Alismataceae (PLATE I, FIGS. 1, 2): —

Alisma plantago. — FIG. 1: grain from Tosterö, Sweden; 25 μ. — Grains spheroidal to polyhedral, cribellate, psilate. Pores 12–15 (FISCHER 1890), according to WODEHOUSE (1936) 17–30. In the material from Tosterö, the number of pores ranges from about 12 to about 30. Grains with a greater number of pores are usually larger than those with fewer pores.

The pores have poorly defined margins and are spanned by pore membranes which blend without interruption into the surrounding exine (WODEHOUSE *l.c.*). The grains have a characteristic granular texture, chiefly confined to the slightly raised exine between the different pore areas.

Sagittaria sagittifolia. — FIG. 2: Grain from Valsberga, Sweden, 1938; 28 μ. — Grains 25.2 by 24.2 μ (ZANDER 1935), spheroidal, cribellate, subechinate. Pores 12–15, equally spaced, frequently poorly defined and, as it seems, of much the same construction as in *Alisma.* Exine provided with numerous small sharp-conical spines and a faint reticulate texture. In many specimens (*cf.* WODEHOUSE) grains appear to have *one* poorly defined pore.

Butomaceae (PLATE I, FIGS. 3, 4): —

Butomus umbellatus. — FIG. 3: lateral view of dyad from a slightly unripe anther; Västerås, Sweden, 1936. FIG. 4: distal side of grain; 37 by 29 μ. — Grains 38.3 by 35 μ (ZANDER 1935), monocolpate, with reticular ornamentation, which, on higher magnification, presents a beaded appearance. The reticulation fades away towards the margins of the furrow.

Centrolepidaceae (PLATE I, FIG. 5): —

Gaimardia setacea. — FIG. 5: distal side of grain; 35 by 31 μ; New Zealand. — Grains spheroidal to more or less irregular, provided with a single irregular " exit "; ektexine finely pitted. In other species (*e.g. Centrolepis aristata, Desvauxia billardieri,* and *D. strigosa*) some of the pits are connected by shallow grooves. There is evident a gradual transition from the irregular exit of this pollen type to the well defined pore in the pollen grains of some plants belonging to the *Restionaceae.*

Cyperaceae (PLATE I, FIGS. 6–9): —

The only pollen type considered here is the *Scirpus-Carex*-type. In this type the pollen grains are psilate, more or less tetrahedral, ranging from high, narrow tetrahedrons (*e.g.* in *Cladium mariscus; cf.* Meinke 1927, p. 435) to short, more or less rounded types, as in *Rhynchospora alba*. Pores four — three lateral, one basal — as a rule poorly defined, particularly the lateral ones; exine with a faint reticular or granular texture. In many species (*cf.* Wodehouse) grains appear to have *one* poorly defined pore.

Carex digitata. — Fig. 6: lateral view of grain; 41 by 32 μ; Västerås 1938.

Eriophorum vaginatum. — Fig. 7: lateral view of grain; 39 by 29 μ; Åker, Sweden, 1938.

Rhynchospora alba. — Fig. 9: lateral view of grain; 28 by 27 μ.

Rhynchospora fusca. — Fig. 8: lateral view of grain; 40 by 29 μ; Sexdrega, Sweden.

High pollen frequencies of the *Cyperaceae* (usually expressed as percentages of the tree pollen total) have often been reported from old deposits such as clay and ooze formed previous to the appearance of forests. Thus Firbas (1934, 1935) found up to 3000 per cent cyperaceous pollen in peat deposits in southern Germany. Although this figure may seem high, nearly the same percentages have been found in many other deposits from Italy in the south to the Scandinavian countries in the north (Faegri 1936, Firbas and Zangheri 1936, Oberdorfer 1937, Schmitz 1929, etc.).

The occurrence of cyperaceous pollen in recent pollen spectra has been studied by Ernst (1934), Firbas (1934), Iversen (1934), and others. As a rule, pollen of the *Cyperaceae* is abundant in samples from places with a heavy growth of sedges. Thus, Firbas (*l.c.*) encountered more than 300 per cent of cyperaceous pollen in a sample from a *Schoenus nigricans*-meadow (with *Carex panicea*, *Eriophorum angustifolium* etc., and much *Carex arenaria* in the vicinity) in the unforested isle of Baltrum (East Frisean Islands, about 10 to 20 kilometers from the nearest woods on the mainland).

Eriocaulaceae (Plate I, figs. 10, 11): —

Eriocaulon septangulare. — Fig. 10: grain from Connecticut; 24 by 23 μ. Fig. 11: contour lines of the back of the pollen grain figured in fig. 10. — Grains spheroidal; exine folded into long, low ridges, separated by narrow grooves, which functionally correspond to the furrows in other grains. The grains are subechinate or warty, provided with a faint reticular texture. The pollen grains of *Aphyllanthes monspeliensis* as well as the grains of some species of *Berberis* and *Pinguicula* show certain resemblance to those of *Eriocaulon*.

Gramineae (Plates I, figs. 12–17, and II, figs. 18, 19): —

Grains usually spheroidal or more or less ovoidal. In size, they range from about 22 to a little over 100 μ in diameter. They have a single germ pore, surrounded by a thickened rim which causes the orifice to be slightly raised above the general surface of the grain, and crossed by a delicate membrane bearing a conspicuous operculum at or near its centre (Wodehouse, pp. 304, 305). The rim is composed

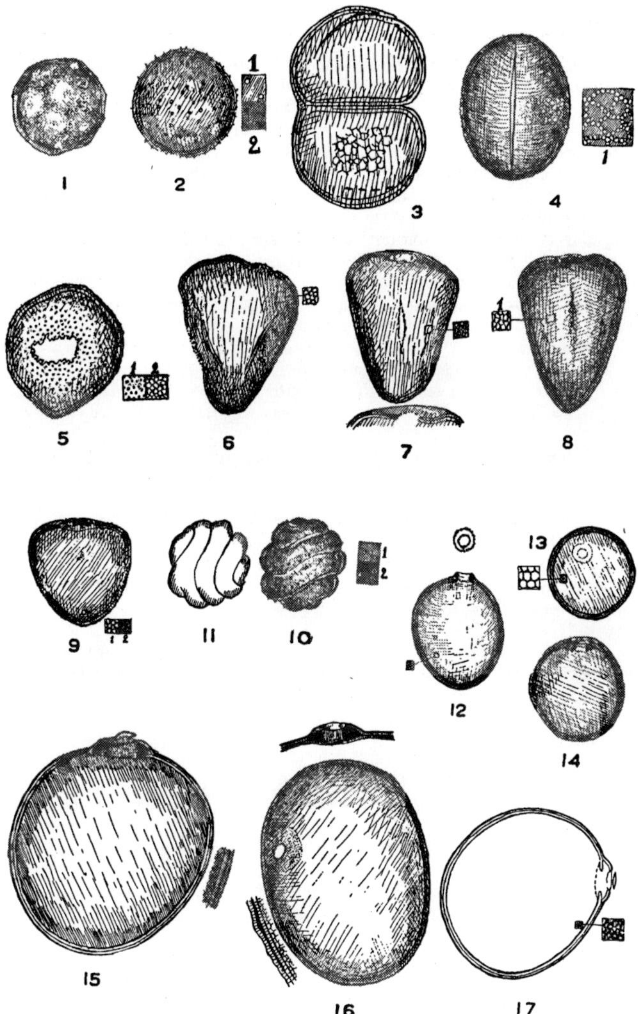

PLATE I (1–17). — *Alismataceae* (1, 2), *Butomaceae* (3, 4), *Centrolepidaceae* (5), *Cyperaceae* (6–9), *Eriocaulaceae* (10, 11), *Gramineae* (12–17). — 1, *Alisma plantago;* 2, *Sagittaria sagittifolia;* 3 and 4, *Butomus umbellatus;* 5, *Gaimardia setacea;* 6, *Carex digitata;* 7, *Eriophorum vaginatum;* 8, *Rhynchospora fusca;* 9, *R. alba;* 10 and 11, *Eriocaulon septangulare;* 12, *Aira flexuosa;* 13 and 14, *Phragmites communis;* 15, *Triticum vulgare;* 16, *Secale cereale;* 17, *Phalaris arundinacea.* — For general remarks concerning the illustrations, *cf.* p. 53 *infra.* Magnification of all plates about 645×.

of three layers: the ektexine and the endexine, which in optical section (compare FIGS. 16–18) deviate from each other as the jaws of an open mouth, and a third, "mesexinous" layer or ring, filling the space between the jaws. The grains may generally be termed psilate. In some species, however, the surface of the grains is slightly rough owing to local separation of the ektexine from the smooth surface of the endexine (FIG. 16, lower left corner). If examined under a high resolving power, the exines are nearly always seen to possess a minute granular or reticulate texture: at high adjustment, dark meshes surrounding bright areas; at low adjustment, bright meshes encircling darker material (compare FIG. 19).

Aira flexuosa. — FIG. 12: lateral view; Östergötland, Sweden; 30 by 25 μ. — Average size 21.4 μ (FIRBAS 1937).

Elymus arenarius. — FIG. 19: grain from Höganäs, Sweden, 1938; 53 by 45 μ. — Average size about 41.9 μ (FIRBAS 1937).

Phalaris arundinacea. — FIG. 17: lateral view; Västerås, Sweden, 1937; 46 by 42 μ. — Size 27 μ (FIRBAS 1937), 37–50 μ (YAMASAKI 1933).

Phragmites communis. — FIGS. 13, 14: grains from Tosterön, Sweden; FIG. 13, 24 μ, FIG. 14, 27 by 25 μ. — Average size about 23.5 μ (FIRBAS 1937).

Secale cereale. — FIG. 16: grains from Åker, Sweden, 1935; 59 by 45 μ. — Size of grains 42–44.5 μ (FERRARI 1927), 41.9 μ (FIRBAS 1937), 62 by 40 μ (WODEHOUSE), 51 by 38.2 μ (ZANDER 1935).

Triticum vulgare. — FIG. 15: lateral view; Västerås, Sweden, 1937; 55 μ. — Size of grains 47.7 μ (FIRBAS 1937), 52 by 46.6 μ (ZANDER 1935).

Zea mays. — FIG. 18: grain from Västerås, Sweden, 1936; 111 μ; in the centre of the figure is an enlargement showing the construction of the pore in detail.

Pollen analysis has been tried as a means of tracing the history of the cultivated cereals. Possibilities along these lines were actually shown to exist in 1933 (ERDTMAN 1938). Systematic investigations in this matter were started by FIRBAS in 1935 and the results published two years later ("Pollenanalytischer Nachweis des Getreidebaus"). This paper contains a review of the pollen morphology of 215 species of European grasses. Two pollen types may be distinguished: the wild grass type and the "cereal" or cultivated grass type ("Getreidetyp"). The maximum diameter of the grains of the former type measures 20 to 25 μ, more seldom 30, and in exceptional cases 35, even 40 μ. They are spheroidal to ellipsoidal; in the latter case with a polar or lateral exit. The exine is thin, generally less than 1 μ in thickness, with or without a faint texture. Diameter of the pore usually less than 2 μ. The grains of the latter type are larger, usually from 35 to 50 μ, rarely as large as 60 μ or smaller than 35 μ. They may be spheroidal, but are more frequently spheroidal to ellipsoidal or decidedly ellipsoidal to ovoidal. The germ pore is often placed laterally and provided with a conspicuous rim; pore diameter about 2 to 7 μ.

184 Species were found to possess pollen of wild grass types. To this group belong nearly all central European species of wild grasses,

together with *Panicum miliaceum*, *Setaria italica* and, probably, *Avena brevis*.

31 Species were provided with grains of the cereal type in a wide sense. They may be divided into:

A. Species with pollen grains fully corresponding to the description of the cereal type (average diameter of the grains generally surpassing 40 μ): *Avena intermedia* (partially), *A. nuda*, *A. orientalis*, *A. sativa*, *Elymus arenarius*, *Hordeum murinum*, *Secale cereale*, *Triticum compactum*, *T. dicoccum* (partially), *T. spelta*, *T. vulgare*, *Zea mays*.

B. Species with pollen grains corresponding to the smaller grains of the cereal type (mean size ranging from 35 to 40 μ): *Agropyron intermedium* (partially), *A. repens* (partially), *Avena fatua*, *A. strigosa*, *Hordeum distichon*, *H. polystichon* ssp. *vulgare* and ssp. *hexastichon*, *H. secalinum* (partially).

C. Species with pollen grains which, according to their shape, should be referred to the cereal type, but, according to their size, to the wild grass type (mean size ranging from 32 to 35 μ): *Agropyron caninum*, *A. junceum*, *A. repens* (partially), *Bromus erectus*, *B. inermis*, *B. ramosus* ssp. *euramosus*, *Glyceria fluitans*, *G. plicata*, *Hordeum maritimum* (partially), *H. secalinum* (partially), *Triticum monococcum* (partially).

Having acquired more detailed information as to exine texture, etc. pollen morphologists will eventually be in a position to recognize some of the genera, or even, in some cases, species, by means of pollen grain characters alone. With this in mind, FIRBAS recommends further investigations (including also the study of pollen grains of different races within the same species) in *Secale*, *Triticum compactum*, and *T. vulgare*. Among the grass pollen grains depicted in this book, those of *Aira* and *Phragmites* (FIGS. 12–14) belong to the wild grass types, the others — *Elymus*, *Phalaris*, *Secale*, *Triticum*, and *Zea* — to the cereal type.

As to *Phalaris arundinacea*, there is an obvious difference between FIRBAS's results and the observations made by YAMASAKI (1933) and by the author. FIRBAS refers the pollen grains of this species to the wild-grass type. Their mean size, expressed as the average length of the largest diameter, is 27.0 ± 0.30 μ; the extreme values 23.75 and 38 μ respectively. Among 100 grains measured, the lowest value occurred in 40 cases, the highest in only one case. According to YAMASAKI the diameter ranges from 37 to 50 μ (preparations probably made as described by MEINKE 1927). Based on herbarium specimens as well as on fresh material from different localities in Sweden, the author's observations show that the grains must be referred to the cereal type, both as to size and appearance. This is also true of the pollen grains of *Phalaris canariensis* (from a plant grown in Västerås; FIRBAS gives the mean size of the grains of this species as 31.9 μ). Nor are FIRBAS's general results in accord with those obtained by WODEHOUSE as shown by TAB. 6.

On the whole there seems to exist a general similarity between the values obtained by WODEHOUSE and those obtained by ERDTMAN, although different methods of preparing the pollen grains were used. On the other hand, FIRBAS followed a method which, as shown by the control experiments which he himself carried out, gives nearly the same results as the acetolysis method employed by ERDTMAN. Nevertheless, the figures given by FIRBAS are sometimes strikingly different from those obtained by ERDTMAN. Some discrepancies may be due to the appearance within the species of forms with different chromosome numbers, while others may be explained by the influence of the mounting medium used for the pollen preparations, etc. There is, however, no doubt that the diagrams published by FIRBAS and others indicate great promise for the use of pollen analysis in tracing the history of cultivated cereals. But great care must be exercised in order to preclude the errors of hasty conclusions. Attention has also been drawn to another aspect of the influence of man upon the occurrence of grasses, *viz.* by IVERSEN (1934), who correlated a marked increase in the frequency of grass pollen in a Greenland bog with the meadows connected with the settlements of Norse immigrants.

Table 6: SIZE OF GRASS POLLEN: —

	DIAMETER IN μ ACCORDING TO		
	FIRBAS (averages quoted in the second column)		WODEHOUSE
Agropyron repens	28.5–34.8	42.8	47 –52
Anthoxantum odoratum	19 –28.3	42.8	37.6–45.6
Arrhenaterum elatius	14.3–25.2	38	34 –39
Avena strigosa	33.3–38.8	47.5	68
Dactylis glomerata	19 –27.2	33.3	28.5–36
Poa trivialis	14.3–20.8	28.5	22.8–25.1
Sorghum halepense	28.5–33.3	42.8	40 –55
Zea mays	52 –66.5	76	90 –100

We now leave the specific problem of the history of the cultivated cereals etc. to consider the general problem of the occurrence of fossil grass pollen. High grass pollen frequencies have often been found in late glacial deposits (FIRBAS 1935, etc.). They testify, as do high frequencies of cyperaceous pollen etc., to the unforested conditions which prevailed during those periods. According to LEWIS and COCKE (1929) the occurrence of open meadows is indicated by high frequencies of pollen of the *Cyperaceae* and *Gramineae* in certain Dismal Swamp peats.

Referring to diagrams from Tierra del Fuego, VON POST (1929) has stressed the possibility of tracing changes from steppe to forest, and vice versa, by means of pollen statistics. Similar problems have been studied by FIRBAS (*l.c.*, pp. 141–143):

„In Mitteleuropa wird vor allem die Entwicklungsgeschichte der Wiesen und der Nachweis steppenartiger Vegetation im Vordergrund des Interesses stehen. Die bisher vorliegenden Oberflächenproben und das sonstige paläontologische Material lassen aber erkennen, dass auch hier besondere Schwierigkeiten bestehen. Zwar wird Gramineenpollen in erheblicher Menge gebildet, und auch über grössere Strecken, von einigen 100 m, reichlich verweht. Oberflächenproben von so gut wie grasfreien Hochmooren (Harz, Rhön) ergaben bis zu 40% Gramineenpollen, der von den umliegenden Bergwiesen bzw. aus grasreichen Fichtenwäldern stammen muss. Ähnliches berichten BRINCKMANN und ERNST aus Grenzgebieten von Heide, Moor und Marsch, und auch SCHUBERT 1933, S. 65 führt den hohen Gramineengehalt von Oberflächenproben auf Wiesenkultur im Bereich der Hochmoore zurück, während er, in Übereinstimmung mit KOCH (1930) und OVERBECK (1931)

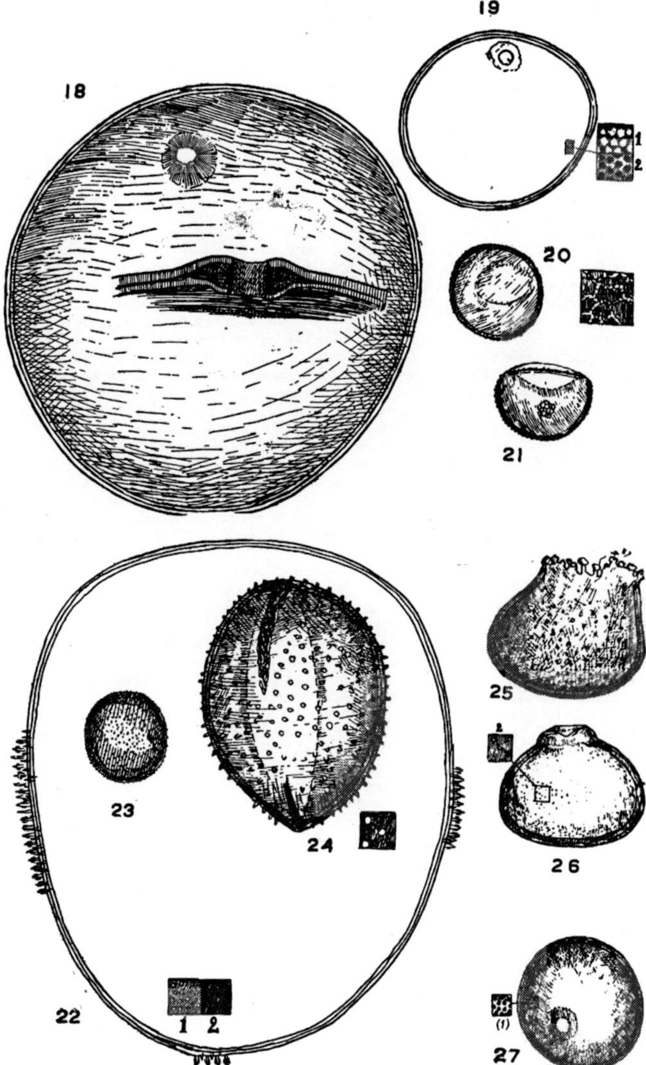

Plate II (18–27). — *Gramineae* (18, 19), *Hydrocharitaceae* (22–24), *Potamogetonaceae* (20, 21), *Restionaceae* (25–27). — 18, *Zea mays;* 19, *Elymus arenarius;* 20 and 21, *Potamogeton perfoliatus;* 22, *Elodea matthewsii;* 23, *Hydrocharis morsus-ranae;* 24, *Stratiodes aloides;* 25, *Lepyrodia anarthria;* 26, *Hypodiscus aristatus;* 27, *Ecdeiocolea monostachya.*

in Hochmoortorfen nur sehr geringe Werte, durchschnittlich etwa 2% vorfand. Aber schon hierin zeigt sich wieder, dass zunächst auf anderem Wege, durch die Höhe der gesamten Nichtbaumpollenwerte, der Nachweis geringer Walddichte geführt werden muss, dann erst — allenfalls — hohe Gramineenwerte als Zeiger freien Graslandes gewertet werden können. In diesem Sinne sprechen dann etwa Werte von 112–910% oder 68–1565% Gramineenpollen, wie sie Brinckmann und Ernst in weiten Marschwiesen und in Schilfbeständen fanden, ebenso wie die angeführten Werte von einer lappländischen Wiese eine deutliche Sprache" . . . „Die grössten Schwierigkeiten aber liegen auch hier wiederum in der Ausschaltung bzw. richtigen Bewertung lokaler Einflüsse. Wenn wir in grasreichen Fichtenwäldern im Harz bis 64% Gramineen finden, wenn in der Altenwalder Heide ein kleiner Molinia-Bestand eine völlige Umkehrung des Verhältnisses von Gramineen- und Ericaceenpollen hervorruft, wenn von paläontologischer Seite auf enge Beziehungen hoher Gramineen- und Cyperaceenwerte zur Torfart hingewiesen wird (z.B. Schubert 1933, S. 65), tritt dies deutlich hervor. Noch stärker scheint der lokale Einfluss des Cyperaceenpollens zu sein, der selbst in kleinen Seggensümpfen der Waldgebiete über 100% ansteigen kann. Wenn daher z.B. von Post (1930, S. 557) aus Werten von etwa 2000% Gramineen, 2700% Cyperaceen eines feuerländischen Moores auf Waldlosigkeit schliesst, ist dieser Schluss sicher berechtigt". (This seems to refer to the bottom sample from the central part of Lago Fagnano. Von Post has expressed the pollen frequencies as percentages of the total of *Nothofagus*, *Gramineae* and (*cf.*) *Cyperaceae* pollen. The bottom sample contains 3 per cent *Nothofagus*, 41 per cent *Gramineae*, and 56 per cent cyperaceous pollen. If the *Nothofagus* pollen frequency is considered to be equal to 100 per cent, the other frequencies would represent 1637 and 1867 per cent respectively.) Firbas continues: „Wenn er hingegen meint, dass damals an Stelle der *Nothofagus*-Wälder eine Vegetation getreten sei deren Charakterpflanzen Gramineen und Cyperaceen gewesen wären, scheint mir die Möglichkeit lokalen Einflusses zu gering eingeschätzt. Denn schon der Umstand, dass das Verhältnis von Cyperaceen- zu Gramineenpollen in entsprechenden Horizonten in dem einen Profil etwa 15 zu 1, in dem anderen etwa 1.5 zu 1 sein kann, spricht für starke lokale Beeinflussung.

Bei dem Reichtum der Verlandungsvegetation an Gräsern und Cyperaceen wird man daher auch aus der Untersuchung von Mudden nicht viel Aufschluss zur Klärung der Steppenfrage erwarten dürfen, und die in methodischer Hinsicht geeigneten Torfe ombrogener Hochmoore werden aus klimatischen Gründen hierfür kaum in Frage kommen. Die Verfolgung der Gramineen, Cyperaceen und der übrigen Kräuterpollen wird daher wohl vorwiegend der eingehenderen Aufklärung der lokalen Sukzessionen dienen müssen, wie dies Ernst überhaupt als besonderen Vorteil der Beachtung der Nichtbaumpollen hervorhebt. Hier vermag sie aber wichtige Ergänzungen zu liefern".

Lane (1931), in an interesting study, used grass pollen as evidence of prairie conditions in Iowa. Other American authors have sometimes uncritically used — Dr. Cain writes me — the occurrence of appreciable quantities of grass pollen at certain bog levels as an indication of a xerothermic period, and an eastward extension of the prairie peninsula. Keller (1943) reports a grass representation of 43 per cent at the 17-foot level of one Indiana bog and of 57 per cent at the 11-foot level of another. There being no other indication of a prairie invasion, he made a size-frequency study of 421 of these fossil grains and found them to have sizes comparable to *Calamagrostis canadensis*. The eastern United States prairie grass dominants (*Andropogon scoparius*, *A. furcatus*, *Sorghastrum nutans*, *Agropyron smithii*, and *Bouteloua curtipendula*), also studied by the size-frequency method by Keller, all have grains that are consistently larger than those of the fossils and *Calamagrostis*. Obviously, a high representation of grass pollen, *per se*, has no value as an indicator of a xerothermic period in northern Indiana.

Hydrocharitaceae (Plate II, figs. 22–24): —

Elodea matthewsii (Plach.) St. John. — Fig. 22: Grain from near La Paz (Asplund no. 3499); 134 by 102 μ.

Hydrocharis morsus-ranae. — Fig. 23: grain from Åker, Sweden, 1935; 23 μ.

Stratiotes aloides. — Fig. 24: proximal side of grain; Bremen; 65 by 49 μ.

The grains of these species are echinate (*Elodea*, *Stratiotes*), or subechinate (*Hydrocharis*). In *Elodea matthewsii*, the spines appear to be interconnected by narrow muri forming a delicate reticulum. The grains of *Elodea* and *Hydrocharis* are apparently acolpate, while those of *Stratiotes* are monocolpate, provided with a long germinal furrow. Those of *Elodea matthewsii* are among the largest pollen grains thus far encountered in the monocotyledons.

Iridaceae (PLATE III, FIG. 28): —

Iris pseudacorus. — FIG. 28: lateral view; 91 by 63 μ; Västerås 1938. — Grains monocolpate, exine reticulate. Reticulum very conspicuous on the proximal side of the grain, gradually becoming less prominent and finally almost disappearing towards the margins of the furrow. If examined with a high resolving power, the muri of the reticulum present a beaded appearance. Size, according to FERRARI (1927) 79.5–84 μ, according to ZANDER (1941), averaging 89.4 by 81 μ.

Potamogetonaceae (PLATE II, FIGS. 20, 21): —

Potamogeton perfoliatus. — FIGS. 20, 21: grains from Åker, Sweden, 1938; 25 μ. — Grains more or less spheroidal with a single oblong or circular depression. Exine reticulate. Reticulum of a more or less beaded appearance.

Restionaceae (PLATE II, FIGS. 25–27): —

Ecdeiocolea monostachya F. M. — FIG. 27: oblique polar view; 34 μ; Australia (PRITZEL no. 611).

Hypodiscus aristatus. — FIG. 26: lateral view; 39 by 29 μ; ex herb. Holm.

Lepyrodia anarthria. — FIG. 25: lateral view; 41 by 32 μ; New South Wales.

In pollen morphology the *Restionaceae* family presents a transitory stage from pollen of the *Centrolepidaceae* type (*cf.* FIG. 25) with a large, more or less irregular exit, to the grass pollen type with a contracted and well-defined pore (*cf.* FIG. 27; FIG. 26 represents an intermediate type). The exine is pitted. The pits are larger in grains approaching the *Centrolepidaceae* type, smaller or sometimes hardly discernible in grains of the *Gramineae* type.

Fossil pollen of *Hypolaena lateriflora* has been identified in New Zealand peat by L. CRANWELL (Geografiska Annaler 1936). There is no detailed description, but the grains are said to be " very striking ".

Scheuchzeriaceae (PLATE III, FIG. 29): —

Scheuchzeria palustris. — FIG. 29: lateral view of dyad; Hallstahammar, Sweden, 1938; size of dyad 44 by 30 μ, of single grain 30 by 22 μ. — Pollen grains acolpate, connected in pairs. Exine reticulate; reticulum continuous from cell to cell, uninfluenced by the suture between them. The transverse wall separating the two grains apparently consists of endexine only.

Sparganiaceae (PLATE III, FIGS. 30, 31): —

Sparganium minimum. — FIG. 30: distal side of grain; 27 by 22 μ; Åker, Sweden, 1938.

Sparganium ramosum. — FIG. 31: lateral view; 34 by 29 μ; Gotland, Sweden. — Grains rounded, with one germ pore, approximately circular in outline, appearing as a jagged hole broken through the exine. Exine reticulate; in *Sparganium ramosum* (FIG. 31, detail figures) the ornamentation seems to be more complicated than in *S. minimum.* — Pollen grains of a similar type occur in *Typha angustifolia.*

Typhaceae (PLATE III, FIGS. 32, 33): —

Typha angustifolia. — FIG. 32: proximal side of grain; 31 by 23 μ; Ekerö, Sweden. — Grains single.

Typha latifolia. — FIG. 33: pollen tetrad; 42 μ; equatorial diameter of the single grains about 23.5 μ; Sala, Sweden. — Grains usually united in tetrads; according to WODEHOUSE irregularly spheroidal or, if united in tetrads, variously modified in shape as a result of their mutual contacts. Germ pore single; its position on the surface of those grains, which are shed united in tetrads, may be anywhere on the distal side, not necessarily at the distal pole. There seems to be a tendency among the grains of those tetrads which are flat for the pores of all four grains to be on the same side of the tetrad. Exine reticulate, reticulum ending at the margins of the pore with open lacunae. The reticulum is continuous throughout the tetrad, passing from cell to cell in the way previously mentioned in *Scheuchzeria.*

The four grains occur in many arrangements, among which the square and rhomboidal predominate (WODEHOUSE). Pollen tetrads also occur in *Typha minima* Funk and *T. shuttleworthii* Koch and Sonder (FIRBAS 1934).

References: —

BRINCKMANN, P., 1934: Zur Geschichte der Moore, Marschen und Wälder Nordwestdeutschlands, III. Das Gebiet der Jade (Bot. Jahrb., vol. LXVI).

ERDTMAN, G., 1937: Literature on pollen-statistics and related topics published 1935 and 1936 (Geol. Fören. Förhandl., vol. 59).

ERNST, O., 1934: Zur Geschichte der Moore, Marschen und Wälder Nordwestdeutschlands, IV. Untersuchungen in Nordfriesland (Schr. Naturwiss. Ver. Schlesw.-Holst., vol. XX).

FAEGRI, K., 1936: Quartärgeologische Untersuchungen im westlichen Norwegen, 1. Über zwei präboreale Klimaschwankungen im südwestlichsten Teil (Bergens Mus. Årbok 1935, Naturvidensk. rekke, no. 8).

FERRARI, A., 1927: Osservazioni di biometria sul polline delle Angiosperme (Atti R. Inst. Bot. Univ. Pavia).

FIRBAS, F., 1934: Zur spät- und nacheiszeitlichen Vegetationsgeschichte der Rheinpfalz (Beih. Bot. Centralbl., Abt. B, vol. LII).

—— —— 1935: Die Vegetationsentwicklung des mitteleuropäischen Spätglazials (Bibl. Bot., H. 112).

—— —— 1937: Der pollenanalytische Nachweis des Getreidebaus (Ztschr. f. Bot., vol. 31).

—— —— und ZANGHERI, P., 1936: Eine glaziale Flora von Forli, südlich Ravenna (Veröff. Geobot. Inst. Rübel, 12. Heft, Zürich).

FISCHER, H., 1890: Beiträge zur vergleichenden Morphologie der Pollenkörner (Breslau).

IVERSEN, J., 1934: Moorgeologische Untersuchungen auf Grönland. Ein Beitrag zur Beleuchtung der Ursachen des Unterganges der mittelalterlichen Nordmännerkultur (Medd. Dansk Geol. Foren., vol. 8).

KELLER, C. O., 1943: A comparative pollen study of three Indiana bogs (Butler Univ. Bot. Studies 6:65).
KOCH, H., 1930: Stratigraphische und pollenfloristische Studien an drei nordwestdeutschen Mooren (Planta, vol. 11).
LANE, G. H., 1931: A preliminary pollen analysis of the East McCulloch peat bed (Ohio J. Sci. 31).
LEWIS, I. F. and COCKE, E. C., 1929: Pollen analysis of Dismal Swamp peat (Journ. Elisha Mitchell Sci. Soc., vol. 45).
MEINKE, H., 1927: Atlas und Bestimmungsschlüssel zur Pollenanalytik (Bot. Arch.).
OBERDORFER, E., 1937: Zur spät- und nacheiszeitlichen Vegetationsgeschichte des Oberelsasses und der Vogesen (Zeitschr. f. Bot., vol. 30).
VON POST, L., 1930: Die Zeichenschrift der Pollenstatistik (Geol. Fören. Förhandl., vol. 51, 1929).
SCHMITZ, H., 1929: Beiträge zur Waldgeschichte des Vogelsbergs (Planta, vol. 7).
SCHUBERT, E., 1933: Zur Geschichte der Moore, Marschen und Wälder Nordwestdeutschlands, II. Das Gebiet an der Oste und Niederelbe (Mitt. Prov.-st. f. Naturdenkmalpfl. Hannover, H. 4).
WODEHOUSE, R., 1935: Pollen grains. Their structure, identification and significance in science and medicine (New York, McGraw-Hill).
—— —— 1936: Pollen grains in the identification and classification of plants, VIII. The *Alismataceae* (Amer. Journ. Bot., vol. 23).
YAMASAKI, T., 1933: Morphology of pollen grains and spores (Rep. Exper. Forest., Kyoto Imp. Univ., no. 5; Japanese).
ZANDER, E., 1935: Beiträge zur Herkunftsbestimmung bei Honig, I (Berlin).
—— —— 1941: Beiträge zur Herkunftsbestimmung bei Honig, III (Leipzig).

Chapter VII

POLLEN MORPHOLOGY — DICOTYLEDONS

(*Aceraceae — Ericaceae*)

Aceraceae (PLATES III, FIGS. 34–41, and IV, FIGS. 42, 43): —

Acer campestris. — FIG. 34: polar view; 37 μ; Lund, Sweden. FIG. 35: equatorial view; 40 by 26 μ; *ibid.* — Size, according to ZANDER (1935), 33 by 31.2 μ.

Acer negundo. — FIG. 36: equatorial view; 37 by 27 μ; Connecticut. FIG. 37: polar view; 27 μ; *ibid.* — Size, according to ZANDER (*l.c.*), 30 by 28 μ.

Acer platanoides. — FIGS. 38, 39: polar and equatorial view respectively, both oblique; about 34 μ; Västerås 1935. — Size, according to ZANDER (*l.c.*) 33.6 by 31.4 μ.

Acer pseudoplatanus. — FIG. 40: oblique polar view; Västerås. FIG. 41: equatorial view; 45 by 34 μ; *ibid.* — Size 35.1–37.4 μ (FERRARI 1927); 33.8 by 29.4 μ (ZANDER *l.c.*).

Acer saccharum. — FIG. 42: equatorial view; 43 by 32 μ; Connecticut. FIG. 43: polar view (outline); 37 μ; *ibid* .— Size 29.8 by 28.4 μ (ZANDER *l.c.*).

The pollen grains of the maples are prevailingly tricolpate, with meridional furrows equally spaced around the equator. When expanded, the grains are often noticeably flattened with the meridional furrows gaping widely open. The furrow membranes are smooth or slightly flecked (WODEHOUSE). The grains are reticulate (*A. negundo*, *A. saccharum*) or granulate with the granules arranged more or less in rows, giving the exine a striated appearance. In certain cases, the patterns of the striae are an important diagnostic character.

With the exception of *Acer negundo*, the maples are primarily insect pollinated. Nevertheless, pollen of several species (besides those of *A. negundo*) were frequently caught on atmospheric pollen plates exposed in Yonkers, N. Y., at considerable distances from the trees (WODEHOUSE). At Västerås, on the other hand, where *A. platanoides* grows abundantly, no pollen grains were caught during the rich blossom season of 1937 in spite of the fact that a more effective method than that of exposing atmospheric pollen plates was used (*cf.* p. 188).

In Sweden, only occasional discoveries of fossil *Acer* pollen are said to have been made. In the U. S. A. (Wisconsin) up to 14 per cent of maple pollen has been reported from a bog in the vicinity of which *A. saccharinum*, *A. saccharum*, and *A. spicatum* grew (HANSEN 1937).

Anacardiaceae (PLATE IV, FIGS. 44, 45): —

Rhus typhina. — FIG. 44: equatorial view; 53 by 35 μ; Connecticut. FIG. 45: polar view (outline). — Grains prolate, tricolpate,

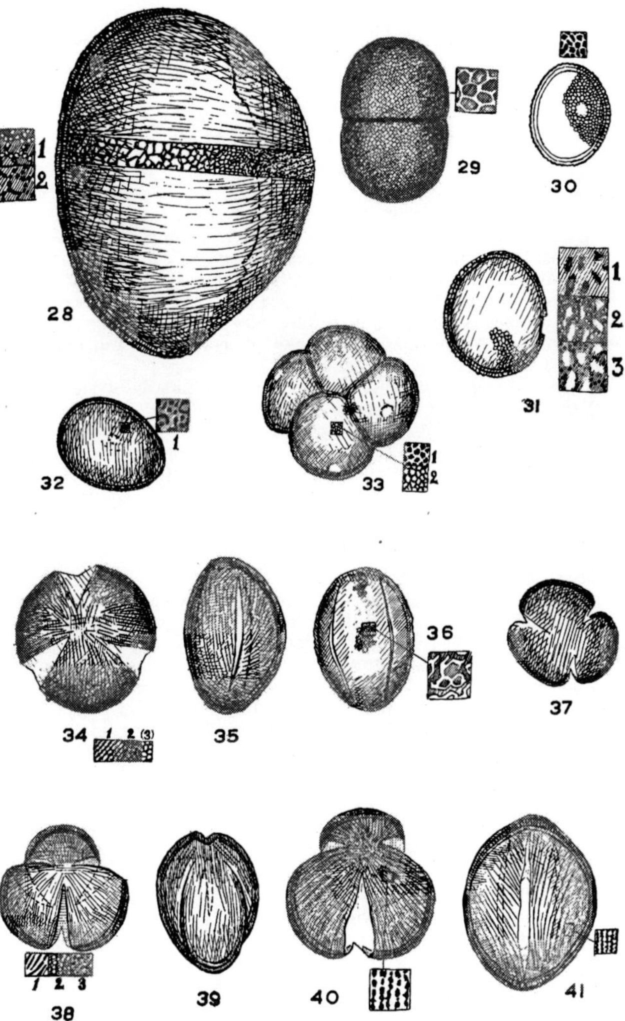

PLATE III (28–41). — *Iridaceae* (28), *Scheuchzeriaceae* (29), *Sparganiaceae* (30, 31), *Typhaceae* (32, 33), *Aceraceae* (34–41). — 28, *Iris pseudacorus;* 29, *Scheuchzeria palustris;* 30, *Sparganium minimum;* 31, *S. ramosum;* 32, *Typha angustifolia;* 33, *T. latifolia;* 34 and 35, *Acer campestris;* 36 and 37, *A. negundo;* 38 and 39, *A. platanoides;* 40 and 41, *A. pseudoplatanus.*

finely reticulate, with long tapering, sharply defined furrows, each with a large germ pore the equatorial diameter of which is about twice as long as the meridional diameter. Ektexine at the pores slightly raised above the general surface of the grain.

Aquifoliaceae (PLATE IV, FIGS. 46, 47): —

Ilex verticillata. — FIG. 46: oblique polar view; 35 μ; Michigan. FIG. 47: equatorial view; 34 by 27 μ. — Grains subprolate to suboblate, tricolpate. Furrows broad and long, their membranes heavy and conspicuously flecked with coarse granules; germ pores represented by hyaline thickenings, one below the centre of each furrow, causing a rounded bulge which may or may not break through the furrow membrane when the grain is moistened. Exine thick and rigid, presenting a coarsely pebbled appearance in surface view (WODEHOUSE). Similar grains occur in *Nemopanthes* and somewhat similar types in the *Cruciferae, Oleaceae* (*Ligustrum*), etc.

Holly (*Ilex aquifolium*) has a wide distribution in western and southern Europe, but fossil grains of this species have, as a rule, been found only occasionally. Higher frequencies (up to 10 per cent) have been reported from bogs in southern Ireland (ERDTMAN 1924) and certain interglacial peats (FIRBAS 1928). Grains of the *Ilex* type have been found on several occasions in Tertiary deposits.

Araliaceae (PLATE IV, FIGS. 48, 49): —

Hedera helix. — FIG. 48; polar view;. 34 μ; Kville, Sweden. FIG. 49: oblique view. — Grains tricolpate with comparatively short furrows tapering to sharp points. Exine thick, provided with an intricate reticulum of somewhat similar type as in *Sparganium* (*cf.* PL. III, FIG. 31): at high adjustment dark lumina surrounded by bright muri appear; at slightly lower adjustment, the muri are replaced by a system of bright points. At still lower adjustment, these points turn dark and the lumina bright. Size of grains, according to ZANDER (1935), 23.8 μ; *cf.* also IVERSEN (1941).

Betulaceae (PLATES IV, FIGS. 50–61, V, and VI, FIGS. 90–93): —

Alnus (PLATE IV, FIGS. 50–61). — Grains oblate spheroidal to suboblate; in polar view triangular, square, pentagonal, or hexagonal according to the number of pores. Pores situated in the angles, aspidate, narrowly elliptical or slit-shaped, meridionally extended (FIG. 58), their length (according to WODEHOUSE) 2.5 to 4.5 μ. Under each exit, a vestibule with an endexinous floor (TEXTFIG. 7, p. 76).

According to ERDTMAN (1936), diagnostic value may be attached to the relative size of the aspides. This is expressed by the ratio IA:A, where IA is the total extension of the interaspidar exine contour and A the sum of the equatorial diameter of the aspides. By means of camera lucida drawings, particularly careful measurements can be made. The IA- and A-lengths should be set off on the line of junction between the ekt- and endexine, not at the outer surface of the ektexine (*cf.* PL. IV, FIGS. 51, 53, 55, and 57).

A characteristic feature is the presence of arci. Their real nature is a matter of debate. According to WODEHOUSE, they are a part of the texture of the grains: *i.e.*, they represent real thickenings which, however, do not protrude above the general surface of the grain. POTONIÉ considers that they belong to the ornamentation of the grain and are to be regarded less as thickenings than as curved margins caused by a slight outward bending of the exine [POTONIÉ 1934*a*, p. 20; however, when speaking of subfossil *Alnus* pollen, he remarks (1934*b*, p. 59): „Längs der Arci scheint die Exine schwach verdickt zu sein, so dass sie der Aussteifung des Pollenkorns dienen "].

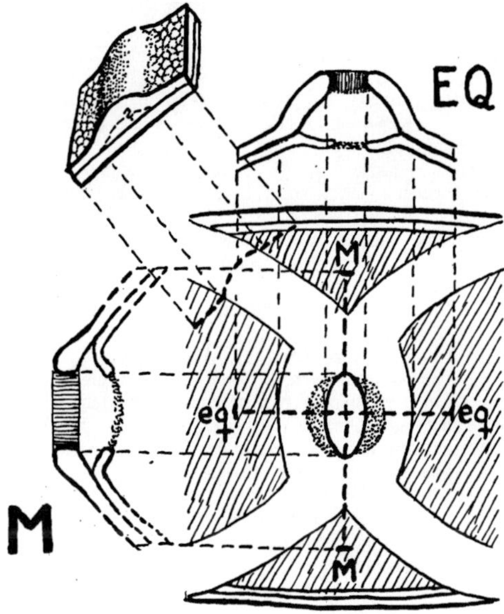

TEXTFIGURE 7. — CONSTRUCTION OF PORE AND ARCI IN *Alnus* POLLEN (schematic). — Eq.–eq, equator; M–M, meridian; EQ, equatorial; M, meridional section of pore. Upper left corner, construction of an arcus.

Basing his opinion only on a study of unsectioned pollen grains, the author feels inclined to consider the arci as an ornamental feature, caused by local thickening of the ektexine or, eventually, by intercalation of mesexinous strands between the ekt- and endexine. Such strands would correspond to the central layer of the exine stratification of five strata in the pollen grains of *Betula*, *Corylus*, and *Myrica*, as described by JENTYS–SZAFER (1928).

If examined with highest power, the exines (the arci possibly excepted) present a finely reticulate texture (TEXTFIG. 7).

Alnus glutinosa. — FIG. 50: five-pored grain (polar view); 29 μ; Visby 1933. FIG. 51: outline of FIG. 50 showing the diameters of the aspides (dotted lines) and the length of the intervening parts of the exine (full lines). FIGS. 52, 53: four-pored grains (polar view); 27 μ; Visby 1933. — Size of expanded grains, according to ZANDER (1935) 21.2 μ (polar axis) by 23 μ (equatorial diameter). According to POTONIÉ (1934), the equatorial diameter ranges from 24 to 28 μ (herbarium material treated with sulphuric acid); according to LÜDI (1932), from 22.8 to 27.1 μ, the average measuring 24.3 μ.

In FIG. 51 (a five-pored grain), the ratio IA:A is 1.17; in FIG. 53 (a four-pored grain), it is 1.38. Drawings of four-pored grains published by HESMER (1929, FIG. 9; IA:A = 1.35) and POTONIÉ (1934, PL. 2, FIG. 34) apparently belong to *A. glutinosa;* this also applies to a five-pored grain photographed by WASSINK (1932, PL. II, FIG. 2; IA:A = 1.16).

Earlier records of fossil pollen of *A. glutinosa* (*e.g.* by STOLLER, SUNDELIN, and C. A. WEBER) are not supported by specific determination of the pollen grains. The finds were made in beds with megascopical remains of *A. glutinosa* or else in beds, where the presence of *A. incana* could reasonably be supposed to be excluded.

Alnus incana. — FIGS. 54, 55: five-pored grains (polar view); 27 μ; Salmis, Sweden. FIGS. 56, 57: four-pored grains (polar view); 24 μ; Strängnäs, Sweden. FIG. 58: three-pored grain (polar view); 21 μ; Connecticut. — Equatorial diameter 20.0 to 28.5 μ, generally about 24 μ (LÜDI 1932, POTONIÉ 1934*b*, WODEHOUSE). IA:A generally <1 (0.94 in FIG. 55, 0.96 in FIG. 57). Grains drawn by DOKTUROWSKY and KUDRJASCHOW (1924, FIG. 9*a* and *b*), belong apparently to this species (IA:A = 0.89).

In the three-pored grain, FIG. 58, IA:A exceeds 1. Three-pored grains are only seldom met with, at least in the typical European *A. incana.* The grain presented in FIG. 58 may eventually prove to have come from a somewhat different form. On the other hand, the grains in a pollen preparation of *A. incana* from Wisconsin were of the same morphological type as those of the European form.

The frequency of pollen grains with four pores varies but usually seems (contrary to a statement by POTONIÉ) to be greater than in *A. glutinosa.* The following percentages were obtained from specimens from Salmis and Härnön (values from the Härnö specimen in parentheses); 3 pores: 0(1); 4 pores: 29(57); 5 pores: 70(42); 6 pores: 1(0).

Alnus rugosa. — FIG. 59: four-pored grain, polar view; 21 μ; Connecticut. — Size about 17.8 by 21.5 μ. Pores four or five, rarely three or six, generally extremely narrow, often slit-shaped (WODEHOUSE). In the preparation from which FIG. 59 was drawn, 97 per cent of the grains were provided with four pores, 2 per cent with 3, and 1 per cent with 5 pores.

Alnus viridis. — FIGS. 60, 61: five-pored grains, polar view; 22 and 21 μ; Wallis, Switzerland. —Equatorial diameter 18.5 to 25.7 μ, averaging 21.7 μ (LÜDI 1932), according to POTONIÉ (1934*b*) 21–23 μ. IA:A>1. Arci generally less conspicuous, aspides less protruding, and exine thinner than in *A. glutinosa* and *A. incana.* By means of these characters (which, incidentally, are not too clear in FIG. 60) and the small size, this pollen type can easily be distinguished from pollen grains of *A. glutinosa* and *A. incana.*

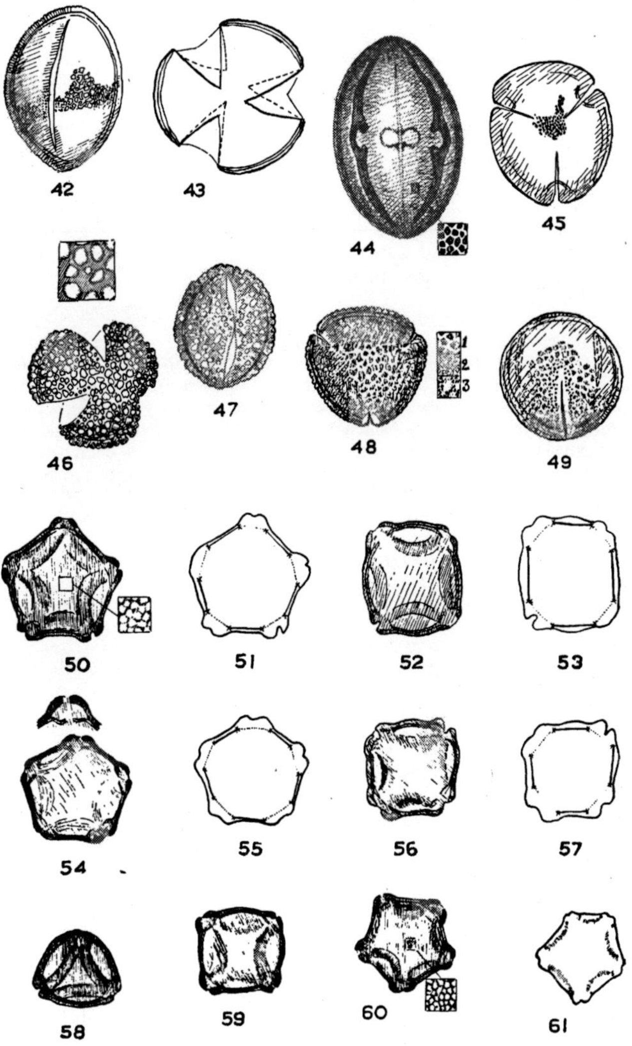

Plate IV (42–61). — *Aceraceae* (42, 43), *Anacardiaceae* (44, 45), *Aquifoliaceae* (46, 47), *Araliaceae* (48, 49), *Betulaceae* (*Alnus*) (50–61). — 42 and 43, *Acer saccharum;* 44 and 45, *Rhus typhina;* 46 and 47, *Ilex verticillata;* 48 and 49, *Hedera helix;* 50–53, *Alnus glutinosa;* 54–58, *A. incana;* 59, *A. rugosa;* 60–61, *A. viridis.*

Swamp-forest peat containing a large number of stumps, roots, etc. of alder and birch is usually very rich in pollen grains from these trees. It is, therefore, evident, that the influence of local pollen is usually rather extensive. In interpreting pollen diagrams from deposits with swamp-forest peat (carr peat, Bruchwaldtorf), due consideration should be given to the fact that in some cases swamp-forest peat indicates a transitional stage in the natural succession of the vegetation, whilst in other cases it may indicate a climatic change. Such a change may be due either to drier conditions, as when "Bruchwaldtorf" is directly formed on wet, paludose mineral soils [*cf.* FIRBAS 1923; the term paludose soils has been suggested by SERNANDER (Geol. Fören. Förhandl., p. 248, 1939) as substitute for the German „anmoorige Böden"].

In districts investigated by HALDEN (1917) and SUNDELIN (1919), the alders exhibit a predilection for shores and do not compete with the true forest trees. The alder pollen grains, therefore, were not included in the forest tree pollen total, but their frequency was expressed in the same way as the hazel pollen frequency: *i.e.* as a percentage of the forest tree pollen total. Furthermore, the alder pollen frequencies were often so high as to distort the pollen diagrams, if expressed in the usual way. The example given by HALDEN and SUNDELIN has been followed by several authors (*e.g.* PAUL and LUTZ 1939) although, for the sake of uniformity, it would be better to abandon it.

Betula (PLATE V, FIGS. 62–75). — Grains suboblate to spheroidal, less flattened than in the grains of *Alnus;* in polar view, more or less angular owing to the aspidate pores; equatorial diameter about 16 to 30 μ, seldom, as in *Betula utilis* (according to WODEHOUSE) 36.5–40 μ. Pores three, less frequently four. In *B. lutea,* occasional grains with five to seven pores have been found (WODEHOUSE). Pores circular, elliptical, or slit-shaped; when elongate, they are meridionally oriented (FIGS. 64, 73) or, when more than three, with their major axes converging in pairs (WODEHOUSE). Exine faintly reticulate.

As a rule, arci have been considered to be characteristic almost exclusively of alder pollen. POTONIÉ (1934*b*, p. 58), however, mentions that the grains of *Betula verrucosa* are provided with arci; and the author's own observations tend to confirm this statement. It seems that arci do in fact occur not only in *Betula,* where they are usually well developed, but also throughout the *Betulaceae,* even though they are sometimes only faintly noticeable. In polar view, the arci follow the grain contour rather closely contrary to the case with alder pollen. Consequently, they tend to conceal the marginal construction of the interaspidar exine and make it difficult, or even impossible, to follow the ekt-endexine suture from pore to pore.

FIGS. 62–75 were drawn before the general occurrence of arci in *Betula* had been fully realised. They are apparently idealised as the ekt- and exdexine contours are clearly shown in the interaspidar exine margin. On the other hand, arci happened to be indicated in FIGS. 71, 73, and 75. In equatorial view, the varying thickness of the exine, due to the presence of arci, is usually apparent. Thus, the exine in FIG. 73 is decidedly thicker at those points where the arci reach the margin of the figure. Arci in *Betula* pollen have also appeared in figures in earlier publications, *e.g.* BURRELL 1934, PL. X, FIG. *k*, but no critical comment on their occurrence has been made.

According to JENTYS–SZAFER (1928), the exine is composed of five layers:

« L'exine comprend cinq couches différentes, dont trois très minces et deux autres plus épaisses. Celles-ci s'étendent entre les couches minces et se dissolvent plus rapidement dans de l'acide chromique.... De toutes les couches formant l'exine, c'est la mince couche extérieure qui est la plus résistante et qui joue pour ainsi dire le rôle d'une cuticule entourant le grain de pollen. Elle résiste le plus longtemps à la dissolution dans l'acide chromique et à la décomposition du pollen fossile dans la tourbe. La couche extérieure est un peu plus épaisse à proximité des pores... et forme ici une sorte d'anneau encadrant ces petites ouvertures ».

The variations in the size of birch pollen may possibly be used under certain conditions as a means of determining the ratio of certain birch species in a given pollen spectrum. JENTYS–SZAFER (*l.c.*) has given a detailed outline of this possibility, summing up her experiences as follows: —

« *1.* Les grains de pollen de *Betula nana* L., *B. verrucosa* Ehrh. et *B. pubescens* Ehrh., se distinguent nettement entre eux par leur grosseur. La différence entre la grosseur des grains de pollen de *Betula nana* et de *Betula verrucosa* correspond à environ 3 microns en moyenne. On observe la même différence entre les grains de *B. verrucosa* et de *B. pubescens.*

2. Les grains de pollen de *Betula humilis* Schrank sont aussi gros que ceux de *B. verrucosa.*

3. La grosseur des grains de pollen de chaque espèce est charactérisée par une courbe dont l'étendue correspond à environs 8 microns.

4. La grosseur des grains de pollen fossiles équivaut à celle des grains actuels bouillis dans de la potasse caustique ou traités par du H_2SO_4 concentré.

5. La courbe correspondant à un mélange de deux espèces de grains de pollen dont la grosseur présente en moyenne une différence de 3 microns, est caractérisée par une étendue qui équivaut elle-même au total de l'étendue des courbes indiquant les pollens purs. Elle se distingue également par un sommet qui coïncide avec le milieu de son trajet, lorsque les deux espèces de pollen sont représentées en proportions égales dans le mélange.

6. A mesure que la quantité exprimée en pour-cents d'une espèce de pollen l'emporte sur la quantité de l'autre, on voit le sommet de la courbe se déplacer dans le sens du sommet caractéristique pour la première espèce.

7. Les courbes, qui correspondent à un mélange de trois espèces de pollen dont les grains diffèrent par la grosseur, sont surtout caractérisées par leur étendue.

8. Les différences morphologiques entre les grains de pollen de différentes espèces de Bouleaux, se maintiennent dans le matériel fossile, de sorte qu'on peut les reconnaître dans des grains typiques.

9. Nous pouvons conclure à la présence de grains de pollen de *Betula humilis* dans un échantillon de tourbe, non seulement d'après la structure de l'exine, mais aussi d'après le caractère de la couche qui révèle certaines conditions climatiques ».

Betula glandulosa. — FIG. 71: polar view; 20 μ; ex herb. Vancouver. — Grains provided with comparatively thick exine, large aspides, and well defined arci. Size of grains (according to YAMASAKI 1933) 27–30 μ.

Betula humilis. — FIG. 62: polar view; 25 μ; Federsee, Germany. — Equatorial diameter (in grains treated with caustic potash or concentrated sulphuric acid) 18.6–25.7 μ, averaging 21.5 μ (JENTYS–SZAFER 1928). Aspides comparatively small (DOKTUROWSKY and KUDRJASCHOW 1924).

Betula lenta. — FIGS. 72, 73: polar and equatorial view respectively; 25 μ; New Jersey. — Equatorial diameter 21.6–25.9 μ (WODEHOUSE).

Betula nana. — FIG. 63: polar view; 20 μ; Storuman, Sweden, 1938. FIG. 64: equatorial view; polar axis 19 μ; Hallstahammar, Sweden, 1938. FIG. 65: polar view; 20 μ; *ibid.* 1938. — Average equatorial diameter according to JENTYS-SZAFER (1928) about 18.6 μ; according to ENEROTH (FAEGRI 1936) 19.1 μ. Maximum size 22.5 μ; aspides relatively large (DOKTUROWSKY and KUDRJASCHOW 1924).

Betula nigra. — FIGS. 74, 75: polar and equatorial view respectively; 24 μ; Pennsylvania.

Betula pubescens. — FIG. 66: polar view; 27 μ; Uppsala, Sweden. — Equatorial diameter 20–28.6 μ, averaging about 24.5 μ (JENTYS-SZAFER 1928; ENEROTH in FAEGRI 1936).

Betula tortuosa. — FIG. 67: three-pored grain, polar view; 30 μ; Pesisvare, Sweden. FIG. 68: four-pored grain, polar view; 32 μ; Enafors, Sweden. — Average equatorial diameter according to ENEROTH (FAEGRI 1936) about 27.3 μ.

Betula verrucosa. — FIGS. 69, 70: polar and oblique equatorial view respectively; 25 μ; Västerås 1937. — Equatorial diameter 18.6–24.3 μ, averaging about 21.8 μ (JENTYS-SZAFER *l.c.;* according to ENEROTH in FAEGRI 1936 22.8 μ).

JENTYS-SZAFER has tried to apply the results of her statistical studies of recent birch pollen to investigations of fossil material. Variation statistics have also been tried by BERTSCH (1928). Because of the results obtained by JAESCHKE (1935), however, he has since doubted the value of variation statistics as an aid in birch pollen determinations.

In bogs in northwestern Germany, SCHUBERT (1933; *cf.* FIGS. 28–33) traced a parallel occurrence of *Betula nana* megafossils and pollens of *B. nana*-type. In the oldest ("pre-boreal") layers, pollen of *B. pubescens-* and *B. verrucosa*-type were found, and probably also a number of pollen grains of birch hybrids. The fluctuations in the frequency of these pollen types, however, were apparently capricious and hardly to be connected with any climatic changes.

According to ENEROTH (FAEGRI 1936), the maximum value of the pollen variation curves is 19.11 (18.6) in *Betula nana*, 22.84 (21.5) in *B. verrucosa*, 24.57 (24.3) in *B. pubescens-* and 27.3 μ in *B. tortuosa* (the figures in parentheses are given by JENTYS-SZAFER). ENEROTH's figures correspond to 7, 8, 9, and 10 scale fractions, respectively, of a Leitz measuring eye-piece, 10 ×, combined with objective no. 5 of the same make. The figures communicated by ENEROTH have been used by FAEGRI (*l.c.*), VON POST, and others. FAEGRI mentions changes from spectra with *B. nana-tortuosa* pollen to such with *B. pubescens*, and further to such with *B. nana-tortuosa, B. verrucosa-pubescens, B. verrucosa* pollen, etc. These changes appear in the same way in the diagrams of two different deposits and effectively substantiate the correlation based on curves of the pollen of other species.

On the other hand, F. and I. FIRBAS (1935) found a parallel decrease of size in *Betula* as well as in *Corylus* pollen in the older layers of two deposits in the Federsee region. The mean size of the birch pollen grains decreased from 20.9 μ (in the birch period; "*verrucosa-*

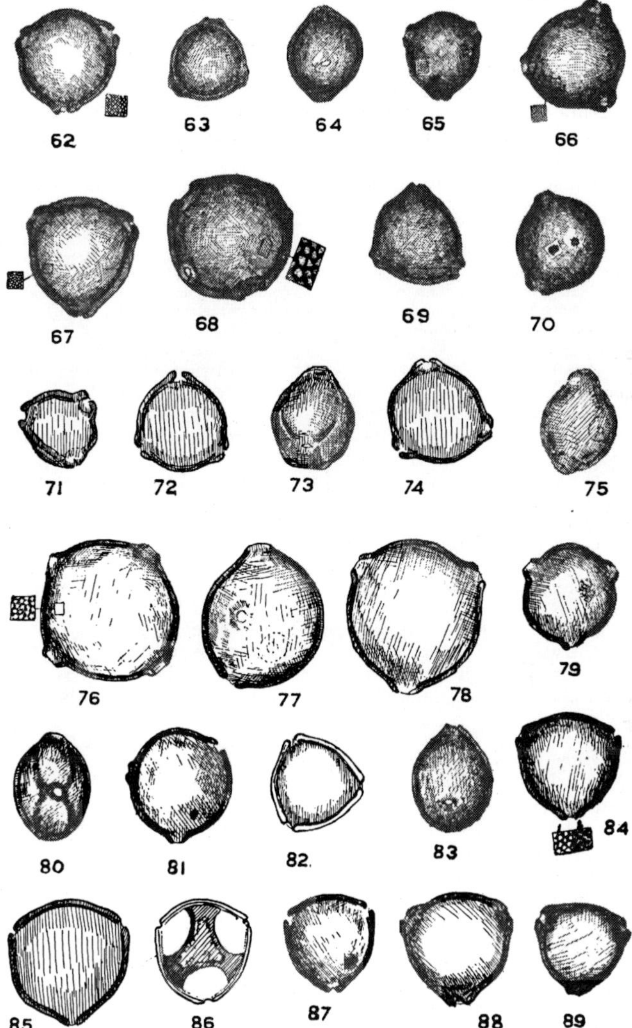

Plate V (62–89). — *Betulaceae* (62–89). — 62, *Betula humilis;* 63–65, *B. nana;* 66, *B. pubescens;* 67 and 68, *B. tortuosa;* 69 and 70, *B. verrucosa;* 71, *B. glandulosa;* 72 and 73, *B. lenta;* 74 and 75, *B. nigra;* 76–78, *Carpinus betulus;* 79, *C. caroliniana;* 80 and 81, *C. duinensis;* 82, *Corylus americana;* 83–87, *C. avellana;* 88, *C. californica;* 89, *C. chinensis.*

type") to 18.1 μ (in the younger pine period; "*nana*-type"), and finally (at the *Corylus* pollen maximum) to 16.7 μ, which is considerably less than the average given for *B. nana*. The material studied by Dr. and Mrs. FIRBAS was boiled a few minutes with ten per cent KOH, carefully washed with water and kept in glycerine for 8–10 months. The final conclusions arrived at were reported as follows:

„Offenkundliche Gesetzmässigkeiten in der Verteilung der Pollengrössen auf verschiedene Perioden, die auf einen Wechsel verschiedener Arten derselben Gattung zurückgehen könnten, können also durch ganz andere Faktoren bedingt werden, neben verschiedener Aufbereitung durch die verschiedene Einwirkung des umgebenden Mediums im Laufe der Zeit. Da wir wenigstens derzeit kein Kriterium haben, um im Einzelfall zu entscheiden, welche der verschiedenen Möglichkeiten zutrifft, wird man den Versuch, eine verlässliche Artdiagnose von *Pinus* und *Betula* allein auf Grund der Pollengrösse zu geben, bis auf weiteres ganz aufgeben müssen. Die bisherigen Angaben dieser Art sind als unzureichend gesichert zu betrachten".

Carpinus (PLATE V, FIGS. 76–81). — Grains of about the same morphological type as those of *Betula*. They are, however, more rounded and the exine is thinner. In some species, arci are fairly well developed, *e.g.* in *Carpinus americana* var. *tropicalis* and *C. duinensis* (FIG. 80). In other species, they are sometimes hardly discernible.

The grains are psilate, provided with a faint reticulate texture and generally have three or four, rarely five or six, aspidate pores. The pore pattern is somewhat similar to that in *Betula*, but the ektexine surrounding the apertures is not thickened and rises more distinctly from the general surface of the grain. The pores are circular, eventually broadly elliptical and, according to WODEHOUSE, usually operculate.

Carpinus betulus. — FIG. 76: polar view of four-pored grain; 37 μ; Öred, Sweden. FIG. 77: equatorial view of four-pored grain; 36 by 32 μ; *ibid.* FIG. 78: polar view of three-pored grain; 39 μ; *ibid.* — Pores usually four, occasionally three or five; diameter 36–40 μ (WODEHOUSE).

Carpinus caroliniana. — FIG. 79: polar view; 27 μ; Connecticut. — Diameter 22–30 μ (POTONIÉ 1934*b*), 27–31 μ (WODEHOUSE). Pores three or four, occasionally five or six; apertures circular, about 4 μ in diameter (WODEHOUSE).

Carpinus duinensis. — FIG. 80: equatorial view; 28 by 21 μ; Duino, Italy. FIG. 81: polar view; 28 μ; *ibid.*

The development of the postglacial distribution of *Carpinus betulus* in Europe shows a gradual spreading northward and westward from a centre in southeastern Europe. Particularly high *Carpinus* pollen frequencies have been found in older (interglacial) deposits; for example, frequencies as high as 88 per cent have been reported by DOKTUROWSKY (1932) from deposits in White Russia.

Corylus (PLATE V, FIGS. 82–89). — Grains approximately 22 to 31 μ in diameter, psilate, aspidate, usually provided with three pores equally spaced on the equator. In polar view, the grains are more or less triangular; in equatorial view suboblate, the relation between the polar axis and the equatorial diameter being about 8.4:10. Pore apertures slightly elliptical or more or less circular; ektexine gradually only slightly raised above the surface of the grain at the pores; arci reduced, often hardly discernible; texture faintly reticulate.

Corylus americana. — FIG. 82: polar view; 25 μ; Connecticut. — According to WODEHOUSE the grains are uniform in size, 22.8 μ in diameter.

Corylus avellana. — FIG. 83: equatorial view; 28 by 22 μ; Västerås 1938. FIG. 84: polar view; 28 μ; *ibid.* FIG. 85: fossil grain; 32 μ; Cushendun, Ireland (the large size can probably be attributed to the chemical reagents — hydrofluoric acid, chlorine, acetic anhydride, etc. — used in preparing the sample for analysis). FIG. 86: polar view: 26 μ; Västerås 1938 (boiled in 10 per cent NaOH). FIG. 87: polar view; 25 μ; *ibid.* (heated in trichloroacetic acid). — Size about 30 μ (KNOLL 1930), 25.7–30.4 μ (FERRARI 1927; in var. *rubra* according to the same author 21–23 μ), 20.2 by 25 μ (ZANDER 1935). Out of 2000 recent grains 99.8 per cent had three, 0.2 per cent two or four pores (SANDEGREN in VON POST 1924).

Corylus californica. — FIG. 88: polar view; 30 μ; California. — The grains of this species seem to have slightly more protruding pores and with more sharply marked arci than the grains of the species previously mentioned.

Corylus chinensis. — FIG. 89: polar view; 24 μ; Kweichow, China. — Pores and arci essentially as in *C. californica.*

In pollen analysis, the frequency of hazel pollen is customarily expressed as a percentage of the forest tree pollen total because hazel usually belongs to the undergrowth and not to the dominating, competing forest types (LAGERHEIM in HOLST 1909, p. 29).

The result of a few analyses by LAGERHEIM led HOLST to believe that earlier hazel had a greater frequency in southernmost Sweden than at the present time and that its gradual decrease was due to the increase of the true forest trees — particularly the oak — which, through shading, gradually killed off the hazel. This idea was followed up by VON POST, who found up to 55 per cent of hazel pollen in peats covered by the sediments of the Litorina sea (VON POST 1918). He later advanced a theory according to which at the beginning of the postglacial period of warmth hazel, in some parts of southern Sweden, did not appear simply as undergrowth in mixed oak forests and in alder forest of Auwald type but took the place of certain deciduous trees: elm, oak, ash, and linden. In so doing the hazel probably formed groves of its own. This was at least, in VON POST's opinion, the case in such places where the "*Corylus* index", or the ratio between the hazel pollen frequency and the total pollen frequencies of alder and the mixed oak forest constituents, attains particularly high values, as 40–50 or even more [the normal value, at least for southern Sweden, is about 0.5–1.0 (VON POST 1920)].

There are other observations in favour of these ideas; notably the fact that the hazel pollen maximum seems to coincide with a tree pollen minimum (tree pollen frequency expressed as the number of tree pollen grains per sq. cm in a preparation). However, it may also be possible that this minimum, which was first observed by VON

Post (1919), is due to an increase of such trees as aspen and ash, the pollen of which are not found in bogs at all or else only in limited amounts. Later (1924) von Post spoke of a "hazel forest". Much discussion has been aroused among phytogeographers as to the real nature of these forests which seem to have no apparent counterpart at the present time, although von Post has drawn attention to the hazel groves of northern Norway, north of the polar limit of oak, linden, and elm.

In connection with the hazel forest problem Erdtman (1929, 1931) has discussed the relation between the fossil pollen flora of a bog and its surface receptivity, *i.e.* its capacity to absorb and preserve pollen grains. It must be observed that there is a fundamental difference between dead, inactive bogs and living bogs, where peat is still formed. Bogs of the first category do not catch and preserve recent pollen grains, whereas bogs of the second category do so if they are not in a temporarily inactive state. In countries with a comparatively warm and wet climate, the living bogs may not have any inactive periods at all, but may be active practically all the year round; while in countries with a continental climate, the active period may be short. Hazel sheds its pollen very early;. in Portugal (Coimbra) even before the New Year (usually about December 27th); at Aberystwyth in Wales, about January 13th; in Bremen, February 23rd; at Karlskrona in southern Sweden, March 10th; in the neighbourhood of Stockholm, about April 6th. Bogs in countries with a maritime climate will be in an active state on these dates, while bogs in countries with a continental climate would still remain inactive. Granted that equal amounts of hazel pollen are distributed over the surfaces of the bogs, pollen records taken will not give the same figures because the oceanic bogs will show a larger admixture of hazel pollen than the continental bogs.

"Hazel forests" have been reported from Sweden, Britain, France, Switzerland, Roumania, etc. The large hazel pollen frequency may, to some extent, be due to earlier springs, which have given the hazel pollen a better chance of being absorbed and preserved in the bogs than is true at the present time. In a review of Erdtman's paper "The Boreal Hazel Forests and the Theory of Pollen Statistics" Rudolph (Ber. über die wiss. Biologie, vol. 18, 1931) mentions that a climatic change would not only cause eventual changes in the surface receptivity of the bogs, but also bring about a change in the flowering time of hazel. This would tend to make more or less invalid the arguments advanced by Erdtman. A climatic shift, however, would not necessarily produce parallel changes in the flowering time of hazel and in the activity of bogs.

Particularly high hazel pollen frequencies have often been encountered in strata which are supposed to have been formed during a "climatic optimum", postglacial as well as interglacial (Jessen and Milthers 1938, Dokturowsky 1931, 1932). Russian interglacial peats have yielded up to 237 per cent of hazel pollen, exhibiting a remarkable contrast to the post-glacial deposits with their much lower values. In some interglacial bogs frequencies even surpassing 237 per cent are said to have been found (Dokturowsky 1931; *cf. e.g.* fig. 2, p. 255). Some of these figures are, however, of rather uncertain value, inasmuch as they are based only upon a few grains (on one occasion, three tree and four hazel pollens were recorded, and the hazel pollen frequency, accordingly, was diagrammed as 132 per cent).

Ostrya (PLATE VI, FIGS. 90–93). — Grains spheroidal; exine thin, particularly the endexine which is only easily discernible at high magnifications. Pores usually three or four, circular (in *O. virginiana* only with decidedly elliptical apertures according to WODEHOUSE). Exine provided with a faint, reticulate texture.

Ostrya carpinifolia. — FIG. 90: polar view; 29 μ; Tyrol. — „Der Pollen von *Ostrya* ist durch relativ dünne klare Exine, nur schwach verdickte Ausstrittsstellen und stark kugelförmige Gestalt charakterisiert, recht unscharfe Merkmale. Immerhin wohl in den meisten Fällen zu erkennen" (FIRBAS 1923). Size 21.0–23.4 μ (FERRARI 1927); about 24 μ (FIRBAS *l.c.*). The size of fossil grains, belonging probably to this species, averages 30 μ (FEURSTEIN 1933).

Ostrya virginiana. — FIGS. 91, 92: three- and four-pored grains in polar view; 28 and 25 μ respectively; Connecticut. FIG. 93: fossil grain; 29 μ; Itasca, Minn. — About one-third of the grains four-pored; grains 28 by 25 μ in diameter; apertures of the pores 4.5 μ long, pore membranes marked with a slight fleck (WODEHOUSE).

Campanulaceae (PLATE VI, FIGS. 94, 95): —

Campanula rotundifolia. — FIG. 94: polar view; 35 μ; Ramnäs, Sweden. FIG. 95: equatorial view; 37 by 31 μ; *ibid.* — Grains echinate, in polar view circular, in equatorial view often more or less flattened (suboblate); pores usually three or four; exine thick, provided with a reticular texture.

Caprifoliaceae (PLATE VI, FIGS. 96–99): —

Linnaea borealis. — FIG. 96: polar view; 40 μ. — Grains spheroidal to suboblate, subechinate, tricolpate; furrows short. Size of grains of *L. borealis* f. *arctica* 57–63 μ (YAMASAKI 1933). On several occasions, pollen grains of *L. borealis* have been found in raw-humus samples from northern Sweden.

Viburnum opulus. — FIGS. 97, 98: equatorial view; FIG. 97: 28 by 21 μ; FIG. 98: 27 μ; Fårö, Sweden. FIG. 99: polar view; 25 μ; *ibid.* — Grains subprolate to spheroidal, tricolpate. Exine reticulate; reticulum gradually disappearing near the margins of the furrows; muri of a more or less beaded appearance. Size 21.1–23.4 μ (FERRARI 1927).

Caryophyllaceae (PLATES VI, FIGS. 100–106, and VII, FIGS. 107–114): —

Pollen grains, as far as known, of two types: cribellate — the predominating type — and colpate (tricolpate; *cf.* FIGS. 113, 114). Grains of the former type may sometimes easily be mistaken for pollen of the *Chenopodiaceae.* Their pores are usually circular in outline, crossed by a membrane which is flecked with a number of granules and provided with a special marginal area. The texture of the exines is usually striking, often of a more or less complicated appearance. It seems that in certain cases generic and even specific determinations may be made on pollenmorphological characters.

Cerastium alpinum. — FIG. 100: pollen from Kongsvold, Norway; 42 μ.

Honckenya peploides. — FIG. 101: pollen from Visby, Sweden; 43 μ.

Lychnis flos-cuculi. — FIG. 102: pollen from Gotland, Sweden; 34 μ. — Size 32.8 - 37.4 μ (FERRARI 1927), 32 by 30 μ (ZANDER 1935); pores 25–30 (ZANDER *l.c.*).

Melandrium rubrum. — FIG. 103: pollen from Hortus Bergianus, Stockholm, 1935; 40 μ. — Size 31.6 μ (ZANDER *l.c.*).

Sagina intermedia. — FIG. 105: pollen from Storlien, Sweden; 28 μ.

Sagina nodosa. — FIG. 104: pollen from Öland, Sweden; 35 μ.

Scleranthus perennis. — FIGS. 109, 110; pollen grains from Gotland, Sweden; 36 μ.

Silene acaulis. — FIG. 106: pollen from Storlien, Sweden; 28 μ. — Size 25.7–30.4 μ (FERRARI *l.c.*).

Silene maritima. — FIG. 108: pollen from Gotland, Sweden; 54 μ. — Size 43 μ (ZANDER 1941).

Silene nutans. — FIG. 107: pollen from Åker, Sweden; 40 μ. — Size 42 by 40.6 μ (ZANDER 1935).

Spergula marina. — FIG. 113: polar view; 23 μ; Koön, Sweden. FIG. 114: equatorial view; 24 by 18 μ; *ibid.*

Stellaria aquatica. — FIG. 112: pollen from Västerås; 35 μ; about 12 pores.

Stellaria uliginosa. — FIG. 111: pollen from Skagershult, Sweden; 29 μ.

Casuarinaceae (PLATE VII, FIG. 115): —

Casuarina equisetifolia. — FIG. 115: polar view; 29 μ; Egypt. — Pollen grains psilate, aspidate, of a somewhat betuloid appearance. Pores three, equatorial; ektexine thick, slightly raised at the pores; texture fine, reticulate.

Ceratophyllaceae (PLATE VII, FIG. 116): —

Ceratophyllum demersum. — FIG. 116: grain from Uppland, Sweden, 1938; 43 μ. — Grains spheroidal, psilate, acolpate. Exine very thin, consisting, apparently, of only one layer. The drawing presents a grain which has been treated with NaOH. If subjected to acetolysis, the grains are either completely destroyed or more or less seriously damaged.

Chenopodiaceae (PLATE VII, FIGS. 117–120): —

Grains spheroidal, cribellate, resembling the cribellate grains of the *Caryophyllaceae.* Pores usually nearly circular in outline, crossed by a delicate membrane flecked with a number of granules which may be aggregated toward the centre and even fused to form a central mass resembling an operculum (WODEHOUSE). Size of grains as well as number of pores frequently widely varying within the species, making a correct identification difficult.

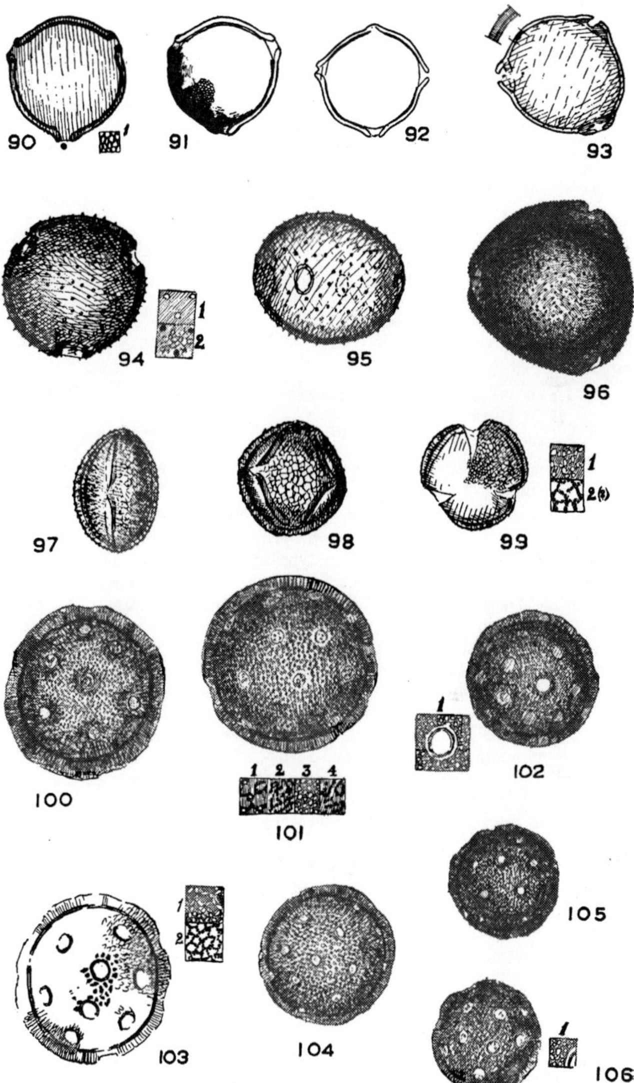

PLATE VI (90–106). — *Betulaceae* (90–93), *Campanulaceae* (94, 95), *Caprifoliaceae* (96–99), *Caryophyllaceae* (100–106). — 90, *Ostrya carpinifolia;* 91 and 92, *O. virginiana;* 94 and 95, *Campanula rotundifolia;* 96, *Linnaea borealis;* 97–99, *Viburnum opulus;* 100, *Cerastium alpinum;* 101, *Honckenya peploides;* 102, *Lychnis flos-cuculi;* 103, *Melandrium rubrum;* 104, *Sagina nodosa;* 105, *S. intermedia;* 106, *Silene acaulis.*

Atriplex latifolium. — FIG. 117: pollen from Gotland, Sweden; 23 μ.

Chenopodium album. — FIG. 118: pollen from Västerås; 34 μ. — Grains with granular texture, apparently caused by long structural elements, radiating from the endexine.

Chenopodium glaucum. — FIG. 119: pollen from Visby, Sweden; 19 μ.

Salicornia herbacea. — FIG. 120: pollen from Minnesota; 24 μ.

Plants belonging to this family are often characteristic of salt marshes and the occurrence of *Chenopodiaceae* pollen in marsh peat and certain sediments may be an indication of changes in the shore line, etc. (ERDTMAN 1921, p. 135; HALDEN 1922, p. 21). From the Oldenbroker Moor in northwestern Germany, OVERBECK and SCHMITZ (1931, p. 90) report the following "*Chenopodiaceae-Alsinoideae*" pollen frequencies: —

175 cm *Sphagnum* peat, slightly humified	0.0–2.0	per cent
75 cm *Eriophorum-Sphagnum* peat	0.0–1.3	per cent
125 cm *Phragmites* peat, upper part	8.0–17.0	per cent
125 cm *Phragmites* peat, lower part, clayey	8.0–30.0	per cent
195 cm clay, with *Phragmites* and marine fossils	8.0–158.0	per cent
40 cm alder brushwood peat	1.0–3.0	per cent

It may be added, however, that the label "*Chenopodiaceae-Alsinoideae* pollen" is somewhat inappropriate since pollen grains of the *Alsinoideae* are of much the same type as those of the *Silenoideae* (*cf.* PLATES VI, VII). Furthermore, *Spergula* and *Spergularia*, often referred to *Alsinoideae*, have divergent (tricolpate) grains [*Spergula* and *Spergularia* are now usually referred to the subfamily *Paronychioideae* of the *Caryophyllaceae* (WETTSTEIN, Handbuch der systematischen Botanik, 4. Aufl., 1935) or to a separate family, the *Illecebraceae* (HUTCHINSON, The Families of Flowering Plants, I, 1926)]. To avoid misunderstandings, the group-name *Chenopodiaceae-Alsinoideae* may be replaced by "*Centrospermae*" (*cf. e.g.* GROSS 1937), a more non-committal yet considerably wider term. H. and M. E. GODWIN (1933) also speak of *Chenopodiaceae-Alsinoideae* pollen in a survey of a peat bed from the Fenland district of eastern England. The pollen flora of this bed is interpreted as indicating a vegetational succession from a salt-marsh with 3 to 7 per cent "*Chenopodiaceae-Alsinoideae*" pollen via brackish-water and willow-alder-brushwood, to oak-wood, and subsequent retrogression via salt-marsh to marine or brackish-water conditions again. Extremely high *Chenopodiaceae* pollen frequencies have been found by ERNST (1934) and BRINCKMANN (1934) in samples from the unforested coasts of northwestern Germany (maximum values 240 and 290 per cent respectively). *Chenopodiaceae* pollen has also been encountered in countries with an extreme continental climate and may eventually come to be used as an indication of aridity (DEEVEY 1937).

Compositae (PLATES VII, FIGS. 121–128, and VIII, FIGS. 129–141): —

The pollen morphology of this family has been partly described, in very great detail, by WODEHOUSE (*l.c.*, pp. 457–540). Further in-

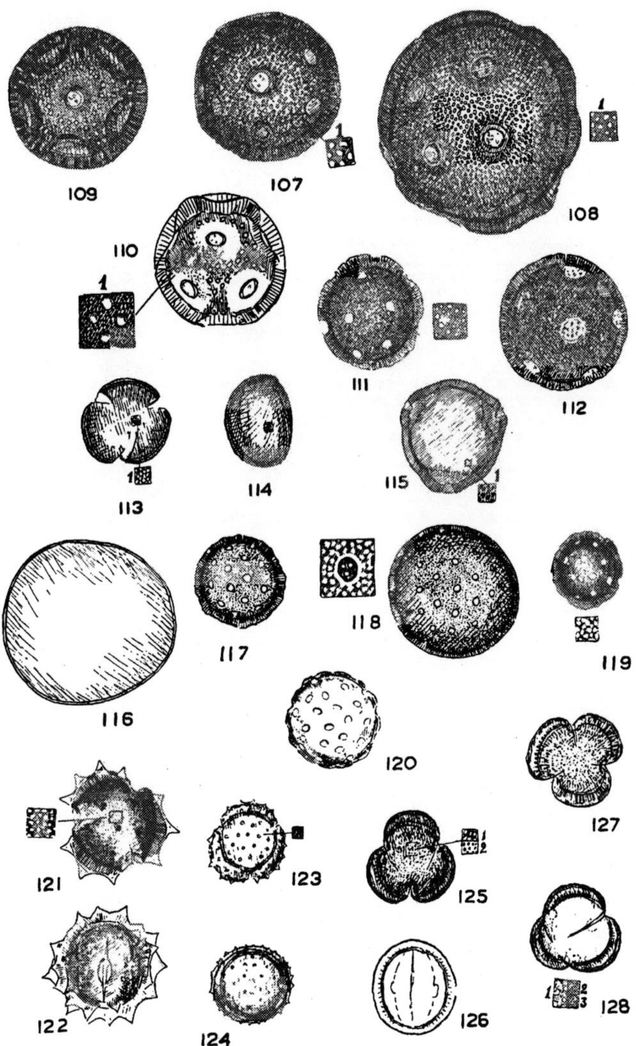

Plate VII (107–128). — *Caryophyllaceae* (107–114), *Casuarinaceae* (115), *Ceratophyllaceae* (116), *Chenopodiaceae* (117–120), *Compositae* (121–128). — 107, *Silene nutans;* 108, *S. maritima;* 109 and 110, *Scleranthus perennis;* 111, *Stellaria uliginosa;* 112, *S. aquatica;* 113 and 114, *Spergula marina;* 115, *Casuarina equisetifolia;* 116, *Ceratophyllum demersum;* 117, *Atriplex latifolium;* 118, *Chenopodium album;* 119, *C. glaucum;* 120, *Salicornia herbacea;* 121 and 122, *Achillea millefolium;* 123 and 124, *Ambrosia maritima;* 125 and 126, *Artemisia borealis* var. *bottnica;* 127, *A. maritima;* 128, *A. vulgaris* f. *tilesii.*

formation is to be found in the works by FISCHER (1890) and ZANDER (1935).

Achillea millefolium. — FIG. 121: polar view; 28 μ; Gotland, Sweden. FIG. 122: equatorial view; 28 μ. — Pollen grains echinate, with spines conspicuous, broadly conical, but sharply pointed. Exine very thick and coarsely granular. Grains 22.5 to 34.2 μ in diameter; spines 2.3 to 4.6 μ long and 4.5 to 1.4 μ apart (WODEHOUSE, p. 500).

Achillea belongs to the tribe *Anthemideae*, the pollen grains of which are characterized by WODEHOUSE (pp. 496, 497) as follows:

"Grains normally tricolpate. Furrows generally long and sharply defined, tapering to pointed ends, their membrane smooth, each provided with a conspicuous germ pore. Exine moderately to exceedingly thick; in insect-pollinated members provided with broad, conical sharp pointed spines; in wind-pollinated members, with spines greatly reduced or entirely absent.

In the echinate grains the exine, if observed in optical section, is seen to consist of two layers. The inner is thicker and appears to be built up of large vertical prisms presenting the appearance of coarse, radial striae. Overlying this is the much thinner layer of more transparent material marked with very fine radial striae. In the non-echinate grains, in which the exine is thinner, the same sort of texture prevails, but usually it is much finer".

Ambrosia maritima. — FIG. 123: polar view.; 20 μ; Crete. FIG. 124: equatorial view; 21 by 20 μ. — The pollen grains of the *Ambrosiinae* (*Ambrosia*, *Iva*, *Xanthium*, etc.) are described by WODEHOUSE (pp. 516, 517) as

"spheroidal or oblately flattened, 16.5–30 μ in diameter, generally tricolpate. Furrows various, long and tapering, of medium length or merely rounded pits only slightly meridionally elongate, almost coinciding with their enclosed germ pores. Exine rather thick, but generally less so than in the grains of entomophilous *Compositae;* generally provided with spines which are short-conical or rounded, or vestigial, less frequently with spines well developed and sharp pointed".

The ektexine is flattened at the poles more than the endexine. In polar view, a single grain resembles somewhat a tetrad with the endexine apparently occupying the position of the top pollen in a tetrahedral tetrad.

Artemisia (FIGS. 125–133). — Grains, when expanded, spheroidal or oblately flattened, 17.5 to 28.5 μ in diameter, normally tricolpate. Furrows long and tapering, their membranes smooth, provided with a germinal aperture. Exine thick; spine vestiges small or absent. In optical section, the grains usually appear rounded-triangular in shape; the exine is thickest in the middle of the lunes, gradually tapering in thickness, in sweeping curves, to the edges of the furrows (WODEHOUSE, pp. 511, 512).

Artemisia borealis var. *bottnica.* — FIG. 125: polar view; 22 μ; Buteå, Sweden. FIG. 126: equatorial view (scheme); 24 by 22 μ.

Artemisia maritima. — FIG. 127: polar view; 21 μ; Gotland, Sweden.

Artemisia vulgaris. — FIG. 130: polar view; 18 μ; Gotland. FIGS. 129, 131, 133: equatorial projections of FIG. 130; polar axe about 22 μ. FIG. 132: meridional section of a pollen grain; orientation as in FIG. 129. — WODEHOUSE (1936, PLATE II) has figured a grain measuring 28.5 μ.

Artemisia vulgaris f. *tilesii.* — FIG. 128: oblique polar view; 21 μ;

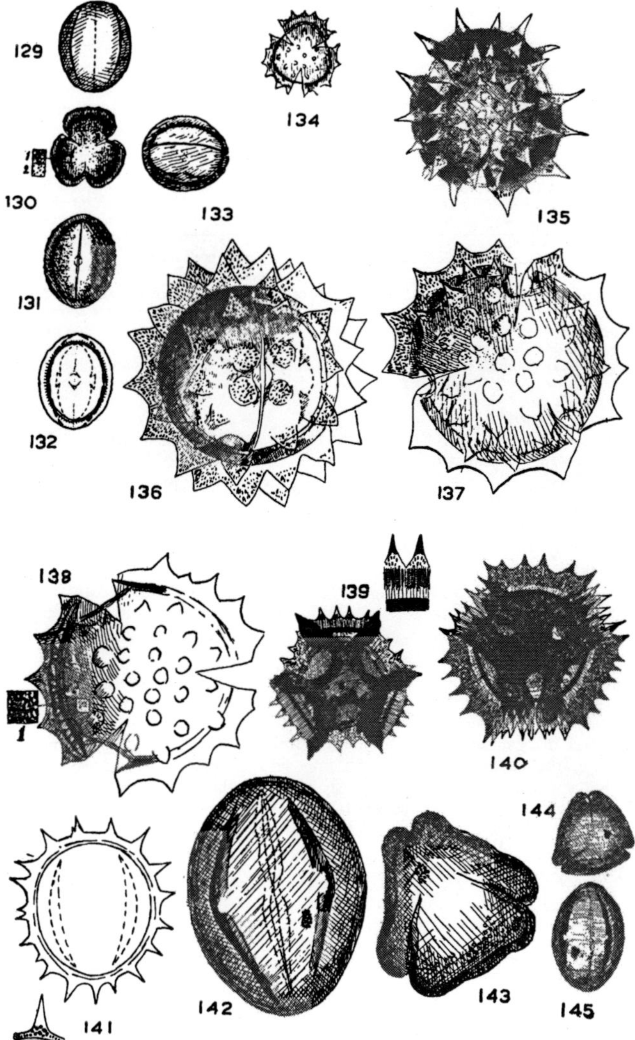

PLATE VIII (129–145). — *Compositae* (129–141), *Cornaceae* (142–145). — 129–133, *Artemisia vulgaris;* 134, *Baccharis halimifolia;* 135, *Bidens tripartita;* 136, *Carduus acanthoides;* 137, *Cirsium palustre;* 138, *C. oleraceum;* 139, *Crepis paludosa;* 140, *Mulgedium lapinum;* 141, *Solidago virgaurea;* 142 and 143, *Cornus amomum;* 144 and 145, *C. suecica.*

Novaya Zemlya. — The white dots in FIG. 128:1 represent vestigial spines, the smaller white and black dots in FIG. 128:2 and 3 texture ("mesexinous elements"). Size 18 to 23 μ (HØEG 1923).

Baccharis halimifolia. — FIG. 134: polar view; 17 μ; Connecticut. — *Baccharis* and *Solidago* (FIG. 141) are referred to the *Astereae.* The grains of this group are spheroidal or slightly flattened.

"In size they range from about 16.5 to about 32 μ in diameter. They are always provided with well developed and characteristic spines which are uniform in size and present the appearance of uniformity of distribution over the surface. The spines are short, broad at the base, and nearly conical in shape. Sometimes they are strictly conical to their apexes, and sometimes they taper slightly into a more or less acuminate tip. The length of the spines and their distance apart are somewhat various in the different species, and there is evidence that measurements of the spine lengths and spine intervals may be used to distinguish some of the genera. The grains of the *Astereae* always have a granular texture, which is rather faint between the spines, but surrounding their bases it is coarser and somewhat more sharply defined; the structure does not extend far up the shaft of the spine (*cf.* FIG. 141), which throughout most of its upper part is quite smooth and homogeneous in appearance" (WODEHOUSE, pp. 488, 489).

Bidens tripartita. — FIG. 135: polar view; 46 μ; Gotland. — *Bidens* is referred to the *Heliantheae,* the relationship of which to the *Ambrosiinae,* according to WODEHOUSE, is unquestionably indicated by the morphology of their pollen grains. Size of grains 30–37 μ (YAMASAKI 1933); 24 μ (ZANDER, 1935).

Carduus acanthoides. — FIG. 136: equatorial view; 66 μ; Gotland. — Size 38 by 37 μ (ZANDER 1935).

Cirsium oleraceum. — FIG. 138: polar view; 56 μ; Degeberga, Sweden, 1938. — The figures clearly indicate the ektexinous nature of the furrows and the endexinous nature of the germ pores. As in *Carduus,* the spines possess a well defined texture from the base to the top. Size 47 by 46 μ (ZANDER 1935).

Cirsium palustre. — FIG. 137: polar view; 61 μ; Gotland.

Crepis paludosa. — FIG. 139: polar view; 37 μ; Svabensverk, Sweden. — Size 30.2 by 29.8 μ (ZANDER 1935). In *Crepis,* as in the *Liguliflorae* in general, the grains are more or less globular, usually tricolpate, the outer surface thrown into high ridges. The ridges are provided with prominent sharp spines, while the floors of the lacunae, enclosed by the ridges, are covered by a thin layer of smooth exine. The pattern of the grains in *Crepis,* as well as in *Mulgedium* (FIG. 140), is essentially the same as in *Taraxacum,* where, according to WODEHOUSE (pp. 458–460):

"the normal pattern consists of 15 lacunae which are of definite shape and arrangement. The three lacunae which encompass the three pores are hexagonal in form, but with two (meridional) gaps in their ridges. If the grains be observed in polar view, the poral lacunae will be in side view, so that only one bounding ridge of each is seen, but they are easily recognized by the gaps just mentioned. In polar view there will also be seen six more lacunae, three of which are adjacent to and in meridional line with the poral lacunae. These are the "abporal" lacunae. Alternating with the pores, and between the abporal lacunae, are the three broad "paraporal" lacunae. Over the pole is seen a rather large triangular or hexagonal area of thickened exine of variable extent — the polar thickening. Since both hemispheres are alike, the whole pattern of the normal *Taraxacum* pollen grain comprises 3 poral, 6 abporal, and 6 paraporal lacunae, marking a total of 15".

Mulgedium alpinum. — FIG. 140: polar view; 48 μ; Duved,

Sweden. — Dry grains 37.2 by 36 μ (ZANDER 1935); *cf.* also the description in connection with *Crepis paludosa*.

Solidago virgaurea. — FIG. 141: equatorial view; 50 by 42 μ; Gotland. — Size 21–27.5 μ (FERRARI (1927), 20–23 μ (YAMASAKI 1933); *cf.* also the description given above concerning *Baccharis*.

As a rule, the pollen grains in the *Compositae* are transported by insects. Nevertheless, they are often found embedded in peat. In many cases, the genus, or even the species, can be determined by means of pollen grain characters. Thus, pollen grains of *Mulgedium alpinum* and *Saussurea alpina* have been identified in peats in northern Sweden. As yet, however, *Compositae* pollen has not received much attention, probably because fossil grains of this family (in accordance with the varied ecological habits) do not appear in the same way as, for example, the pollen grains of *Cyperaceae*, *Ericaceae*, and *Gramineae*.

Particularly high *Compositae* pollen frequencies have been recorded from bogs in Ohio by DRAPER (1929) and from Germany by ERNST (1934; 176 per cent in a sample of recent *Sphagnum*, 440 per cent in shore sand; both samples from Föhr, North Frisean Islands).

The pollen grains of *Artemisia* are shed in large quantities and are widely distributed by the wind. The sages frequently grow in steppe or shore regions; hence the tracing of fossil *Artemisia* pollen may help elucidate certain climatic and geological relations of the past (*cf.* also WODEHOUSE 1935*b*). The author has found sage pollen in bogs in northwestern Europe, from old layers, such as laminated clays, all the way up to the youngest strata. However, more information is much needed, particularly since misleading illustrations have repeatedly appeared and have contributed to a wide-spread confusion regarding "salicoid" pollen, a group to which *Artemisia* pollen often, although without reason, has been referred.

Cornaceae (PLATE VIII, FIGS. 142–145): —

Grains tricolpate, furrows long, tapering. Exine with fine reticulate, or granular, texture. Furrow membranes nearly smooth or slightly flecked, each with a well-defined aperture. Perhaps the most characteristic feature about the grains is the slight inwardly projecting thickening which borders the furrows, more pronounced around the pores (WODEHOUSE).

Cornus amomum. — FIG. 142: equatorial view; 63 by 48 μ; Connecticut. FIG. 143: oblique polar view; 43 μ.

Cornus suecica. — FIG. 144: polar view; Åre, Sweden. FIG. 145: equatorial view; 29 by 19 μ; *ibid.* — Size of grains, according to YAMASAKI (1933), 30–33 μ.

Cruciferae (PLATE IX, FIGS. 146, 147): —

Crambe maritima. — FIG. 146: oblique polar view; 29 μ; cult., Västerås 1938. FIG. 147: equatorial view; 30 μ. — Grains spheroidal, tricoplate; muri of a more or less beaded appearance.

Desfontaineaceae (PLATE IX, FIGS. 148, 149): —

Desfontainea spinosa. — FIG. 148: equatorial view; 28 by 34 μ; Chile. FIG. 149: polar view. — Grains suboblate, tricolpate; furrows short; exine pitted.

Dipsacaceae (PLATE IX, FIGS. 150, 151): —

Scabiosa succisa. — FIG. 150: equatorial view; 99 by 91 μ; Åker, Sweden, 1938. FIG. 151: part of a grain; oblique polar view. — Grains spheroidal, echinate, usually (if correctly? *Cf.* ZANDER 1935, p. 298) referred to as tricolpate. Furrows short, closed by small opercula. Exine provided with a thick "palisade layer". Size 80–96 by 70–90 μ (FIRBAS 1931), 84.2 by 82.4 μ (ZANDER *l.c.*).

Droseraceae (PLATE IX, FIGS. 152–154): —

Drosera. — Grains united in tetrads, generally tetrahedral in arrangement. Exine of the distal surface of the grain thick, rigid, and provided with spines; that of the proximal surface soft and flexible and thrown into plaits which converge toward the innermost point of the grain where it joins with its three neighbours at the centre of the tetrad (WODEHOUSE). The flowers are sometimes cleistogamous. Pollen tetrads are occasionally found in peat and are easily identifiable.

Drosera intermedia. — FIG. 152: tetrad about 57 μ; equatorial diameter of single grain about 42 μ; Uppland, Sweden. — Spines short, blunt.

Drosera longifolia. — FIG. 153: tetrad about 72 μ; equatorial diameter of single grain about 50 μ; Uppland, Sweden. — Large tetrads with characteristic granular texture on the distal side of the grains, between the spines.

Drosera rotundifolia. — FIG. 154: tetrad about 55 μ; equatorial diameter of single grain about 34 μ. — Tetrads much smaller than in *D. longifolia.* Spines comparatively longer and more sharply pointed than in *D. intermedia.*

Elaeagnaceae (PLATE IX, FIGS. 155–158): —

Hippophaë rhamnoides. — FIG. 156: oblique polar view; 29 μ; Ålvkarleby, Sweden. FIGS. 157, 158; grains in equatorial view; 25 by 27 and 25 by 29 μ respectively. — Grains spheroidal to suboblate, tricolpate, with narrow, tapering furrows, each with a single germ pore. In surface view, the pore resembles the figure 8 laid sideways. Exine with reticulate texture; at the germ pores the ektexine is somewhat raised above the general surface of the grain.

A beautiful species of virgin land, *Hippophaë* is climatically rather indifferent. It disappears under increasing competition from other shrubs and trees. In Scandinavia, it is confined almost exclusively to sea-shores, continuously rejuvenated by isostatic land upheaval. In only one place (Junker Valley in Norway, near the Arctic Circle) it lingers on the mountains, possibly a relict from those times when it was widely spread over the Scandinavian inland as well. Its pollen output

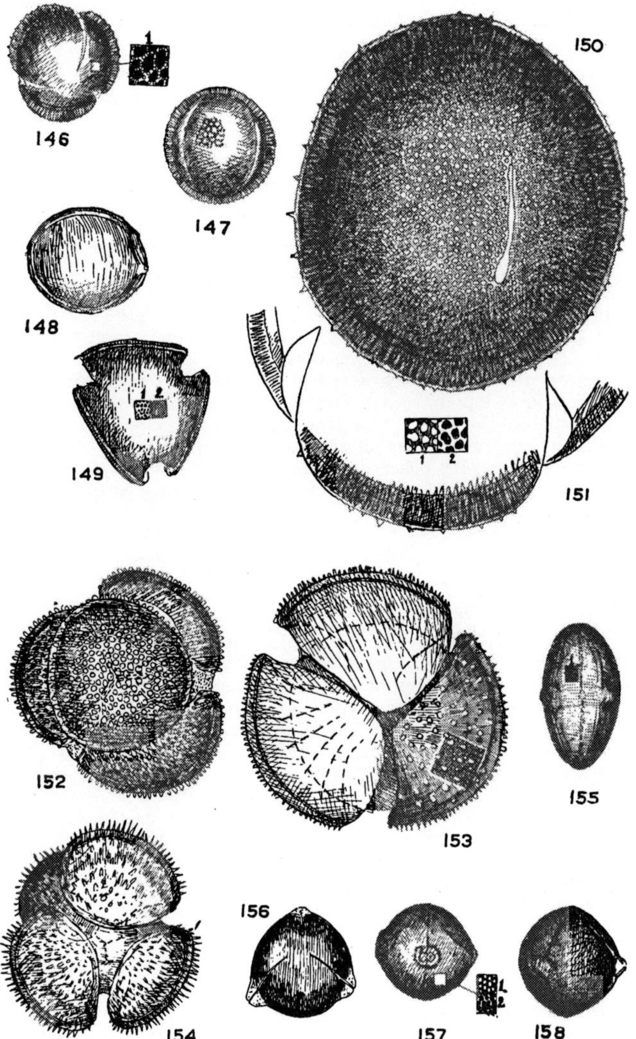

Plate IX (146–158). — *Cruciferae* (146, 147), *Desfontaineaceae* (148, 149), *Dipsacaceae* (150, 151), *Droseraceae* (152–154), *Elaeagnaceae* (155–158). — 146 and 147, *Crambe maritima;* 148 and 149, *Desfontainea spinosa;* 150 and 151, *Scabiosa succisa;* 152, *Drosera intermedia;* 153, *D. longifolia;* 154, *D. rotundifolia;* 155, *Shepherdia canadensis;* 156–158, *Hippophaë rhamnoides.*

may not be considerable; according to FIRBAS (1934) it may possibly be of the same magnitude as that of *Salix.*

The pollen frequency is usually expressed in the same way as the frequency of hazel pollen: *i.e.* as a percentage of the forest tree pollen total. FAEGRI (1936), however, contends that this is illogical, since *Hippophaë* does not appear as an undergrowth but represents special areas. Low frequencies (usually less than 2 per cent) of *Hippophaë* pollen have been encountered in many old postglacial sediments: *e.g.* in southern Sweden (VON POST 1924, NILSSON 1935 etc.) and northern Germany (SCHÜTRUMPF 1935, BOEHM-HARTMANN 1937, GROSS 1937 etc.), under conditions from which it might be inferred that *Hippophaë* was growing in these countries during the period of transition from tundra to subarctic birch or pine forests. The author has found up to 40 per cent *Hippophaë* pollen (also stellate hairs in great profusion) in old lake sediments near Stor-Uman in southern Lapland; 32 to 48 per cent *Hippophaë* pollen were found by BENRATH and JONAS (1937) in certain soil profiles from northern Germany. FIRBAS (1934) has also studied the pollen flora of recent samples from Baltrum (East Frisean Isles), which has an abundant growth of *Hippophaë*, but no forest. *Hippophaë* pollen frequencies from 27 to 50 per cent are quoted, but they are somewhat dubious, since they are based only upon 5 to 19 grains.

Shepherdia canadensis. — FIG. 155: equatorial view; 39 by 21 μ; Wyoming. — Grains prolate, tricolpate; each furrow with a single germ pore.

Empetraceae (PLATE X, FIGS. 159–162): —

Empetrum. — Grains similar to those in the *Ericaceae,* united in tetrahedral tetrads, tricolpate, furrows fairly short, narrow, tapering. Each is flattened against its three neighbours in the tetrad, the flattening extending very near to the equator of the grains. The furrows of each cell contiguous and continuous with those of its three neighbours directly across the edges of their faces of contact. Each furrow encloses a single germinal aperture near its point of contact with the furrow of the neighbouring grain. This is indicated in the detailed figure in the lower right hand corner of FIG. 160. A narrow, ektexinous furrow is seen crossing the contact surfaces of two grains. Its tapering ends are surrounded by what seems to be endexinous thickenings. These formations taper slightly towards the ends of the furrow and are abruptly cut at the equators. The germinal apertures, therefore, appear as narrow channels between the endexinous contact walls on one side and the endexinous thickenings on the other. Comparison may be made with *Ericaceae* pollen (*e.g. Erica tetralix,* FIG. 173), where the germinal apertures sometimes look like equatorial transverse furrows. The ektexine appears to be confined to the general surface of the tetrad, forming a thin, smooth coat, cut by the twelve colpae, which unite to form six composite furrows with 12 germinal apertures in all.

Empetrum hermaphroditum. — FIG. 159: tetrad, Enafors, Sweden; diameter about 47 μ; equatorial diameter of single grains about 34 μ. — Diameter of the tetrads 34.0 to 47.6 μ (*cf.* TAB. 7), mean value 40.68 ± 0.20 μ. Corresponding figures in *Empetrum nigrum:* 22.1 to

34.0, mean value 27.86 ± 0.14 μ. The tetrads of the tetraploid *E. hermaphroditum* are usually much larger than those of the diploid *E. nigrum*. Tetrads of *E. nigrum* are, as a rule, easily recognizable, but it may sometimes be difficult to make a safe distinction between tetrads of *E. hermaphroditum* and of diverse ericaceous plants, *e.g. Ledum palustre*. Tetrads described and illustrated by SKOTTSBERG (1901) probably belong to *E. hermaphroditum*.

Table 7: DIAMETER RANGE IN 200 TETRADS OF *Empetrum hermaphroditum* AND THE SAME NUMBER OF TETRADS OF *E. nigrum:* —

Diameter in μ	22.1	23.8	25.5	27.2	28.9	30.6	32.3	34.0	35.7	37.4	39.1	40.8	42.5	44.2	45.9	47.6
Emp. nigrum	1	8	34	**72**	47	31	6	1	—	—	—	—	—	—	—	—
E. hermaphroditum	—	—	—	—	—	—	—	6	15	19	27	**60**	37	26	9	1

Empetrum nigrum. — FIGS. 160, 161: tetrad (scheme). FIG. 162: tetrad; diameter about 29 μ; equatorial diameter of single grains about 21 μ. — Diameter of tetrads 22.1–34.0 μ (*cf.* TAB. 7); mean value 27.86 ± 0.14 μ. Diameter according to OVERBECK (1934) 23–32 μ. Outline more angular and wall thicker than in the tetrads of *E. hermaphroditum.*

The tetrads of *Empetrum rubrum* (diameter about 29 μ; specimen from W. Falkland) and *Ceratiola ericoides* (diameter about 30 μ; specimen from Florida) are of quite similar shape.

As yet little is known about the occurrence of *Empetrum* in European late-quaternary deposits. Systematic attempts to determine the tetrads of both species, *E. hermaphroditum* and *E. nigrum*, have not been made, although this possibility was incidentally mentioned by FAEGRI in 1936 and later more fully demonstrated by ERDTMAN (1938).

Subrecent samples from the arctic heath of the isle of Heinäsaari off Petsamo in northern Finland yielded 400–800 per cent *Empetrum* tetrads (FIRBAS 1934). In late glacial sediments at Dannenberg near Bremen *Empetrum* tetrads occur fairly regularly, attaining a maximum frequency of 25 per cent (OVERBECK and SCHNEIDER 1938). Samples from sediments of similar age collected at Roundstone, western Ireland, by JESSEN, yielded about 100 per cent *Empetrum* pollen, all or most of them of the nigrum-type (78 *Empetrum* tetrads and 76 " tree " pollen in all were recorded; of the last were 37 *Betula*, 4 *Pinus*, and 35 *Salix;* analysis by ERDTMAN). " *Ericaceae* tetrads " encountered in some bogs in Tierra del Fuego are probably tetrads of *Empetrum rubrum* (VON POST 1930).

Ericaceae (PLATE X, FIGS. 163–177): —

Pollen grains as a rule united in tetrads of the same type as those of the *Empetraceae.*

Andromeda polifolia f. *acerosa.* — FIGS. 164, 165: tetrads from Björkliden, Sweden; diameter 39 μ. — A tetrad of *A. polifolia*, figured by BEIJERINCK (1935), measures 65 by 60 μ. OVERBECK (1934) describes the tetrads of the same species as follows: „Oberflächenstruktur sehr variabel, meist unregelmässig wabig-netzig. Keimfalten lang, scharf begrenzt, und spitz auslaufend. Grösse der Tetrade 33 bis 45 μ, häufigster Wert 40 μ. Die kleineren Varianten dieses Pollens sind wohl kaum mit Sicherheit gegen *Vaccinium*-typen abzugrenzen, wenn nicht rauhe Exemplare mit stark wabig-netziger Struktur vorliegen".

Arbutus unedo. — FIG. 163: tetrad (Palermo): diameter 48 μ. — Tetrads rounded, almost spherical. Exine with very faint, reticular or granular, texture. Outline very smooth.

Arctostaphylos uva-ursi. — FIG. 166: tetrad (Gotland 1935); diameter 49 μ. — Tetrads rounded; diameter, according to OVERBECK (1934), about 42 μ (38–45 μ), according to ZANDER (1935), about 46 μ. Exine with faint, reticular texture. Outline not absolutely smooth if examined with highest resolving power.

Bruckenthalia spiculifolia. — FIG. 167: polar view (Transylvania); 22 μ.

Calluna vulgaris. — FIG. 168: tetrahedral tetrad (Gotland); 40 μ. FIG. 169: rhomboidal tetrad (Lönshult, Sweden, 1938); 52 by 44 μ. — Easily recognizable tetrads with thick exine and more or less uneven surface. Furrows comparatively short, poorly defined. Size of tetrads 32–44 μ (FIRBAS 1931), 27–37 μ (OVERBECK 1934; mean size about 32 μ), 32 by 38 μ (ZANDER 1935). The heather sheds pollen in great profusion [according to POHL (Beih. Bot. Centralbl., Abt. A, vol. LVI, 1937) about 4000 millions of tetrads per square meter in areas covered with dense *Calluna* growth]. As to the transportability of the tetrads opinions are divided. According to HESMER (Zeitschr. Jagd- u. Forstwesen, vol. LXV, 1933) the pollen settles on the spot or in the immediate vicinity; according to JONASSEN (Bot. Tidsskr., vol. 43, Copenhagen 1935, p. 190) it may be carried in great amounts by the wind.

Erica arborea. — FIG. 170: tetrad (Morocco); 27 μ. — Tetrads comparatively small and thick-walled; exine with well defined granular texture. Size 28–33 μ (FIRBAS 1927), 20–30 μ (FIRBAS 1931).

Erica carnea. — FIG. 171: tetrad (cult., Lund, Sweden); 32 μ. — Size of tetrads 28 μ (ARMBRUSTER and OENIKE 1929); 31.8 by 26.2 μ (ZANDER 1935).

Erica cinerea. — Tetrads 36 by 50 μ (FIRBAS 1931; OVERBECK 1934).

Erica stricta. — FIG. 172: equatorial view; 22 μ; Catania. — Grains single (SAMUELSSON, Svensk Bot. Tidskr., vol. 7, 1913), tricolpate; furrows long; germinal pores elliptical, transverse; exine flecked.

Erica tetralix. — FIG. 173: tetrad (Lönshult, Sweden, 1938); 45 μ. — Tetrads somewhat similar to those of *Calluna* — particularly as to exine structure — but as a rule tetrahedral and provided with comparatively longer furrows with more sharply defined margins. Size 32–42 μ (FIRBAS 1931), 30 to 40 μ, averaging 35 μ (OVERBECK 1934), 37.4 by 35 μ (ZANDER 1935).

Kalmia latifolia. — FIG. 174: tetrad (Connecticut); 45 μ. — Furrows short, not well defined. Size 40–55 μ (HANSGIRG 1897).

Ledum palustre. — FIG. 175: tetrad (Hallstahammar, Sweden, 1938); 35 μ. — Tetrads thick-walled, fairly similar to those of *Empetrum;* diameter according to OVERBECK (1934) 28.4–36.7 μ (averaging

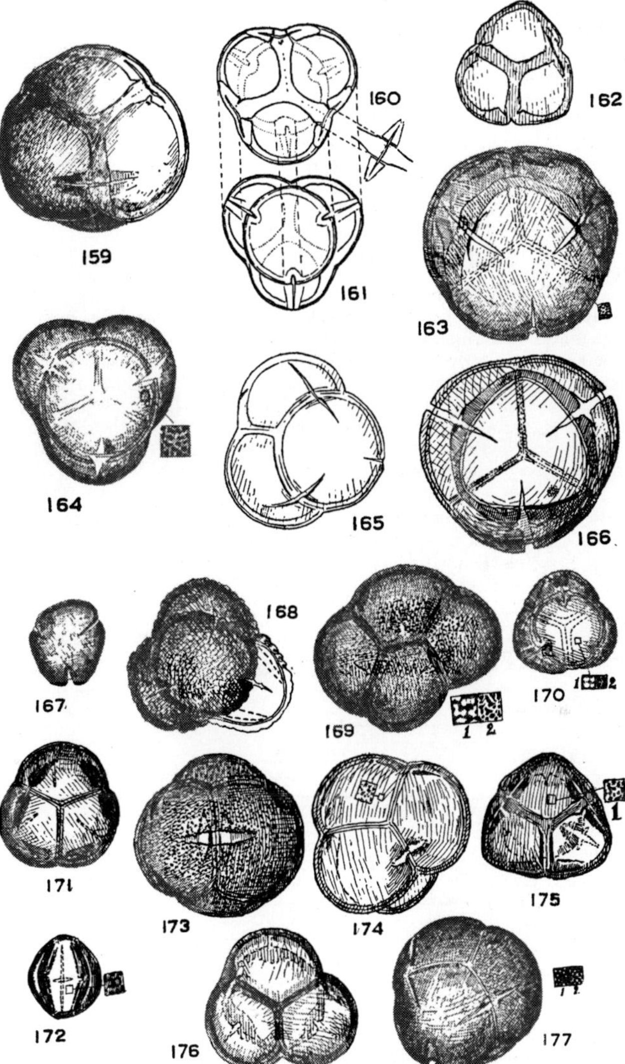

PLATE X (159–177). — *Empetraceae* (159–162), *Ericaceae* (163–177). — 159, *Empetrum hermaphroditum;* 160–162, *E. nigrum;* 163, *Arbutus unedo;* 164 and 165, *Andromeda polifolia* f. *acerosa;* 166, *Arctostaphylos uva-ursi;* 167, *Bruckenthalia spiculifolia;* 168 and 169, *Calluna vulgaris;* 170, *Erica arborea;* 171, *E. carnea;* 172, *E. stricta;* 173, *E. tetralix;* 174, *Kalmia latifolia;* 175, *Ledum palustre;* 176, *Loiseleuria procumbens;* 177, *Vaccinium oxycoccus.*

33.4 μ). The tetrads of *L. latifolium* are very similar. Size 32.2 by 28.6 μ (ZANDER 1935).

Loiseleuria procumbens. — FIG. 176: tetrad (Enafors, Sweden); 38 μ. — Diameter of tetrads 23.4–35 μ; averaging 28.4 μ (OVERBECK 1934).

Vaccinium oxycoccus. — FIG. 177: tetrad (Hallstahammar, Sweden, 1938); 40 μ. — „Oberfläche fast glatt, nur bei Immersion wird oft eine feine Körnelung sichtbar. Membran zart, Farbe gelblich. Keimfalten lang und nur schwach umrandet. Grösse der Tetrade 28.4 bis 36.7 μ, häufigster Wert 33.4 μ " (OVERBECK 1934). Size 32–40 μ (FIRBAS 1931).

High *Ericaceae* pollen frequencies are usually found in deposits grown over or surrounded by heath. If forests are absent the frequencies are particularly large. In samples from an Orcadian bog, for example, 30 tree pollen and 547 *Ericaceae* tetrads, *i.e.* about 1800 per cent, were found (ERDTMAN, Journ. Linn. Soc. (Bot.), vol. XLVI, 1924). A reexamination of peat samples from the Shetland Islands revealed still higher frequencies — up to 11,000 per cent, if calculated in the usual way — as shown in TAB. 8.

Table 8: TREE POLLEN AND *Ericaceae* TETRAD FREQUENCIES IN NINE SAMPLES FROM A DEPOSIT IN THE SHETLAND ISLANDS. The percentages are based on the total number of pollen grains (tetrads counted as single grains) and pteridophyte spores: —

DEPTH UNDER THE SURFACE (CM)	NUMBER OF POLLEN GRAINS AND PTERIDOPHYTE SPORES	TREE POLLEN (PER CENT)	*Ericaceae* TETRADS (PER CENT)
20 (*Vaginatum*-peat)	1084	0.64	96.9
40	105	0.9	99.1
60	356	2.0	91.5
120 (do., much humified)	2579	4.0	93.0
130	2007	0.8	96.5
190 (sedge peat, much humified)	757	14.0	34.5
200	1044	9.6	76.0
218	82	35.0	13.0
236	193	9.0	5.0

In northwestern Germany JONAS (Fedde's Rep., Beih. LXXXVI, 1936, p. 8) reports up to 5000 per cent of *Ericaceae* tetrads. In samples of late glacial sediments from southwestern Norway, *Ericaceae* tetrads form about 5 to 50 per cent of the non-tree pollen flora and their frequency sometimes even exceeds the highest tree pollen frequencies.

Thus it seems evident that ericaceous heath existed in the neighbourhood during the formation of these sediments [FAEGRI, Bergens Mus. Årsbok 1935 (1936)].

On the other hand FIRBAS found only from 1 to 10 per cent of *Ericaceae* tetrads in young-diluvial peats at Marga, near Dresden, supposed to have been formed under climatic conditions similar to those now prevailing at the polar tree limit (FIRBAS and GRAHMANN, Abh. math.-phys. Kl., Sächs. Akad. Wiss., vol. XL, 1928). Likewise, OVERBECK and SCHMITZ (1931) state that *Ericaceae* tetrads are sparse or lacking in " pre-boreal " layers in northwestern Germany. They also claim that the occurrence of *Ericaceae* tetrads is very erratic, often of an extremely local nature. In pollen diagrams, the *Ericaceae* curves, therefore, often change abruptly from high values to low, and vice versa. In *Sphagnum* peat, such changes are undoubtedly associated with the presence of alternating " Bulten " (tussock) and " Schlenken " (pool) layers. In spite of the well-known instability of the *Ericaceae* curves, OVERBECK and SCHMITZ conclude that the heaths of northwestern Germany are of an anthropogenous origin. There is evidence to show that they date back to the " sub-Boreal " period: *i.e.* roughly contemporary with the Bronze age. Natural heath was probably confined to particularly exposed coastal districts. Submerged peats, dredged in the Weser estuary, have furnished some evidence to show that this kind of heath is of greater antiquity. It may have existed from early " Atlantic ", or even from " Boreal " times.

FIRBAS has often contributed to the heath question from the point of view of pollen statistics. His results „ zwingen zu der Annahme, dass die *Calluna*-Werte selbst dann, wenn das Untersuchungsgebiet vollständig bewaldet wäre, dort, wo das Heidekraut an Ort und Stelle wuchs und reichlich zur Blüte gelangte, offenbar noch eine Höhe zwischen 100 und 200% vielleicht auch noch mehr, erreichen könnten. Wenn also bei pollenanalytischen Untersuchungen von Torfen und Humusböden an der Stelle der Probeentnahme oder in unmittelbarer Nähe *Calluna* gewachsen sein kann, besagen selbst Werte zwischen 100 und 200% *Calluna*-Pollen nichts über ein Vorhandensein offener Heide. Sie sind selbst in geschlossenen Waldgebieten zu erwarten und dürfen daher nicht als Ausdruck geringer Waldbedeckung betrachtet werden. Ist das Gebiet nicht vollständig bewaldet — und in den nordwestdeutschen Küstengebieten beträgt der Anteil natürlicher Waldböden infolge der grossen Ausdehnung waldloser Hochmoore, Rohrsümpfe und Salzwiesen im Umkreis von vielen km oft noch weniger als 50% — dann sind selbst *Calluna*-Werte bis zu 350% (und mehr?) noch kein Beweis für ein Vorkommen waldloser Heide " (FIRBAS 1937, p. 15).

A striking example which shows that high *Ericaceae* pollen frequencies may not always indicate heath has been given by AUER (Acta Geogr., vol. 5:2, Helsinki), who encountered great amounts of *Ericaceae* pollen (*cf. Pernettya*) in peats from the rain-forest area of Tierra del Fuego (up to 98 per cent of the total number of pollen grains and spores).

References: —

ARMBRUSTER, L. und OENIKE, G., 1929: Die Pollenformen als Mittel zur Honigherkunftsbestimmung (Bücherei f. Bienenkunde, vol. X).

BEIJERINCK, W., 1933: De oorsprong onzer Heidevelden (Nederl. Kruidkundig Arch., vol. 43).

BENRATH, W. and JONAS, F., 1937: Zur Entstehung der Ortstein-Bleichsandschichten an der Ostseeküste (Planta, vol. 26).

BERTSCH, K., 1928: Wald- und Florengeschichte der Schwäbischen Alb (Veröff. Staatl. Stelle f. Natursch., Württemb. Landesamt f. Denkmalpfl., H. 5).

BOEHM-HARTMANN, H., 1937: Spät- und postglaziale Süsswasserablagerungen auf Rügen, I (Arch. Hydrobiol., vol. XXXI).

BRINCKMANN, P., 1934: Zur Geschichte der Moore, Marschen und Wälder Nordwestdeutschlands, III. Das Gebiet der Jade (Bot. Jahrb., vol. LXVI).

BURRELL, W. H., 1924: Pennine peat (The Naturalist).

DEEVEY, E. S., 1937: Pollen from interglacial beds in the Pang-gong valley and its climatic interpretation (Amer. Journ. Sci., vol. XXXIII).

DOKTUROWSKY, W. S., 1931: Zur Flora der Inter- und Postglazialen Ablagerungen der USSR. (All Ukrain. Acad. Sci., vol. II).

—— —— 1932: Neue Angaben über die interglaziale Flora in der USSR. (Abh. Nat. Ver. Brem., vol. XXVIII).

—— —— und KUDRJASCHOW, W., 1924: Schlüssel zur Bestimmung der Baumpollen im Torf (Geol. Arch., vol. 3).

DRAPER, P., 1929: A comparison of pollen spectra of old and young bogs in the Erie basin (Proc. Okla. Acad. Sci., vol. 9).

ERDTMAN, G., 1921: Pollenanalytische Untersuchungen von Torfmooren und marinen Sedimenten in Südwest-Schweden (Ark. Bot., vol. 17).

—— —— 1924: Mitteilungen über einige irische Moore (Svensk Bot. Tidskr., vol. 18).

—— —— 1929: Review of FIRBAS and GRAHMANN: Über jungdiluviale und alluviale Torflager in der Grube Marga bei Senftenberg (*Ibid.*, vol. 23).

—— —— 1931: The boreal hazel forests and the theory of pollen statistics (Journ. Ecology, vol. XIX).

—— —— 1936: New methods in pollen statistics (Svensk Bot. Tidskr., vol. 30).

—— —— 1938: Pollenanalys och pollenmorfologi (*Ibid.*, vol. 32).

ERNST, O., 1934: Zur Geschichte der Moore, Marschen und Wälder Nordwestdeutschlands, IV. Untersuchungen in Nordfriesland (Schr. Naturwiss. Ver. Schlesw.-Holst., vol. XX).

FAEGRI, K., 1936: Quartärgeologische Untersuchungen im westlichen Norwegen, I. Über zwei präboreale Klimaschwankungen im südwestlichsten Teil (Bergens Mus. Årbok 1935, Nat.-vet. rekke, no. 8).

FERRARI, A., 1927: Osservazioni di biometria sul polline delle Angiosperme (Atti Ist. Bot. Univ. Pavia).

FEURSTEIN, P., 1933: Geschichte des Viller Moores und des Seerosenweihers an den Lanser Köpfen bei Innsbruck (Beih. Bot. Centralbl., vol. LI).

FIRBAS, F., 1923: Pollenanalytische Untersuchungen einiger Moore der Ostalpen (Lotos, vol. 71).

—— —— 1928: Beiträge zur Kenntnis der Schieferkohlen des Inntals und der interglazialen Waldgeschichte der Ostalpen (Ztschr. Gletscherkunde, vol. XV).

—— —— 1931: Über die Waldgeschichte der Süd-Sevennen und über die Bedeutung der Einwanderungszeit für die nacheiszeitliche Waldentwicklung der Auvergne (Planta, vol. 13).

—— —— 1934: Über die Bestimmung der Walddichte und der Vegetation waldloser Gebiete mit Hilfe der Pollenanalyse (*Ibid.*, vol. 22).

—— —— 1935: Die Vegetationsentwicklung des mitteleuropäischen Spätglazials (Bibl. Bot., Heft 112).

—— —— 1937: Ein nordböhmischer Beitrag zur pollenanalytischen Behandlung der Heidefrage (Natur u. Heimat, 8. Jahrg.).

—— —— und FIRBAS, I., 1935: Zur Frage der grössenstatistischen Pollendiagnosen (Beih. Bot. Centralbl., vol. LIV).

FISCHER, H., 1890: Beiträge zur vergleichenden Morphologie der Pollenkörner (Breslau).

GODWIN, H. and M. E., 1933: Pollen analyses of Fenland peats at St. Germans, near King's Lynn (Geol. Mag., vol. LXX).

GROSS, H., 1937: Nachweis der Allerödschwankung im süd- und ostbaltischen Gebiet (Beih. Bot. Centralbl., vol. LVII).

HALDEN, B., 1917: Om torvmossar och marina sediment inom norra Hälsinglands litorinaområde (Sveriges Geol. Unders., ser. C, no. 280).

—— —— 1922: Tvänne intramarina torvbildningar i norra Halland (*Ibid.*, ser. C, no. 310).

HANSEN, H. P., 1937: Pollenanalysis of two Wisconsin bogs of different age (Ecology, vol. 18).

HANSGIRG, A., 1897: Beiträge zur Biologie und Morphologie des Pollens (Sitz.-Ber. Böhm. Ges. Wiss., Prag, Math.-Nat. Cl., vol. 39).
HESMER, H., 1929: Mikrofossilien in Torfen (Palaeont. Ztschr., vol. 11).
HØEG, O. A., 1924: Pollen on bumble-bees from Novaya Zemlya (Rep. scient. res. Norv. Exp. N. Z. 1921, no. 27).
HOLST, N. O., 1909: Postglaciala tidsbestämningar (Sveriges Geol. Unders., ser. C, no. 216).
IVERSEN, J., 1941: Land occupation in Denmark's Stone Age (Danm. Geol. Unders., II. R., no. 66).
JAESCHKE, J., 1935: Zur Frage der Artdiagnose der *Pinus silvestris*, *Pinus montana* und *Pinus cembra* durch variationsstatistische Pollenmessungen (Beih. Bot. Centralbl., vol. LII).
JENTYS-SZAFER, J., 1928: La structure des membranes du pollen de *Corylus*, de *Myrica* et des espèces européennes de *Betula* et leur détermination à l'état fossile (Bull. Acad. Polon. Sc. Lettr., Cl. Sc. Mat., Sér. B).
JESSEN, K. and MILTHERS, V., 1928: Stratigraphical and Paleontological Studies of Interglacial Fresh-water Deposits in Jutland and Northwest Germany (Danmarks Geol. Unders., II. R., no. 48).
KNOLL, F., 1930: Über Pollenkitt und Bestäubungsart (Ztschr. f. Bot., vol. 23).
LÜDI, W., 1932: Die Waldgeschichte der Grimsel (Beih. Bot. Centralbl. vol. XLIX).
NILSSON, T., 1935: Die pollenanalytische Zonengliederung der spät- und postglazialen Bildungen Schonens (Geol. Fören. Förhandl., vol. 57).
OVERBECK, F., 1934: Zur Kenntnis der Pollen mittel- und nordeuropäischer *Ericales* (Beih. Bot. Centralbl., vol. LI).
—— —— und SCHMITZ, H., 1931: Zur Geschichte der Moore, Marschen und Wälder Nordwestdeutschlands, I. Das Gebiet von der Niederweser bis zur unteren Ems (Mitt. Prov.-St. Naturdenkmalpfl. Hannover, H. 3).
—— —— und SCHNEIDER, S., 1938: Mooruntersuchungen bei Lüneburg und bei Bremen und die Reliktnatur von *Betula nana* L. in Nordwestdeutschland (Ztschr. f. Bot., vol. 33).
PAUL, H. and LUTZ, J., 1939: Zur Kenntnis der Moore des Oberpfälzer Mittellandes (*Ibid.*, vol. 34).
VON POST, L., 1918: Skogsträdspollen i sydsvenska torvmosselagerföljder (Forhandl. 16. Skand. Naturforskermøte, Kristiania).
—— —— 1919: Ett par offerdammar från Skånes bronsålder (Rig, vol. 2).
—— —— 1920: Postarktiska klimattyper i södra Sverige (Geol. Fören. Förhandl., vol. 42).
—— —— 1924: Some features of the regional history of the forests of southern Sweden in post-arctic time (*Ibid.*, vol. 46).
—— —— 1930: Die Zeichenschrift der Pollenstatistik (*Ibid.*, vol. 46).
POTONIÉ, R., 1934*a*: Zur Mikrobotanik der Kohlen und ihrer Verwandten, I. Zur Morphologie der fossilen Pollen und Sporen (Arb. Inst. Paläobot. Petrogr. Brennst., vol. IV).
—— —— 1934*b*: Zur Mikrobotanik der Kohlen und ihrer Verwandten, II (*Ibid.*, vol. IV).
SCHUBERT, E., 1933: Zur Geschichte der Moore, Marschen und Wälder Nordwestdeutschlands, II. Das Gebiet an der Oste und Niederelbe (Mitt. Prov.-St. f. Naturdenkmalpfl. (Hannover), H. 4).
SCHÜTRUMPF, R., 1935: Pollenanalytische Unterschungen der Magdalenien- und Lyngby-Kulturschichten der Grabung Stellmoor (Nachrichtenbl. f. Deutsche Vorzeit, 11. Jahrg.).
SKOTTSBERG, C., 1901: Einige blütenbiologische Beobachtungen im arktischen Teil von Schwedisch Lappland 1900 (Bih. Vet.-Akad. Handl., vol. 27).
SUNDELIN, U., 1919: Über die spätquartäre Geschichte der Küstengegenden Östergötlands und Smålands (Bull. Geol. Inst. Upsala, vol. XVI).
WASSINK, E. C., 1932: Abbildungen von Baumpollen aus dem Torf (*In* FLORSCHÜTZ, F.: Resultate von Untersuchungen an einigen niederländischen Mooren. Rec. Trav. bot. néerl., vol. 29).
WODEHOUSE, R., 1935: Pollen grains (New York, McGraw-Hill).
YAMASAKI, T., 1933: Morphology of pollens and spores (Rep. Exper. Forest, Kyoto Imp. Univ., no. 5; Japanese).
ZANDER, E., 1935: Beiträge zur Herkunftsbestimmung bei Honig (Berlin).
—— —— 1941: Beiträge zur Herkunftsbestimmung bei Honig, III (Leipzig).

Chapter VIII

POLLEN MORPHOLOGY — DICOTYLEDONS

(*Fagaceae — Violaceae*)

Fagaceae (PLATES XI, FIGS. 178–195, and XX, FIG. 388): —

Castanea dentata. — FIG. 178: equatorial view; 16 by 10.6 μ; Tennessee. — Grains prolate to subprolate, tricolpate, 14.7 by 11.2 μ. Furrows long and tapering, almost meeting at the poles, each enclosing a well-marked germinal aperture and a distinct transverse furrow with a thickened rim. The latter is conspicuous in optical section and, in fossil material, serves as a useful means of identification. Exine perfectly smooth (WODEHOUSE).

Castanea sativa (*C. vesca*). — FIG. 179: equatorial view; 20 by 14 μ. FIG. 180: polar view. — Germ pores well developed, more or less rounded; exine with a faint reticulate texture. Size 9.4–14.0 μ (FERRARI 1927); 11–17 by 8–14 μ (FIRBAS 1931); 12–16 μ (POTONIÉ 1934*b*); 9.5 by 13.5 μ (ZANDER 1935). Attention was first directed to the occurrence of fossil chestnut pollen by KELLER (Jahrb. St. Gall. Naturwiss. Ges., vol. 64, 1929, p. 78). DOKTUROWSKY (Ber. Deutsch. Bot. Ges., vol. XLIX, 1931) has found up to 12–30 per cent of *Castanea* pollen in peats of western Caucasus.

Castanopsis chrysophylla. — FIG. 388 (PLATE XX): two grains in lateral and one grain in polar view; 17.5 by 12 and 15 by 11 μ respectively; Oregon. — It may be difficult or even impossible to decide whether a pollen grain of the *Castanopsis* type should be referred to *Castanopsis* or to *Castanea* (*cf. Castanea dentata!*). "*Castanopsis*" pollen has been encountered in Tertiary deposits (p. 221).

Fagus silvatica. — FIG. 181: equatorial view; 42 μ; Ryttern, Sweden. FIG. 182: polar view; 41 μ. FIG. 183: destruction of pollen by oxidation by heating with $HClO_4$ (60 per cent). Exine fragment with ektexinous furrow and endexinous germ pore. — Grains spheroidal, tricolpate; diameter (25–)30–34(–38) μ (RUDOLPH and FIRBAS 1924), 26–46 μ (ČERNJAVSKI 1935), or, according to ZANDER (1935) approximately 44 μ. The furrows are proportionately shorter than in *Castanea* and *Quercus*. Each furrow has a distinct, equatorial germ pore. The grains of *F. orientalis* are nearly of the same type; ČERNJAVSKI (*l.c.*) quotes the following sizes: 29–51 μ (pollen from the Balkan peninsula), (31–)37.5(–46) μ (pollen from Caucasus). Pollen grains, somewhat similar to those of *Fagus*, occur in several unrelated plants, as for example *Cedrela*.

Nothofagus antarctica. — FIGS. 184, 185: polar and equatorial view; 26 and 29 by 23 μ respectively; Chile. — Grains suboblate with about five meridionally elongated furrows with thickened rims. Diameter, according to CRANWELL (1939), usually about 30 μ. The number of furrows of the pollen grains of this and two other species has been

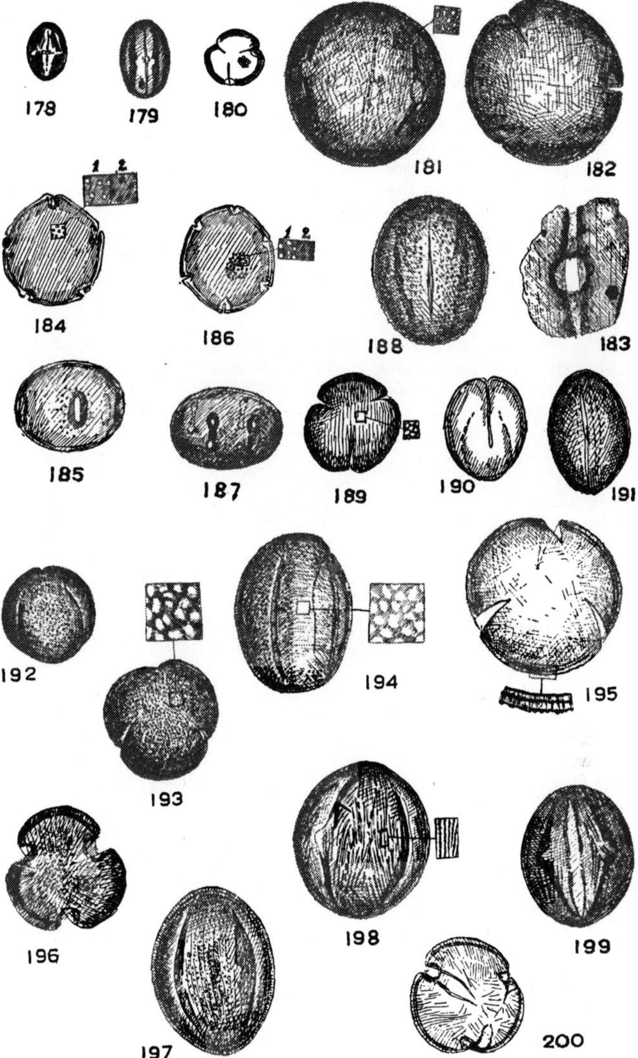

PLATE XI (178–200). — *Fagaceae* (178–195, *see also* PLATE XX: 388), *Gentianaceae* (196–200). — 178, *Castanea dentata;* 179 and 180, *C. sativa* (*vesca*); 181–183, *Fagus silvatica;* 184 and 185, *Nothofagus antarctica;* 186 and 187, *N. cliffortioides;* 188, *Quercus borealis;* 189–191, *Q. ilex;* 192 and 193, *Q. robur;* 194 and 195, *Q. sessiliflora;* 196 and 197, *Gentiana pneumonanthe;* 198–200, *Menyanthes trifoliata.*

investigated by VON POST (1930). The results are outlined in TAB. 9. Pollen diagrams from Tierra del Fuego have been provided with auxiliary diagrams showing the variation in the number of furrows of *Nothofagus* pollen at different levels (VON POST *l.c.*). From the evidence derived from these diagrams it may be assumed that changes have occurred, and that these changes comprise a shift from forests with *Nothofagus antarctica* predominating to forests with *N. pumilio* dominating and back again to domination by *N. antarctica*. However, the evidence may not be definite, since statistics based upon furrow and pore number tend to be somewhat misleading, unless they are supported by particularly substantial proof (*cf.* pore numbers in *Alnus incana*, p. 70).

Table 9: NUMBER OF FURROWS IN *Nothofagus* POLLEN (ACCORDING TO VON POST 1930; *cf.* ALSO CRANWELL 1939): —

	NUMBER OF FURROWS					
	2	3	4	5	6	7
N. antarctica (4 specimens)	1	1	2	71	25	1 per cent
N. betuloides (3 specimens)	1	1	14	74	10	0 " "
N. pumilio (3 specimens)	0	0	1	22	65	13 " "

Nothofagus cliffortioides. — FIGS. 186, 187: polar and equatorial view; 27 and 29 by 21 μ respectively; New Zealand. — Furrows frequently six, meridionally elongated; their equatorial part often pinched. A detailed description of the southern beech pollens has recently been given by CRANWELL (1939).

Quercus borealis Michx. (*Q. rubra* Du Roi). — FIG. 188: equatorial view; 37 by 31 μ; Connecticut. — Grains subprolate to spheroidal, generally tricolpate. Furrows narrow, of medium length. As in pollen grains of other species of oak hitherto investigated, there are no defined germ pores, although there may be traces of such. Size 36.5 by 25.6 μ (WODEHOUSE).

Quercus ilex. — FIG. 189: polar view; 25 μ; Florence. FIG. 190: oblique polar view. FIG. 191: equatorial view; 31 by 22 μ. — A small type with sharply defined furrows and a nearly smooth exine. Size, according to ČERNJAVSKI (1935), 16.1–28.5 μ.

Quercus robur. — FIGS. 192, 193: grains in oblique polar view; 24 and 29 μ respectively; Västerås 1937. — Exine distinctly granular. Size according to ČERNJAVSKI (1935) (18–)25(–28.5) μ.

There exist various quercoid pollen types. These may, sometimes, be difficult to distinguish from oak pollen (*cf. e.g. Caltha palustris*, FIGS. 310, 311).

Quercus sessiliflora. — FIG. 194: equatorial view; 39 by 29 μ; Oskarshamn, Sweden. FIG. 195: polar view; 39 μ; Cottian Alps. — Size according to ČERNJAVSKI [1935] (18–)22.5(–28.5) μ. In connection with pollen analyses the postglacial history of *Quercus sessiliflora* has been discussed by ERDTMAN (Ark. f. Bot., vol. 17, 1921) and VON POST (1924). ISBERG (Ark. f. zool., vol. 21a, 1929) has tried, although with doubtful success, to separate the pollen frequencies of *Quercus robur* from those of *Q. sessiliflora.*

Gentianaceae (PLATE XI, FIGS. 196–200): —

Gentiana pneumonanthe. — FIGS. 196, 197: polar and equatorial view; 30 and 41 by 31 μ respectively. — Grains subprolate, tricolpate. Each furrow with a single germ pore; exine with a well defined granular texture.

Menyanthes trifoliata. — FIGS. 198, 199: grains in equatorial view; 39 by 34 and 34 by 28 μ respectively; Emtervik, Sweden. FIG. 200: oblique polar view; 29 μ. — General plan of the grain similar to that of *Gentiana pneumonanthe.* The exine has a well defined striated appearance which is due to a great number of low and densely, though somewhat irregularly, spaced ridges. Size 43–50 μ (YAMASAKI 1933); 32.2 by 29.4 (ZANDER 1935).

Geraniaceae (PLATE XII, FIGS. 201, 202): —

Geranium robertianum. — FIG. 201: polar view; 74 μ; Eskilstuna, Sweden, 1938. FIG. 202: equatorial view; 79 by 72 μ; *ibid.* — Grains large, usually oblate spheroidal, tricolpate; furrows very short. Exine with prominent ornamentation, which consists of densely arranged rods about 7.5 μ long.

Haloragidaceae (PLATE XII, FIGS. 203–215): —

Gunnera bracteata. — FIG. 203: polar view; 27 μ; Masatierra. FIG. 204: equatorial view; 28 by 30 μ; *ibid.* — Grains tricolpate, in polar view three-lobated, in equatorial view more rounded. Furrows in the meridional depressions between the lobes. Polar caps smooth, unaffected by furrows and intercolpar bulges. Exine reticulate. Pollen grains of *Gunnera* are rather common (up to 70 per cent of the total number of pollen grains and fern spores) in certain deposits in Tierra del Fuego (AUER 1933).

Myriophyllum. — Grains aspidate (*cf.* FIGS. 207, 212, 214), generally rather flattened, angular in outline, with pores at the angles. Pores provided with elongate apertures, generally equally spaced around the equator; when three, meridionally arranged; when four or five, with their axes biconvergent (*cf.* FIG. 214). Texture slightly rough, especially around the pores. The formation of aspides (*cf. e.g.* FIG. 207) is apparently due to a mesexinous filling between the ektexine and the endexine. There may be a slight similarity between *Myriophyllum* pollen and the pollen of *Alnus* (*cf.* FIGS. 211 and 215) or *Betula* (*cf.* FIG. 209). *Myriophyllum* pollen, however, lacks arci, nor is its pore pattern the same as in *Alnus* and *Betula.*

In samples from European deposits, pollen of *Myriophyllum* is often encountered. It has also been found in samples from Tierra del Fuego, where it forms up to 50 per cent of the total number of pollen grains and spores (AUER 1933).

Myriophyllum alterniflorum. — FIGS. 205–207: grains from Botkyrka, Sweden; diameter 31, 35 by 25, and 29 μ respectively. — Easily recognizable grains (the "X-pollen" of SUNDELIN 1919, fig. 1, p. 218) with four or five, sometimes even three or six, usually irregularly spaced pores, each enclosing a germinal aperture with a high and steep wall. Exine faintly reticulate.

High frequencies of *M. alterniflorum* pollen — up to several hundred per cent — have been found in old post-glacial sediments in wide areas of northwestern Europe. L. VON POST (Geol. Fören. Förhandl., vol. 57, 1935) has also found the grains of this species in young sediments of at least two deposits in southern Sweden. He attributes a certain significance to the parallel behaviour of *M. alterniflorum* and the northern coniferous forest. The great distribution of these forests in early post-glacial times was followed by a retreat due to the combined pressure of mixed oak forest, Auwald, and other elements. On account of ensuing climatic deterioration, the retreat was followed by an advance, an advance which is possibly still in progress. These ideas of post-glacial climatic development, broadly outlined by VON POST, have been widely accepted. Yet the history of *Myriophyllum alterniflorum* cannot properly be fitted into this climatological picture until more complete and reliable information relative to the ecology of this species has been obtained.

Myriophyllum heterophyllum. — FIG. 208: polar view; 27 μ; Michigan. — Grains about 32 μ in diameter; nearly always with four pores, rarely with five or three (WODEHOUSE).

Myriophyllum spicatum. — FIG. 209: three-pored grain in polar view; 25 μ; Bohemia. FIG. 210: four-pored grain in polar view; 29 μ; Cottian Alps. FIGS. 211, 212: polar and equatorial view; 28 and 23 by 29 μ respectively; Lidingö, Sweden. — Pores generally equally spaced around the equator; ektexine rising rather steeply from the general surface of the grain to form the outer wall of the aspides.

Myriophyllum verticillatum. — FIG. 213: polar view; 35 μ; Gotland, Sweden. FIGS. 214, 215: equatorial and polar view; 24 by 34 and 25 μ respectively; Lidingö, Sweden. — Grains essentially as in *M. spicatum*, but ektexine rising at the pores only gradually. Hence the aspides are lower, wider, and rather more poorly defined than in *M. spicatum* as shown from a comparison between FIG. 214 and FIG. 212. Lacustrine sediments of Lake Bålen, west of Stockholm, have yielded (0–)1–3(–23) per cent pollen of this species (SELLING, Geol. Fören. Förhandl., vol. 60, 1938).

Hamamelidaceae (PLATE XII, FIGS. 216, 217): —

Hamamelis virginiana. — FIG. 216: oblique polar view; 19 μ; Connecticut. — Grains small, spheroidal, tricolpate, reticulate.

Liquidambar styraciflua. — FIG. 217: grain from Pennsylvania; 34 μ. — Grains spheroidal, about 38 μ in diameter, provided with 12 to 20 approximately circular pores; pore membranes conspicuously flecked; ektexine pitted with minute, round pits (WODEHOUSE). Characteristics of the pollen of other genera of this family have been given briefly by SIMPSON (1936, pp. 99, 100).

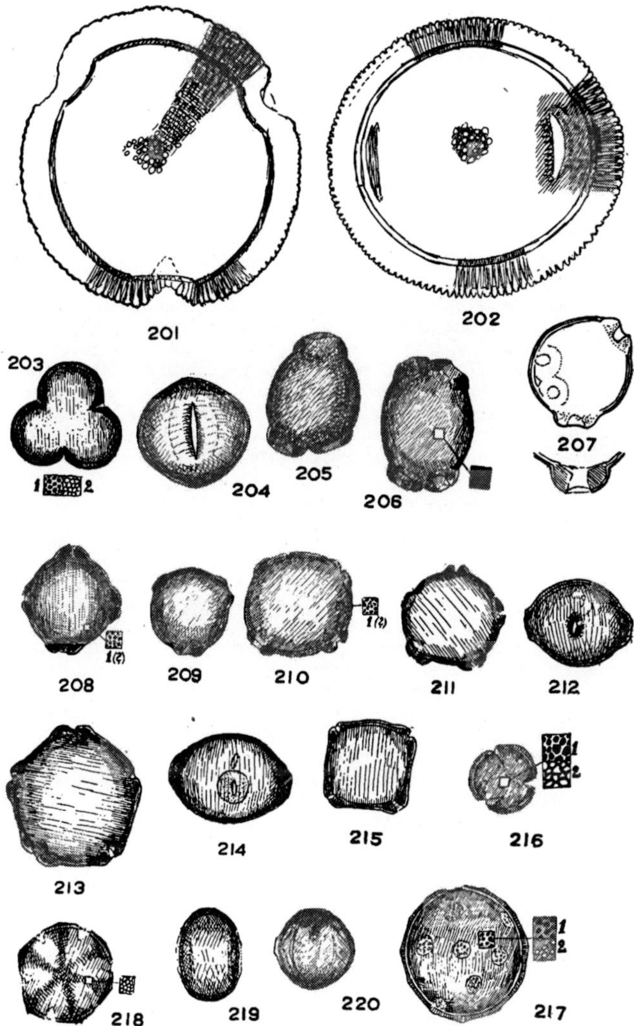

PLATE XII (201–220). — *Geraniaceae* (201, 202), *Haloragidaceae* (203–215), *Hamamelidaceae* (216, 217), *Hippuridaceae* (218–220). — 201 and 202, *Geranium robertianum;* 203 and 204, *Gunnera bracteata;* 205–207, *Myriophyllum alterniflorum;* 208, *M. heterophyllum;* 209–212, *M. spicatum;* 213–215, *M. verticillatum;* 216, *Hamamelis virginiana;* 217, *Liquidambar styraciflua;* 218–220, *Hippuris vulgaris.*

Hippuridaceae (PLATE XII, FIGS. 218–220): —

Hippuris vulgaris. — FIG. 218: polar view; 27 μ; Jonsberg, Sweden. FIGS. 219, 220: equatorial view; 24 by 18 and 21 by 22 μ respectively. — Grains thin-walled, prolate to spheroidal, generally with five or six meridional, poorly defined furrows. Intercolpar strands of thicker exine unite at the pores forming a supporting framework.

Hydrocaryaceae (PLATE XIII, FIGS. 221–223): —

Trapa natans. — FIG. 221: equatorial view; 68 by 54 μ; Immeln, Sweden. FIG. 222: optical meridional section through a pore. FIG. 223: polar view (proximal pole) of young grain showing tetrad scar; 58 μ. — Pollen grains, according to ASSARSSON (1927), 50–70 μ in diameter, provided with three well defined meridional crests, made up of more or less folded ektexinous material and meeting at the poles. If the crests be removed, the pollen grain would appear as a thick lens with three equatorial, equally spaced and meridionally elongate pores. In the vicinity of the pores, the folds of the crests are small; their size increases towards the poles, where they attain a maximum.

Juglandaceae (PLATE XIII, FIGS. 224–232): —

Carya (FIGS. 224–227). — Grains spheroidal to oblate, larger than those of *Juglans*, 40 to 52 μ in diameter. Germ pores usually three (occasionally some grains have four or six), apertures short, elliptical, or nearly circular, 3.4 to 5.7 μ in diameter, equally spaced in a circle, parallel with and not far from the equator (WODEHOUSE). Thus one hemisphere (the proximal?) has all of the pores, while the other has none. The exine of the pore-bearing hemisphere is particularly thick. The increased thickness is due to the outer part of the exine; the endexine is thin throughout. The thickening encroaches somewhat on the other hemisphere, leaving only a polar area of varying size free from the thickening (FIGS. 225 and 227). Apparently the pores should be regarded as some kind of aspides with large horizontal and very minute vertical extensions (as shown by FIGS. 224 and 225 the endexine separates from the ektexine at a considerable distance from the apertures).

Carya cordiformis. — FIG. 224: four-pored grain; polar view, poral hemisphere; 51 μ; Connecticut. FIG. 225: three-pored grain; polar view, abporal hemisphere; 50 μ. — Normal grains oblate-spheroidal to nearly or quite spheroidal, averaging 52.4 by 45.6 μ. Pores three or, in some grains which may or may not be considerably larger than normal, four; apertures circular or broadly elliptical, averaging 5.24 μ in diameter; exine thickenings extending over the whole surface of the grain, except on a small area at the pole of the abporal hemisphere. Texture slightly granular (WODEHOUSE).

Carya glabra. — FIG. 226: three-pored grain; polar view, poral hemisphere; 50 μ; Connecticut. FIG. 227: three-pored grain; equatorial view; 38 by 54 μ. — The grains are somewhat smaller than in *C. cordiformis*, 44.8 by 38.7 μ; a large proportion have four pores, and a

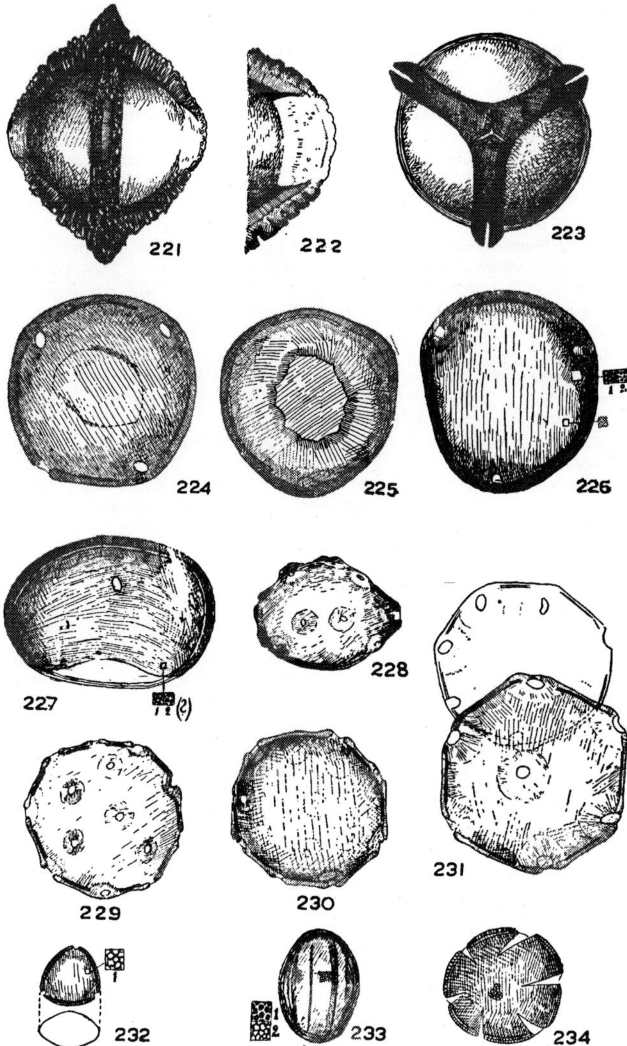

PLATE XIII (221–234). — *Hydrocaryaceae* (221–223), *Juglandaceae* (224–232), *Labiatae* (233–234). — 221–223, *Trapa natans;* 224 and 225, *Carya cordiformis;* 226 and 227, *C. glabra*; 228, *Juglans australis;* 229 and 230, *J. nigra;* 231, *Pterocarya caucasica* (*P. fraxinifolia*); 232, *Engelhardtia chrysolepis;* 233 and 234, *Lycopus europaeus.*

few, which are always giants, have six. Apertures slightly elongated; 4.56 μ in length (WODEHOUSE).

Engelhardtia. — Grains somewhat flattened, triangular in outline; pores three, one at each angle, aspidate. The pollen grains of *E. spicata* Blume are 19.4–21.6 μ in diameter; pores three, their apertures elliptical to circular, 2.6 to 3.4 μ long; pore diagram similar to that of *Myrica;* exine faintly granular (WODEHOUSE). *Pollenites levis* R. Pot. and *P. bituitus* R. Pot. from Tertiary deposits in Germany, should be referred to *Engelhardtia* according to THIERGART (1937).

Engelhardtia chrysolepis. — FIG. 232: polar view and outline of equatorial view; 15.5 and 9.3 by 15.5 μ respectively; Hongkong. — Grains oblate, circular-triangular in outline. Exine faintly reticulate.

Juglans (FIGS. 228–230). — Grains aspidate, decidedly flattened, oblate-spheroidal, 34 to 42 μ in diameter. There are always more than three pores (about 6 to 15) mostly confined to one hemisphere and, though encroaching somewhat upon the other hemisphere, always leaving the greater part of it free. Openings of the pores elliptical to circular, 2.3 to 3.3 μ in length (WODEHOUSE).

Juglans australis. — FIG. 228: oblique equatorial view; 28 by 38 μ; Argentine.

Juglans nigra. — FIG. 229: polar view, poral hemisphere; 40 μ; Connecticut. Fig. 230: polar view, abporal hemisphere; 40 μ. — Grains 34.2 to 30.8 μ in diameter (WODEHOUSE); germ pores 12 to 15.

Juglans regia. — Diameter of pollen grains 32–48 μ; pores generally 12–15 (FIRBAS 1931).

Pterocarya. — Grains aspidate, much flattened, 27 to 34 μ in diameter. Pores 5 to 7, occasionally three or four, one at each angle of the grain, their apertures elliptical, 3 to 4.6 μ long, arranged around the equator and meridionally oriented or converging in pairs alternately above and below the equator, occasionally irregular in arrangement (WODEHOUSE).

Pterocarya caucasica (*syn. P. fraxinifolia*). — FIG. 231: two grains in polar view; 43 and 47 μ respectively; Visby, Sweden (*cult.*). — The lower grain shows a tendency of the pores to assemble on one hemisphere as in the grains of *Carya* and *Juglans*. Size of grains 23–32 μ (BAAS 1932).

Fossil grains occurred but sparingly (less than 5 per cent) in a bog in western Caucasus, although *Pterocarya* was a common associate of the alder swamps surrounding the bog (DOKTUROWSKY, Ber. Deutsch. Bot. Ges., vol. XLIX, 1931). In early diluvial deposits in Germany up to 20 per cent of *Pterocarya* pollen has been found (BAAS *l.c.*).

Labiatae (PLATES XIII, FIGS. 233, 234, and XIV, FIGS. 235, 236): —

Lycopus europaeus. — FIG. 233: equatorial view; 28 by 21 μ; Gotland, Sweden. FIG. 234: oblique polar view; 31 μ; *ibid.* — Grains hexacolpate as in the "M" (*Majorana*) form described by ZANDER (1935, p. 272); exine reticulate.

Stenogyne kamehamehae. — FIG. 235: polar view; 42 μ; Maui, T. H., 1938. FIG. 236: equatorial view, somewhat oblique; 49 by 42 μ. — Grains tricolpate, as in the "L" (*Lamium*) form of ZANDER (1935).

The pollen grains of about 40 species belonging to the *Labiatae* family are described in the work by ZANDER. *Cf.* also MEINKE 1927, p. 446 and J. LEITNER 1942.

Lentibulariaceae (PLATE XIV, FIGS. 240–242): —

Utricularia minor. — FIG. 240: equatorial view; 22 by 30 μ; Råstasjön, Sweden. FIG. 241: polar cap. — According to FISCHER (1890, p. 54), the grains of *U. intermedia*, *U. minor*, *U. neglecta*, and *U. vulgaris* possess about 15 to 20 furrows, which frequently are confluent at their polar ends, as shown in FIGS. 240 and 241.

Utricularia vulgaris. — FIG. 242: equatorial view; 45 by 36 μ; Stavkyrka, Sweden. — In pollen analyses, *Utricularia* pollen is seldom encountered, except as stray grains. Nevertheless, HALDEN (Sveriges Geol. Unders., Ser. C, no. 280, 1917) once found as much as 51 per cent *Utricularia* pollen.

Loganiaceae (PLATE XIV, FIG. 237): —

Gelsemium sempervirens. — FIG. 237: oblique polar view; 39 μ. — Grains tricolpate; each furrow with a conspicuous germ pore; exine reticulate.

Loranthaceae (PLATE XIV, FIGS. 238, 239): —

Viscum album. — FIG. 238: polar view; 42 μ; Barkarö, Sweden. FIG. 239: equatorial view; 34 by 42 μ. — Grains suboblate, tricolpate, echinate. Furrows short, their membranes provided with a granular ornamentation. Polar caps and intercolpar parts of the exine with irregularly spaced blunt spines. The exine attains considerable thickness, but there is no clear division between ekt- and endexine (*cf.* also IVERSEN 1941, PL. IX, FIG. 4).

Lythraceae (PLATE XIV, FIG. 243): —

Peplis portula. — FIG. 243: polar view; about 17 μ; Ängelsberg, Sweden. Grains probably tricolpate, in polar view hexagonal (three long, three shorter sides) with germ pores in the centre of the shorter sides.

Magnoliaceae (PLATE XIV, FIGS. 244–246): —

Drimys axillaris. — FIG. 244: tetrad; 39 μ. FIG. 245: exine pattern at the angles where the outer surfaces of three grains meet. — Grains united in tetrahedral tetrads. Individual grains circular in outline, each with a single furrow occurring as a large, roundish thin area in the exine at the distal side. The exposed surface of the grain, except the furrow and its margin, is covered with a reticulate system of ridges enclosing angular lacunae. The pattern of the reticulum is not continuous across the sutures between the grains of the tetrad (FIG. 245).

In deposits of the rain forest area of western Tierra del Fuego the frequency of *Drimys* pollen comprises up to 20 per cent of the total amount of pollen grains and fern spores (AUER 1933).

Liriodendron tulipifera. — Fig. 246: distal side; 67 by 47 μ; Connecticut. — Grains monocolpate, about 62 μ long and 49 μ broad, with a deep longitudinal furrow. "Texture of the outside . . . finely pitted, with large, wart-like nodules superimposed in irregular fashion. The rugged, coarse appearance that these grains present is unique and serves to distinguish them from all other monocolpate grains that I have seen" (Wodehouse).

Myricaceae (Plate XIV, fig. 247): —

Myrica gale. — Fig. 247: polar view; 32 μ; Svanå, Sweden, 1938. — Grains usually provided with three pores, aspidate, somewhat similar to those of *Betula*. Size about 27 by 23.5 μ (Wodehouse); equatorial diameter, according to Jentys–Szafer (1928), 21.5–27.2 μ averaging 23.9–24.7 μ. Of 2000 recent grains studied, 98.4 per cent were provided with three and 1.6 per cent with two or four pores (Sandegren *in* von Post 1924). The ektexine is considerably thickened within the aspides (in optical section, according to Wodehouse, this thickening would suggest in appearance the terminal joint, or tarsus, of the hind legs of some insects).

According to Jentys–Szafer there are three layers in the exines of *Myrica* pollen: two thin layers enclosing a thick one. However, most of the illustrations published by Jentys–Szafer show only two layers, which are apparently identical with ektexine and endexine respectively. The supposed third layer appears to represent the particularly resistant surface layer of the ektexine (*cf.* Jentys–Szafer *l.c.*, pl. 10, fig. 13*b*).

If well preserved, the pollen grains of *Myrica* are readily distinguished from those of *Betula* and *Corylus* on account of the pore construction. The pollen grains of *Betula* and *Corylus*, as mentioned previously, are provided with arci, swinging pairwise from pore to pore, tending — in grains in polar view — to obscure the interaspide wall construction and — in grains in equatorial view — to render the outline of the pollen grain more or less angular. Acetolysed pollen grains of *Cerothamnus cerifera*, *C. inodorus*, *Myrica asplenifolia*, *M. carolinensis*, and *M. gale* have been studied and no undoubted traces of arci have been found. Grains of *Myrica gale*, treated with concentrated sulphuric acid for six days, were particularly transparent and left no doubt whatsoever as to the absence of arci. In polar view the strongly refractive endexine stood out prominently from aspis to aspis. In equatorial view, the wall construction appeared equally sharp, exhibiting smooth exine layers of uniform thickness. In *Corylus*, however, the endexine never appears equally clear; it is even more obscure in *Betula*.

Myrtaceae (Plate XIV, figs. 248–250): —

Eucalyptus nandiniana. — Fig. 248: polar view; 19 μ; India. — The grains in *Eucalyptus* are generally flattened, with triangular outline in polar view; diameter about 19–25 μ (Wodehouse 1932). Germinal furrows three — one at each angle — each crossed by a transverse

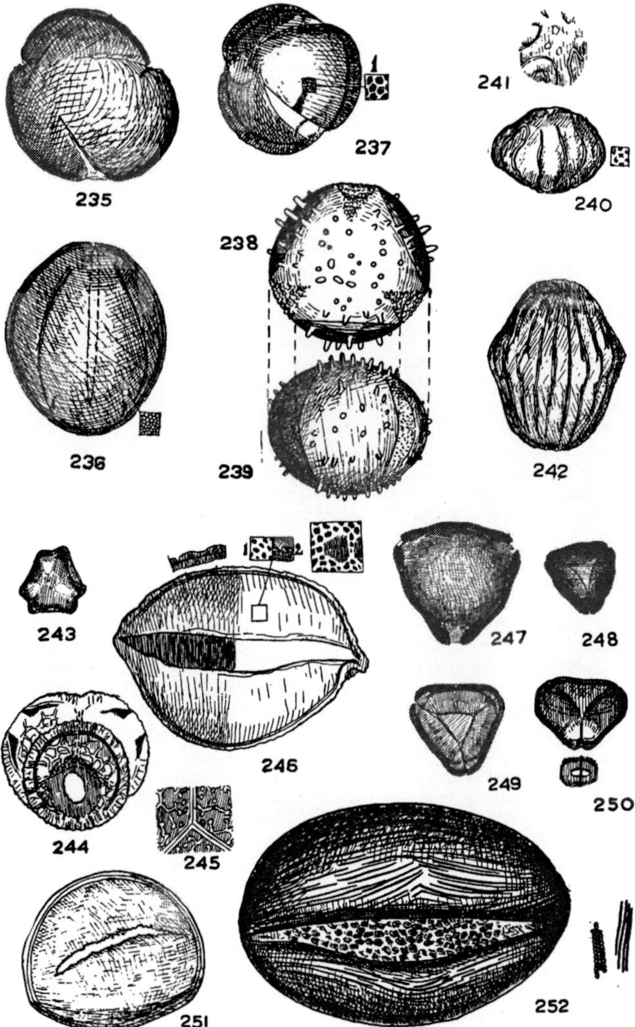

PLATE XIV (235–252). — *Labiatae* (235, 236), *Lentibulariaceae* (240–242), *Loganiaceae* (237), *Loranthaceae* (238, 239), *Lythraceae* (243), *Magnoliaceae* (244–246), *Myricaceae* (247), *Myrtaceae* (248–250), *Nymphaeaceae* (251–252). — 235 and 236, *Stenogyne kamehamehae;* 237, *Gelsemium sempervirens;* 238 and 239, *Viscum album;* 240 and 241, *Utricularia minor;* 242, *U. vulgaris;* 243, *Peplis portula;* 244 and 245, *Drimys axillaris;* 246, *Liriodendron tulipifera;* 247, *Myrica gale;* 248, *Eucalyptus nandiniana;* 249 and 250, *Metrosideros polymorpha;* 251, *Brasenia purpurea;* 252, *Cabomba aquatica.*

endexinous furrow or pore; exine smooth. The flattened polar surfaces are arched.

Metrosideros polymorpha. — Fig. 249: polar view; 24 μ; Hawaii 1938. Fig. 250: oblique equatorial view; detail figure: equatorial view of pore (aspis?).

Nymphaeaceae (Plates XIV, figs. 251, 252, XV, figs. 253–267, and XVI, fig. 268): —

Barclaya sp. — Fig. 255: lateral view; proximal part uppermost; 41 by 24 μ; Buitenzorg 1938. — Pollen grains bean-shaped, resembling certain monolete fern spores. They are apparently monocolpate and provided with thick, faintly textured exine.

Brasenia purpurea. — Fig. 251: distal part; 48 by 41 μ; Cuba. — Grains monocolpate, psilate.

Cabomba aquatica. — Fig. 252: distal side; 88 by 57 μ; Para. — Grains monocolpate, exine thick, conspicuously striate.

Euryale ferox. — Fig. 253: lateral view; 51 by 37 μ; Hort. Bergianus, 1938. — Grains echinate, morphologically of about the same type as the pollen grains of *Nymphaea.*

Nelumbo jamaicensis. — Fig. 254: oblique polar view; 58 μ; Cuba. — Grains tricolpate, a feature not encountered in other genera of the recent *Nymphaeaceae.*

Nuphar advena Ait. (*Nymphaea advena* Soland.). — Fig. 256: proximal part; 65 by 40 μ; Connecticut. — Grains about 51.3 μ long, monocolpate. Furrow closed by a narrow, linear operculum. Exine heavy, slightly granular, with long, sturdy, and obtusely pointed spines which are longer on the proximal than on the distal side. When dry, the grains are more or less boat-shaped, with the furrow tightly closed and somewhat invaginated. When moistened the furrow becomes evaginated, causing the grain to assume an oblate spheroidal form. In this condition, the furrow is seen to be provided with a narrow strip of exine, which is the operculum, of a texture similar to that of the rest of the grain but bearing slightly shorter spines (Wodehouse).

Nuphar luteum. — Fig. 257: oblique lateral view; distal part down; polar axis about 28 μ; Västerås 1936. — Distinguished from the grains of *N. pumilum* by the greater number and length of the spines.

Nuphar pumilum. — Fig. 258: proximal part; 58 by 39 μ; Sala, Sweden. — Spines shorter and less numerous than in *Nuphar luteum.*

Nymphaea alba. — Fig. 259: side view; distal part uppermost; 47 by about 37 μ; Närke, Sweden. Fig. 260: oblique view, showing extent of operculum; 42 by 37 μ. — Exine with a broadly elliptical or circular operculum occupying most of the distal part of the grain. Exine echinate, the operculum excepted. Spines considerably smaller and more numerous than in *Nuphar.* Operculum attached to the rest of the exine as a cover to a watch (fig. 259) or completely surrounded by a narrow strip of thin, flexible exine (fig. 260).

Nymphaea candida. — Fig. 261: lateral view; distal part (with operculum) uppermost; 30 by 22 μ; Ljusdal, Sweden. Fig. 262: polar view (proximal part of grain); 28 μ. — Of a type similar to the grains of *N. alba*, but smaller and provided with small warts instead of spines.

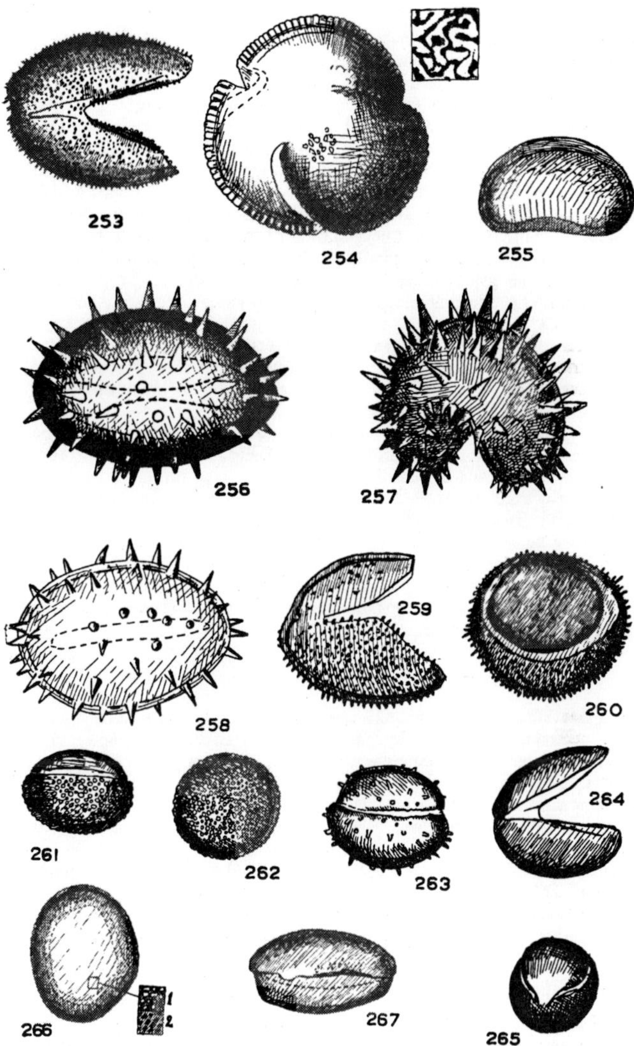

PLATE XV (253–267). — *Nymphaeaceae.* — 253, *Euryale ferox;* 254, *Nelumbo jamaicensis;* 255, *Barclaya* sp.; 256, *Nuphar advena* (*Nymphaea advena*); 257, *N. luteum;* 258, *N. pumilum;* 259 and 260, *Nymphaea alba;* 261–262, *N. candida;* 263, *N. odorata;* 264 and 265, *N. stellata;* 266 and 267, *N. tetragona.*

Nymphaea odorata. — Fig. 263: lateral view; distal part up; 34 by 27 μ; Connecticut. — Grains about 38 by 30 μ. "The entire surface is generally covered with long, conspicuous spines or with rounded wart-like protuberances or both. The spines are unique: they have a resinous appearance and, when they reach their best development, are long, straight, cylindrical rods, not tapering but maintaining their full diameter almost or quite to their tips, which are generally rounded. They are set on the surface of the grain at all possible angles and are often extremely oblique. They are never uniform, in spacing, shape, or length, differing in these respects markedly from the spines encountered in the grains of most other plants. Sometimes they are well developed throughout the grain, imparting to the whole surface, except for the furrow ring (*i.e.* the ring-shaped strip of thin exine connecting the operculum with the rest of the grain) which is always smooth, a bristly appearance, but nearly always they are better developed on the proximal surface than on the distal, and frequently on the operculum they are represented only by little wart-like nodules" (Wodehouse, pp. 345, 346).

Nymphaea stellata Willd. — Fig. 264: side view, distal part uppermost; 34 by 24 μ; Hort. Bergianus 1938. Fig. 265: side view; about the same as Fig. 264 if turned nearly 90°.

Nymphaea tetragona. — Fig. 266: polar view (proximal part up); 36 by 30 μ; Tammela, Finland. Fig. 267: lateral view (proximal part up) 40 by 22 μ. — Exine provided with very small granules.

Victoria cruziana. — Fig. 268: tetrad; diameter about 86 μ; equatorial diameter of single grains about 60 μ; Hort. Bergianus 1938. — Pollen grains united in tetrads; grains similar in structure to those of *Euryale* and *Nymphaea.*

Nyssaceae (Plate XVI, figs. 269, 270): —

Nyssa silvatica. — Fig. 269: polar view; 32 μ; New Jersey. Fig. 270: equatorial view; 33 by 32 μ. — Grains tricolpate, spheroidal, in polar view somewhat triangular in outline; furrows at the angles, long and tapering, each provided with a distinct germinal aperture, rounded or somewhat elliptical in outline, in the latter case with its long axis crossing that of the furrow. Exine finely but distinctly granular, with the granules, as stated by Wodehouse (p. 446), appearing as the ends of fine, vertical rods of one material embedded in a matrix of another material of a different refractive index, appearing radially striate in optical section. Further notes on the pollen morphology of *Nyssa* are found in the papers by Rudolph (1935) and Kirchheimer (1938).

Oenotheraceae (Onagraceae) (Plate XVI, figs. 282–284): —

Chamaenerium angustifolium (*Epilobium angustifolium*). — Fig. 282: polar view; 85 μ; Ångelsberg, Sweden. Fig. 283: aspis, equatorial view; 25 μ in diameter. — Grains aspidate, in polar view triangular in outline with one pore at each angle.

Epilobium palustre. — Fig. 284: polar view; 76 μ; Västerås. — Grains sometimes loosely united in tetrads.

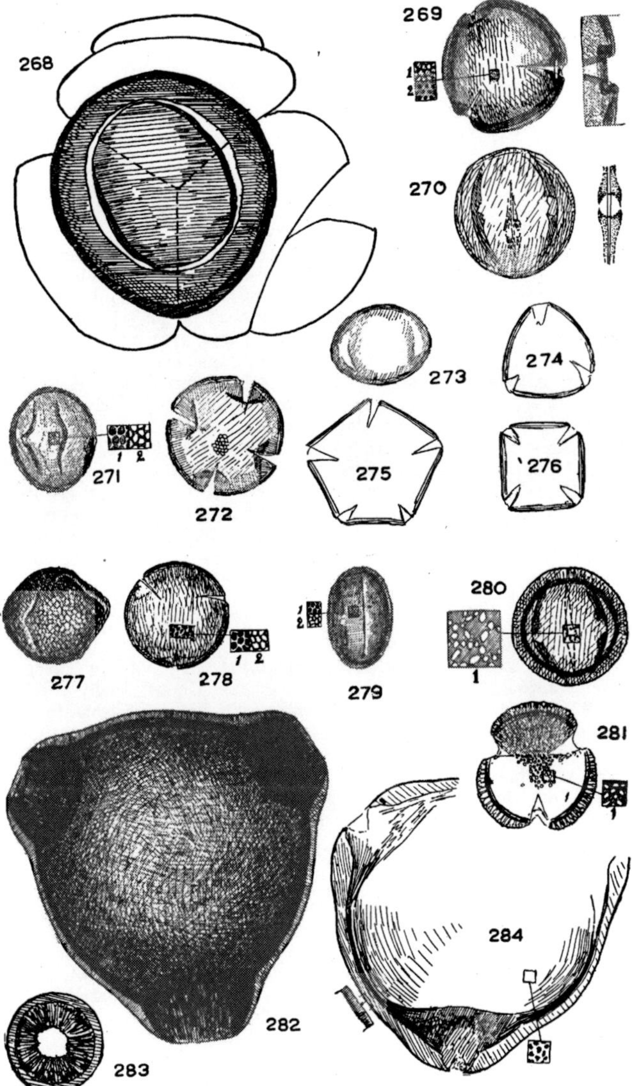

PLATE XVI (268–284). — *Nymphaeaceae* (268), *Nyssaceae* (269, 270), *Oenotheraceae* (282–284), *Oleaceae* (271–281). — 268, *Victoria cruziana;* 269 and 270, *Nyssa silvatica;* 271–276, *Fraxinus americana;* 277 and 278, *Fraxinus excelsior;* 279, *F. ornus;* 280 and 281, *Ligustrum vulgare;* 282 and 283, *Chamaenerium angustifolium;* 284, *Epilobium palustre.*

Oleaceae (PLATE XVI, FIGS. 271–281): —

Fraxinus. — Grains subprolate to suboblate, when fully expanded 20 to 25 μ in diameter, tri- to pentacolpate. Furrows without germpores. Exine reticulate, the net usually simple and weakly developed, constant for all the grains of a given species but of varying degrees of coarseness among the different species (WODEHOUSE).

Fraxinus americana. — FIGS. 271, 272: tetracolpate grains in equatorial and polar view; 27 by 22 and 30 μ respectively; Connecticut. FIGS. 273–276: outline of pollen grains in a preparation made by WODEHOUSE according to the method described on p. 31; FIG. 273: 26 by 20 μ; FIG. 274: 24 μ; FIG. 275: 32 μ; FIG. 276: 23 μ.

Grains, according to WODEHOUSE, when expanded about 24 μ in diameter, flattened and angular in outline. Furrows four, less frequently three, or rarely five, short and broad, gaping widely as the grain expands, giving it an angular appearance. Exine finely reticulate.

The illustrations afford striking examples of the effect of different chemical treatments on the shape of the pollen grains. Thus, while the acetolysed tetracolpate grain, FIG. 272, is well rounded and has its furrows, except at the equator, somewhat depressed below the general surface of the grain, tetracolpate grains from WODEHOUSE's preparation are quadrangular and do not show any traces of depressed furrows. It may also be noticed that the diameter of the grain in FIG. 275 is as much as 32 μ (owing to swelling after preparation) or considerably larger than the maximum size (25 μ) of *Fraxinus* pollen mentioned by WODEHOUSE himself.

Fraxinus excelsior. — FIG. 277: equatorial view; 25 by 29 μ; Västerås 1936. FIG. 278: polar view; 27 μ. — Grains usually tricolpate. Reticulum with lumina of fairly uniform size, larger than in *F. americana* (ERDTMAN 1936). As mentioned by several authors (*e.g.* SUNDELIN, in Sveriges Geol. Unders., Ser. Ca, no. 16, 1917) there is a certain resemblance between the pollen of *Fraxinus excelsior* and that of *Salix*. But while the reticulum of the former extends to the margin of the furrows, the lumina of the latter shrink and finally disappear towards the furrows.

Fraxinus ornus. — FIG. 279: equatorial view; 27 by 17 μ; Slovakia. — Grains conspicuously reticulate; muri apparently of the same or similar type as in *Ligustrum* (*cf.* FIGS. 280, 281).

Ligustrum vulgare. — FIG. 280: equatorial view; 30 by 31 μ; Tanum, Sweden. FIG. 281: polar view; 31 μ. — Grains averaging about 28.5 μ in diameter, tricolpate, with sharply defined furrows, each provided with a fairly well-defined germ pore. Furrow membranes smooth. Exine reticulate, reticulum much coarser than in *Fraxinus*, with lacunae larger and muri higher and buttressed, presenting a beaded appearance. Towards the furrows, the reticulum is slightly finer than elsewhere, and it ends abruptly along the furrow margins with closed lacunae (WODEHOUSE).

Onagraceae (*see* **Oenotheraceae**).

Papaveraceae (PLATE XVII, FIGS. 285–287): —

Papaver orientale. — FIG. 285: hexacolpate grain; 32 μ; Visby, Sweden, 1935. FIG. 286: tetracolpate grain; 30 μ. FIG. 287: tricolpate grain; 31 μ. — Grains spheroidal with furrows of various number and arrangement. Furrow membranes without germ pores, copiously and uniformly flecked with granules. Exine reticular. The arrangement of the furrows in the hexacolpate grains (FIG. 285) may be explained by comparing the six furrow axes to the six edges of a tetrahedron and the points of convergence to the four solid angles of such a figure. Tetracolpate grains (FIG. 286) have four furrows equally spaced on the equator. Their axes are not meridionally arranged but cross the equator obliquely and converge in pairs, at angles of 120°, towards four centres (*cf.* WODEHOUSE, pp. 170–172).

Plantaginaceae (PLATE XVII, FIGS. 288–292, *cf.* also IVERSEN 1941, pp. 39–41).

Litorella americana. — FIG. 288: pollen from Quebec; 34 μ. — Grains cribellate, similar to those of *Plantago maritima*, but with surface more corrugated. For exine structure, *see P. maritima.*

Plantago coronopus. — FIG. 291: pollen from Kullaberg, Sweden; 32 μ. — Grains spheroidal, cribellate-aspidate; pores five to seven, circular in outline, their membranes flecked with minute granules. Exine conspicuously thickened at the pores, encircling them with a low, well defined rim. Exine surface: *cf. P. maritima.*

Plantago juncoides. — FIG. 290: pollen from Greenland; 53 μ.

Plantago major. — FIG. 292: pollen from Västerås 1937; 22 μ. — Grains somewhat various, approximately spheroidal, 16 to 21 μ in diameter. Germ pores 4 to 6, according to the size of the grain, but usually five, irregular in shape; their membranes flecked with granules. Texture of the exine rough and warty (WODEHOUSE).

Plantago maritima. — FIG. 289: pollen from Västerås 1936; 32 μ. — In conformity with the pollen grains of *Litorella*, those of the plantains are provided with a somewhat uneven surface, resembling ripple marks, with thin exine areas — hatched in the figures — merging into and alternating with areas of thicker exine. The exine has a fine granular texture (*cf.* FIGS. 288 — *Litorella* — and 289) which is lacking only in the pore membranes and in the immediate vicinity of the pores.

Fossil *Plantago* pollen has been observed particularly in coastal peats. The frequency may be high, up to 56 per cent having been reported by ERNST (Schr. Naturwiss. Ver. Schlesw.-Holstein, vol. XX, 1934) from northwestern Germany.

Platanaceae (PLATE XVII, FIG. 293): —

Platanus occidentalis. — FIG. 293: polar view; 21 μ; Massachusetts. — Grains spheroidal, about 15–20 μ in diameter, tricolpate, with broad furrows of medium length. Furrow membranes without germ pores, copiously and uniformly flecked with granules (WODEHOUSE). Exine reticulate.

Plumbaginaceae (PLATE XVII, FIGS. 294–297): —

Armeria vulgaris. — FIG. 296: polar view of pollen grain from a plant with cob-like stigmata; 78 μ; Ryttern, Sweden. — As stated by KULCZYŃSKI (1932), this species has two kinds of pollen grains. One of them, produced by specimens with papillose stigmata, is of the same type as that described below under *A. vulgaris* var. *maritima*. The other, still more intricate, is produced by specimens with cob-like stigmata. It is very schematically shown in FIG. 296. The ornamentation consists of long rods, laterally — their pointed apices excepted — united to high muri forming a coarse reticulum with big lacunae (ERDTMAN 1940, IVERSEN 1940).

Armeria vulgaris var. *maritima.* — FIG. 294: equatorial view of pollen grain from a specimen with papillose stigmata; 64 by 56 μ; Thorshavn, Faeroes. FIG. 295: polar view; 68 μ. — Grains suboblate to spheroidal, tricolpate, less frequently tetracolpate. Furrows comparatively short, with tapering ends. Exine provided with a very conspicuous ornamentation consisting of rods with characteristically swollen ends. In surface view, these ends appear roundish, in lateral view pointed and more or less angular. The rods are densely spaced and arranged in rows forming a close reticulum (FIG. 294). Pollen grains produced by specimens with cob-like stigmata are similar to the grain figured in FIG. 296.

Statice limonium. — FIG. 297: polar view of pollen grain produced by a brevistylous plant; 61 μ; Vellinge, Sweden. — As in *Armeria* there are two kinds of pollen grains (MACLEOD, J., Untersuchungen über die Befruchtung der Blumen; Bot. Centralblatt, vol. XXIX, 1887); the grain figured in FIG. 297 comes from a specimen with papillose stigmata.

Polygalaceae (PLATE XVII, FIGS. 298, 299): —

Polygala sepyllacea. — FIGS. 298, 299: equatorial and oblique polar view respectively; 31 μ; Bergen, Norway. — Grains spheroidal, polycolpate, with equatorial germ pores. The furrows, with membranes and germ pores, are enclosed by a framework of thick exine which also encloses a number of circular areas with thin exine in the polar caps. In a peat sample from England the author once found a grain which should be referred to this species. This indicates that in pollen analysis almost anything in the way of pollen grains may be encountered.

Polygonaceae (PLATES XVII, FIGS. 300, 301, and XVIII, FIGS. 302–307): —

Polygonum bistorta. — FIG. 300: equatorial view of tricolpate grain; 64 by 43 μ; Sauerland, Germany. FIG. 301: equatorial view of tetracolpate grain; 55 by 37 μ. — Grains prolate, psilate, tri- or tetracolpate, with narrow, tapering furrows, each of which is provided with an oblong germ pore. Exine thick, granular. A fossil pollen (polar axis about 40 μ), which has been referred to *Polygonum bistorta*,

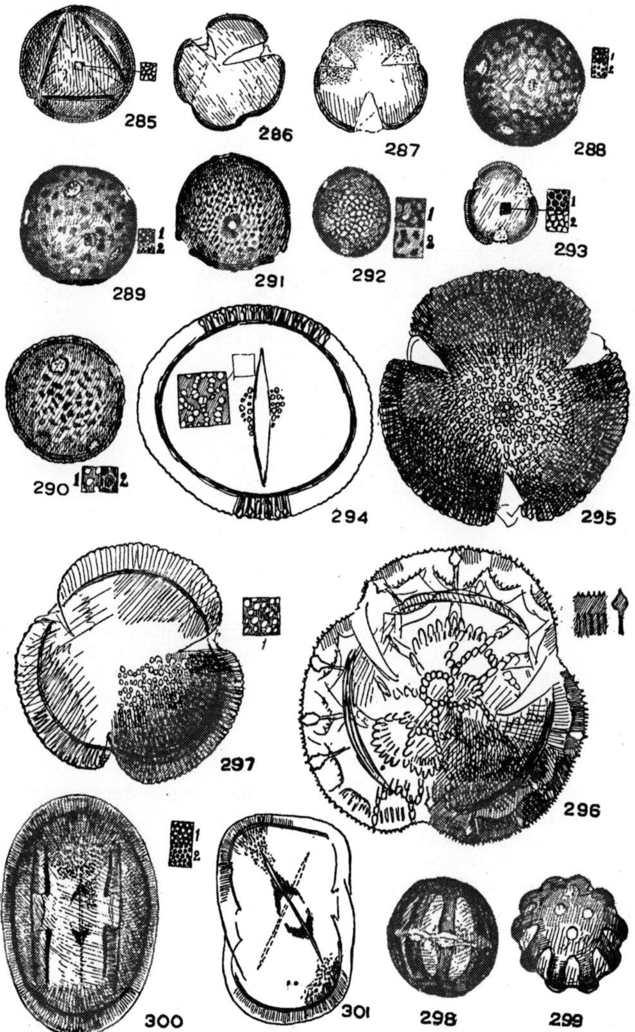

Plate XVII (285–301). — *Papaveraceae* (285–287), *Plantaginaceae* (288–292), *Platanaceae* (293), *Plumbaginaceae* (294–297), *Polygalaceae* (298, 299), *Polygonaceae* (300, 301). — 285–287, *Papaver orientale;* 288, *Litorella americana;* 289, *Plantago maritima;* 290, *P. juncoides;* 291, *P. coronopus;* 292, *P. major;* 293, *Platanus occidentalis;* 294–295, *Armeria vulgaris* var. *maritima;* 296, *A. vulgaris;* 297, *Statice limonium;* 298 and 299, *Polygala sepyllacea;* 300 and 301, *Polygonum bistorta.*

has been figured by FIRBAS (FIRBAS and GRAHMANN 1928). They report a maximum frequency of 105 per cent.

Polygonum convolvulus. — FIGS. 303, 304: grains in equatorial view; 22 by 18 and 24 by 20 μ respectively; Ekerö, Sweden. — Like the grains of *Polygonum viviparum*, but smaller and provided with elongated transverse furrows.

Polygonum viviparum. — FIG. 302: equatorial view; 42 by 37 μ. — Grains spheroidal, tricolpate; at the equator, the furrows are crossed by endexinous transverse germ pores (or furrows). Size, according to HØEG (1924), 55–60 by 35–40 μ.

Rumex acetosella. — FIGS. 305–307: equatorial, oblique, and polar view; FIGS. 305, 306: 21 by 24 μ; FIG. 307: about 20 μ; Gotland, Sweden. — Grains 22–24 μ in diameter, spheroidal or somewhat flattened, slightly bulging between the furrows. Furrows sometimes three, but more frequently four or ocasionally six; long, slender, and pointed at their ends. Each furrow is provided with an elliptical germ pore about 3.4 μ long, with its long axis oriented in the same direction as that of the furrow. Exine coarsely pitted (WODEHOUSE). A pollen type, supposedly of a weed, attaining a relative frequency of up to 513 per cent in the upper layers of a bog in northwestern Germany (SCHROEDER 1939) probably belongs to a species of *Rumex* (probably *R. acetosella*).

Proteaceae: —

No examples of the pollen morphology of this family are given in this book. However, the pollen grains seem, as a rule, to be easy to identify. A few types have been described by WODEHOUSE (1932) and by ZANDER (1941). AUER (1933) has encountered *Proteaceae* grains in peats of the rain-forest area in the western part of Tierra del Fuego (up to 85 per cent of the total number of pollen grains and spores).

Ranunculaceae (PLATE XVIII, FIGS. 308–313): —

The pollen morphology of this family has recently been studied by KUMAZAWA (1936) and WODEHOUSE (1936). Fossil ranunculaceous pollen are seldom mentioned. In some peats in Tierra del Fuego, however, AUER encountered considerable quantities of *Hamadryas* pollen (up to 42 per cent of the total number of pollen grains and spores).

Aconitum septentrionale. — FIG. 308: polar view; 25 μ; Lapland 1936. FIG. 309: equatorial view; 25 by 23 μ. — Grains spheroidal, tricolpate, possessing broad furrows without germ pores. Furrow membranes copiously flecked with more or less irregular large granules. Polar and intercolpar areas smooth, with a faint reticular texture.

Caltha palustris. — FIG. 310: polar view; 25 μ; Gotland, Sweden. FIG. 311: oblique equatorial view; 25 by 23 μ. — Grains tricolpate, furrows long and broad, lightly tapering to broad rounded ends, their membranes finely flecked. Exine moderately thick, granular, and densely coated with minute papillae similar to the granules of the furrow membranes (WODEHOUSE 1936). Grains rather similar to those

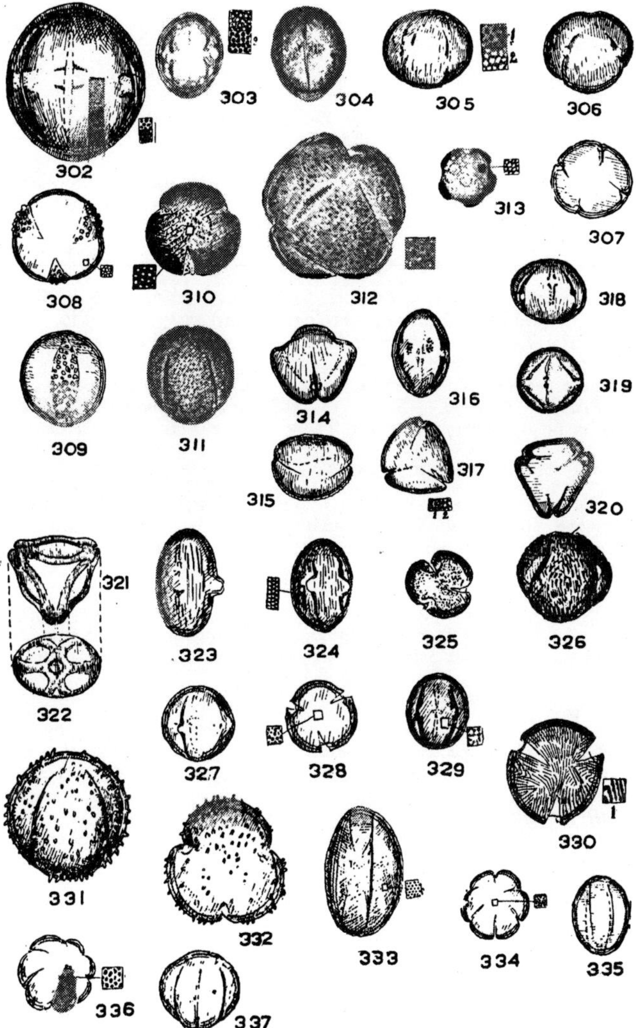

PLATE XVIII (302–337). — *Polygonaceae* (302–307), *Ranunculaceae* (308–313), *Rhamnaceae* (314–320), *Rhoipteleaceae* (321, 322), *Rosaceae* (323–333), *Rubiaceae* (334–337). — 302, *Polygonum viviparum;* 303 and 304, *P. convolvulus;* 305–307, *Rumex acetosella;* 308 and 309, *Aconitum septentrionale;* 310 and 311, *Caltha palustris;* 312, *Ranunculus paucistamineus;* 313, *Thalictrum flavum;* 314–317, *Rhamnus cathartica;* 318 and 319, *R. frangula;* 320, *R. frangula;* 321 and 322, *Rhoiptelea chiliantha;* 323–325, *Comarum palustre;* 326, *Dryas octopetala;* 327 and 328, *Filipendula hexapetala;* 329, *F. ulmaria;* 330, *Rosa rugosa;* 331 and 332, *Rubus chamaemorus;* 333, *Sorbus aucuparia;* 334 and 335, *Asperula odorata;* 336 and 337, *Galium boreale.*

of *Quercus*, but easily distinguished by having flecked furrow membranes.

Ranunculus paucistamineus. — FIG. 312: hexacolpate grain; about 39 μ; Gotland, Sweden. — Grains more or less irregular; number of furrows variable. Exine conspicuously, though not densely, granular.

Thalictrum flavum. — FIG. 313: grains from Västerås, Sweden, 1936; 16 μ. — Grains cribellate, with about 8 pores. Exine with reticulate texture.

Rhamnaceae (PLATE XVIII, FIGS. 314–320): —

Rhamnus cathartica. — FIG. 314: tricolpate grain, oblique view; 22 μ; Connecticut. FIG. 315: dicolpate grain; 17 by 22 μ. FIG. 316: equatorial view of tricolpate grain; 23 by 15.5 μ. FIG. 317: tricolpate grain, polar view; 19 μ. — Grains as a rule tricolpate, with conspicuous germ pores; exine of a delicate reticulate texture.

Rhamnus frangula. — FIGS. 318, 319: grains in equatorial view; 320: in polar view 17 by 20 and 17 by 18 μ respectively; Västerås.

Rhoipteleaceae (PLATE XVIII, FIGS. 321, 322): —

Rhoiptelea chiliantha. — FIG. 321: polar view; 23 μ; Kweichow. FIG. 322: equatorial view (schematic). — Grains flattened, in polar view of triangular outline, tricolpate, with furrows at the angles. Each furrow provided with a large germ pore (vestibule?). As in *Alnus*, the grains are conspicuously arched with arci swinging in pairs (or possibly in threes) from pore to pore. So far this pollen type has been of no interest in pollen analysis. It is mentioned here because of its similarity — in polar view — to the pollen of *Alnus* and *Betula.*

Rosaceae (PLATE XVIII, FIGS. 323–333): —

Comarum palustre. — FIGS. 323, 324: equatorial view; 27 by 16 and 24 by 16 μ respectively; Västerås. FIG. 325: polar view; 17 μ. — Grains prolate, tricolpate, with conspicuous germ pores; exine striate. Size 23–27 μ (YAMASAKI 1933).

Dryas octopetala. — FIG. 326: oblique view; 23 by 25 μ; Åre, Sweden. — Exine striate; striation irregular, consisting of more or less parallel striae. Size about 26–28 μ (FERRARI 1927, ARMBRUSTER-OENIKE 1929).

Filipendula hexapetala. — FIGS. 327, 328: equatorial and polar view respectively; 19 μ; Åker, Sweden, 1938.

Filipendula ulmaria. — FIG. 329: equatorial view; 19 μ by 16 μ; Rogberga, Sweden. — Size 32.8–35.1 (FERRARI 1927), 15 by 13 μ (ZANDER 1935).

Rosa rugosa. — FIG. 330: polar view.

Rubus chamaemorus. — FIG. 331: oblique view; 32 μ. FIG. 332: polar view; 32 μ. — Exine provided with blunt spines of somewhat varying length, often irregularly placed over the surface. Size, according to YAMASAKI (1933) 40–37 by 37–30 μ.

Sorbus aucuparia. — FIG. 333: equatorial view; 32 by 21 μ; Vartofta, Sweden. — Size 27–33 μ (YAMASAKI 1933), 26 by 24.8 μ (ZANDER 1935).

As yet but little is known of the occurrence of fossil rosaceous pollen. However, the pollen of *Rubus chamaemorus* is easily recognized and often found in peats and raw humus in Finland and Sweden. The highest frequency (44–50 per cent) has been reported by BENRATH and JONAS (Planta, vol. 26, 1937, p. 618) from soil profiles in northern Germany. Recently AUER (1933) has found up to 60 per cent *Acaena* pollen in South American peats (frequency calculated in relation to the total number of pollen grains and spores).

Rubiaceae (PLATE XVIII, FIGS. 334–337): —

Asperula odorata. — FIG. 334: polar view; 17 μ. FIG. 335: equatorial view; 20 by 15.5 μ.

Galium boreale. — FIG. 336: oblique polar view; 21 μ. FIG. 337: oblique equatorial view; 20 by 34 μ; Lapland 1937. — Grains, as in *Asperula*, with six or seven furrows without germ pores. Exine finely reticulate.

Salicaceae (PLATE XIX, FIGS. 338–348): —

Populus tremula. — FIG. 338: grain from Västerås 1936; 29 μ. — Grains spheroidal, acolpate; exine thin, faintly granular.

Populus tremula gigas. — FIG. 339: grain from Tynderö, Sweden, 1936; 35 μ. — Grains larger than those of *P. tremula.* Exine provided with a minute sculpturing in the form of small granules rising above the general surface of the exine (cf. the illustr. in the upper left corner).

Salix. — Grains usually prolate, less frequently subprolate, prevailingly tricolpate, possessing long, tapering furrows without germ pores. Exine reticulate, with sharply angular lacunae. Reticulation gradually disappearing towards the margins of the furrows. Whether specific identifications in this genus can be based on pollen grain characteristics, such as size, shape, and sculpturing, is still a question.

Salix caprea. — FIG. 340: equatorial view; 24 by 16 μ; Västerås. — Size of grains, according to FERRARI (1927), 25.7 by 16.4 μ.

Salix cinerea. — FIG. 341: equatorial view; 25 by 16 μ; Gotland.

Salix nigricans. — FIG. 342: equatorial view; 27 by 15.5 μ; Vreta, Sweden.

Salix polaris. — FIG. 343: equatorial view; 31 by 19 μ; Novaya Zemlya. FIG. 344: polar view; 19 μ.

Salix repens. — FIGS. 345, 346: polar and equatorial view respectively; FIG. 345: 17 μ; FIG. 346: 25 by 15.5 μ.

Salix reticulata. — FIG. 347: equatorial view; 22 by 15 μ; Tromsö, Norway.

Salix triandra. — FIG. 348: equatorial view; 24 by 15 μ; Säter, Sweden.

Statements regarding fossil *Salix* pollen may not always be correct. Misleading illustrations have been published and these have led to much uncertainty and confusion. There can be hardly any doubt that, in certain cases, *Artemisia* pollen has been referred to the genus *Salix*, although these pollen types, as may be seen from PLATES XIX and VII, are morphologically very different.

Pollen spectra with extremely high frequencies of *Salix* pollen are considered as indicating a tundra vegetation of low or prostrate willows [FIRBAS, Bibl. Bot., H. 112, 1935; FAEGRI, Bergens Mus. Årsbok 1935 (1936)] Recent and subrecent samples from areas of dense thickets of taller willows, *e.g.* northern Finland and Swedish Lapland, are poor in *Salix* pollen, its frequency being from zero to a few per cent [according to FIRBAS (Planta, vol. 22, 1934) and ERDTMAN]. An exceptionally high *Salix* pollen frequency (95 per cent) was found by FIRBAS (Saalburgjahrbuch 1930) in a peat, containing pollen grains of *Carpinus*, *Fagus* etc. and the remains of willow wood. This may possibly have been due to the embedding of stamens, or even of catkins, in the peat. High frequencies have been reported not only from northern and central Europe. Thus, in Italy, CHIARUGI (N. Giorn. Bot. Ital., n.s., vol. XLIII, 1936) has found up to 122 per cent of *Salix* pollen (frequency expressed in the same way as the hazel pollen frequency).

Sarraceniaceae (PLATE XIX, FIGS. 349, 350): —

Sarracenia purpurea. — FIG. 349: equatorial view; 24 by 18 μ. FIG. 350: polar view; 23 μ. — Grains more or less spheroidal to subprolate, provided with about 8 furrows, each with a single germ pore.

Saxifragaceae (PLATE XIX, FIGS. 351–356): —

Saxifraga cotyledon. — FIG. 351: polar view; 24 μ; Lapland 1936.

Saxifraga hirculus. — FIG. 352: oblique view; 27 μ; Hammerdal, Sweden. FIG. 353: equatorial view: 24 by 31 μ. — Size, according to HØEG (1921), 18–23 by 30–36 μ.

Saxifraga oppositifolia. — FIG. 354: equatorial view; 29 by 21 μ; Åre, Sweden.

Saxifraga stellaris. — FIG. 355: equatorial view; 17 by 14 μ; Åre, Sweden. FIG. 356: oblique polar view; 15 μ. — The pollen grains of *Saxifraga* are subprolate to suboblate, tricolpate. Exine patterns widely varied. *Cf.* also MEINKE (1927).

Scrophulariaceae (PLATE XIX, FIGS. 357, 358): —

Scrophularia nodosa. — FIGS. 357, 358: equatorial views; 26 by 19 and 24 by 21 μ respectively; Åker, Sweden, 1938. — Grains tricolpate, each furrow with an equatorial germ pore; exine thin, finely reticulate.

Symplocaceae (PLATE XX, FIG. 389): —

Symplocos crataegoides. — FIG. 389*a*: polar view; 31 μ; Pindar Valley, India. FIG. 389*b*: equatorial view; 20 by 35 μ. — Grains triangular in polar view; at the angles, short furrows, each underlain by an endexinous, transverse furrow. In a description of the pollen morphology of *S. spicata* (KIRCHHEIMER 1938) the grains are said to measure from 16 to 27 μ and to possess three pores and a smooth exine.

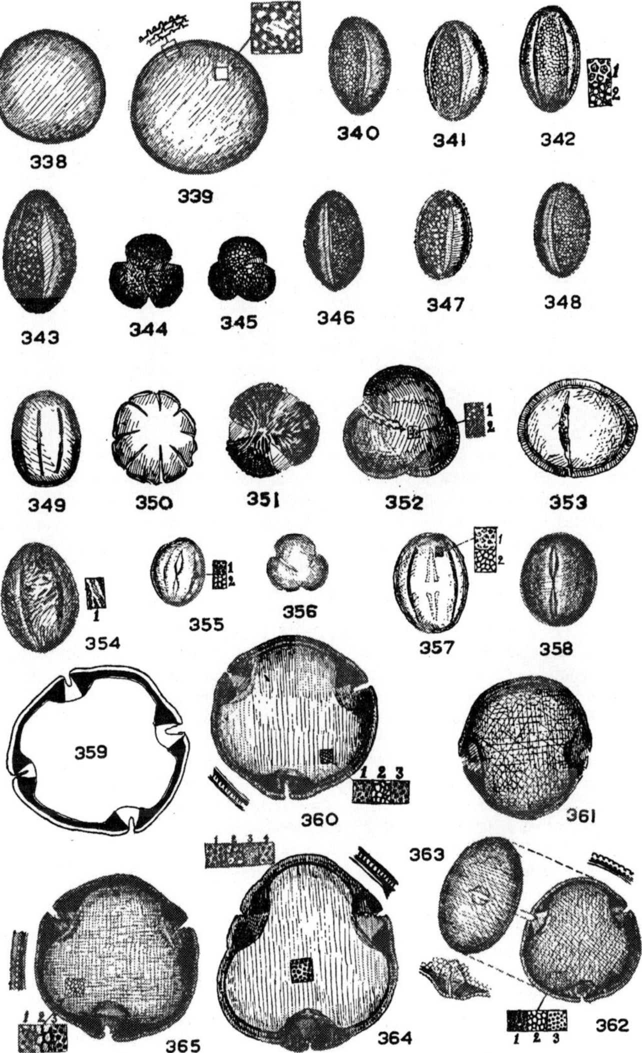

PLATE XIX (338–365). — *Salicaceae* (338–348), *Sarraceniaceae* (349, 350), *Saxifragaceae* (351–356), *Scrophulariaceae* (357–358), *Tiliaceae* (359–365). — 338, *Populus tremula;* 339, *P. tremula gigas;* 340, *Salix caprea;* 341, *S. cinerea;* 342, *S. nigricans;* 343 and 344, *S. polaris;* 345 and 346, *S. repens;* 347, *S. reticulata;* 348, *S. triandra;* 349 and 350, *Sarracenia purpurea;* 351, *Saxifraga cotyledon;* 352 and 353, *S. hirculus;* 354, *S. oppositifolia;* 355 and 356, *S. stellaris;* 357 and 358, *Scrophularia nodosa;* 359 and 360, *Tilia americana;* 361–363, *T. cordata;* 364, *T. platyphyllos;* 365, *T. tomentosa.*

Tiliaceae (PLATE XIX, FIGS. 359–365): —

Tilia. — Grains suboblate to oblate, generally tricolpate. Furrows very short, pitlike depressions enclosing circular germ pores. Ektexine thin, apparently formed by delicate rods, the upper parts of which have coalesced to a continuous cover (FISCHER 1890). Endexine generally of approximately the same thickness as the ektexine. Pore construction: *see* under *T. cordata.*

Tilia americana. — FIG. 359: tetracolpate grain, polar view (optical section; scheme); 45 μ; New Jersey. FIG. 360: tricolpate grain, polar view; 43 μ. — Grains about 36.5 by 28 μ (WODEHOUSE).

Tilia cordata. — FIG. 361: bicolpate grain, polar view; 37 μ. FIG. 362: tricolpate grain, polar view; 32 μ. FIG. 363: tricolpate grain, equatorial view; 21 by 32 μ. — Grains about 25–37 μ in diameter, in average about 31 μ (TRELA 1928, GODWIN 1934); in polar view rounded or, sometimes, faintly triangular. The surface of the grain, if examined with high resolving power, is not quite smooth (FIG. 362, upper detail figure; other traces of the intricate texture are shown in the lower detail figure). In the detail figure of FIG. 363 a furrow and the appendant germ pore are shown in oblique view and in optical section. Under the ektexine, cut by the short furrow, there is what appears to represent a mesexinous filling. It forms the bulk of the pore wall and is underlain by a thin endexine lamella. As stated by FISCHER (1890, p. 24), the lumen of the pores gradually tapers towards the surface of the grain (a somewhat divergent picture of the pore construction is given by POTONIÉ (1934, p. 24).

Tilia platyphyllos. — FIG. 364: polar view; 51 μ; Strömstad, Sweden. — Grains about 37 μ in diameter; in polar view more or less rounded to hexagonal. Exine proportionately thicker than in *T. cordata;* pores slightly bulging (TRELA 1928). The exine texture is still more intricate and the distance from the centre of the grain to the bottom of the pores comparatively shorter than in grains of *T. cordata.* By means of these characters, even single grains may readily be distinguished from those of *T. cordata.* Nevertheless, only a few discoveries of fossil grains of *T. platyphyllos* have been made (*cf.* DOKTUROWSKY, Geol. Fören. Förhandl., vol. 51, 1929; footnote, p. 406).

Tilia tomentosa. — FIG. 365: polar view; 45 μ.

Ulmaceae (PLATE XX, FIGS. 366–370): —

Celtis aculeata. — FIG. 366: polar view; 19 μ; Venezuela. FIG. 367: equatorial view; 17 by 21 μ. — Grains suboblate, in polar view more or less angular, provided with three or four equatorial germ pores, one at each angle of the grain. Ektexine comparatively thick, faintly textured. According to WODEHOUSE, the grains of *C. laevigata* K. Koch (*C. mississippiensis* Spach) are larger (diameter averaging about 40 μ) and provided with decidedly aspidate pores.

Ulmus (FIGS. 368–370). — Grains suboblate, 23 to 38 μ in diameter. Germ pores three to seven, generally five or four, rarely three, elliptical in shape, their aperture 3.5 to 6 μ in length, equatorially arranged, with their long axes converging in pairs (WODEHOUSE). Exine with a faint, undulating sculpturing imparting a more or less reticular appearance to the surface of the grains. According to WODEHOUSE, traces of arci may be found in the pollen grains of *U. crassifolia*, while the pollen grains of the water elm, *Planera aquatica*, are said to have typical arci.

Ulmus laevis. — FIG. 368: polar view; 32 μ; Öland, Sweden. — According to SAURAMO and others, there is evidence to show that the pollen grains of *U. laevis*, at least the typical ones, may be distinguished from the grains of *U. scabra*. The distinctive characters have not been described in detail. However, it is worth while to note that the grains of *U. laevis* frequently are provided with four pores and with thicker exine, coarser sculpturing — suggestive of the surface markings of a peanut shell — and a more angular outline than the grains of *U. scabra*. Suggestions concerning the postglacial history of this interesting species are found in the papers of LINKOLA (Acta Forest. Fenn., no. 40, 1934, p. 38), BENRATH (Inaug.-Diss., Königsberg Pr., 1934, p. 59), JONAS (Fedde's Rep., Beih. LXXXVI, 1936, p. 9), and THOMSON (Schr. phys. ökon. Ges. Königsberg, vol. LXIX, 1937, p. 287).

Ulmus scabra. — FIG. 369: polar view; 33 μ; Västerås 1936. FIG. 370: oblique equatorial view; equator marked by a broken line; 32 by 27 μ. — Grains in polar view more or less rounded, provided with five, or sometimes with four or six pores. Exine comparatively thin; sculpturing usually less pronounced than in the grains of *U. laevis*.

Umbelliferae (PLATE XX, FIGS. 371–380): —

Grains subprolate to perprolate, in polar view more or less triangular. They are usually tricolpate with long, tapering furrows at the angles; each furrow encloses an equatorial germ pore. Ektexine, in most cases, slightly bulging at the pores. The grains are provided with a rather pronounced texture, dotted or reticulate in surface view (FIGS. 371, 380), and, in many cases, apparently due to mesexinous rod-shaped elements (FIG. 376).

Angelica archangelica. — FIG. 371: equatorial view; 32 by 14 μ; Dovre, Norway.

Azorella pedunculāris. — FIG. 372: equatorial view; 21 by 11 μ; Pichincha, Ecuador.

Hydrocotyle vulgaris. — FIG. 373: equatorial view; 21 by 17 μ. FIG. 374: polar view; 15 μ.

Oenanthe aquatica. — FIG. 375: equatorial view; 34 by 15.5 μ; Tosterön, Sweden. FIG. 376: polar view; 18 μ.

Pimpinella saxifraga. — FIG. 377: equatorial view; 33 by 14 μ; Ekerö, Sweden.

Xanthosia ciliata. — FIGS. 378, 379: grains in equatorial view; FIG. 378, 27 by 17, FIG. 379, 22 by 17 μ; Australia. FIG. 380: oblique polar view; 19 μ. — In polar view, the grains are very similar to the pollen of *Artemisia* (*cf.* FIG. 130). Certain pollen types of the *Umbelliferae* and the *Compositae* have several morphological characters in common.

Urticaceae (PLATE XX, FIGS. 381–382): —

Urtica dioica. — FIG. 381: grains from Åker, Sweden, 1938; 15 by 17 and 14 by 15 μ respectively. — Grains spheroidal, diameter, according to WODEHOUSE, about 10.5 μ. The grains usually are provided with three, but sometimes with two or four equatorially arranged pores with circular apertures. Exine smooth, texture very faint or lacking. Fossil pollen and stinging hairs have been found by WEBER (Engler Bot. Jahrb., vol. 54, Beibl. no. 120, 1917).

Urtica urens. — FIG. 382: polar view; 14 μ; Åker, Sweden, 1938. — Grains similar to those of *U. dioica*, exine sometimes slightly bulging at the pores.

Valerianaceae (PLATE XX, FIG. 383): —

Valeriana excelsa. — FIG. 383: oblique polar view; 64 μ; Västerås. — Grains spheroidal, tricolpate (with wide furrows), echinate (with blunt spines uniformly spaced), and of a reticulate texture.

Violaceae (PLATE XX, FIGS. 384–386): —

Viola mauiensis. — FIG. 384: polar view; 40 μ; Hawaii 1938. FIG. 385: equatorial view; 36 by 38 μ. — Grains spheroidal, tricolpate; each furrow with an equatorial germ pore. Exine smooth, provided with a granular texture.

Viola palustris. — FIG. 386: equatorial view; 37 by 27 μ; Västerås 1938. — The pollen grains bear no apparent similarity to *Quercus* pollen, but for a long time they were considered to be indistinguishable.

References: —

ARMBRUSTER, L. und OENIKE, G., 1929: Die Pollenformen als Mittel zur Honigherkunftsbestimmung (Bücherei f. Bienenkunde, vol. X).

ASSARSSON, G., 1927: Fossilt pollen av *Trapa natans* (Geol. Fören. Förhandl., vol. 49, Stockholm).

AUER, V., 1933: Verschiebungen der Wald- und Steppengebieten Feuerlands in postglazialer Zeit (Acta Geographica 5:2, Helsinki).

BAAS, J., 1932: Eine frühdiluviale Flora im Mainzer Becken (Ztschr. f. Bot., vol. 25).

ČERNJAVSKI, P., 1935: Über die rezenten Pollen einiger Waldbäume in Jugoslavien (Beih. Bot. Centralbl., vol. LIV).

CRANWELL, L., 1939: Southern-Beech Pollens (Rec. Auckl. Inst. Mus., vol. 2).

ERDTMAN, G., 1936: New methods in pollen analysis (Svensk Bot. Tidskr., vol. 30).

—— —— 1940: Flower dimorphism in *Statice Armeria* L. (*Ibid.*, vol. 34).

FERRARI, A., 1927: Osservazioni di biometria sul polline delle angiosperme (Atti Ist. Bot. Pavia).

FIRBAS, F., 1927: Beiträge zur Geschichte der Moorbildungen und Gebirgswälder Korsikas (Beih. Bot. Centralbl., vol. XLIV).

—— —— 1931: Über die Waldgeschichte der Süd-Sevennen und über die Bedeutung der Einwanderungszeit für die nacheiszeitliche Waldentwicklung der Auvergne (Planta, vol. 13).

—— —— und GRAHMANN, R., 1928: Über jungdiluviale und alluviale Torflager in der Grube Marga (Abhandl. math.-phys. Kl. sächs. Akad. Wiss., vol. XL).

FISCHER, H., 1890: Beiträge zur vergleichenden Morphologie der Pollenkörner (Breslau).

GODWIN, H., 1934: Pollen analysis. An outline of the problems and potentialities of the method (New Phytologist, vol. XXXIII).

IVERSEN, J., 1940: Blütenbiologische Studien, I. Dimorphie und Monomorphie bei *Armeria* (Kgl. Danske Vidensk. Selsk., Biol. Meddel. XV:8).

—— —— 1941: Land occupation in Denmark's Stone Age (Danm. Geol. Unders., II. R., no. 66).

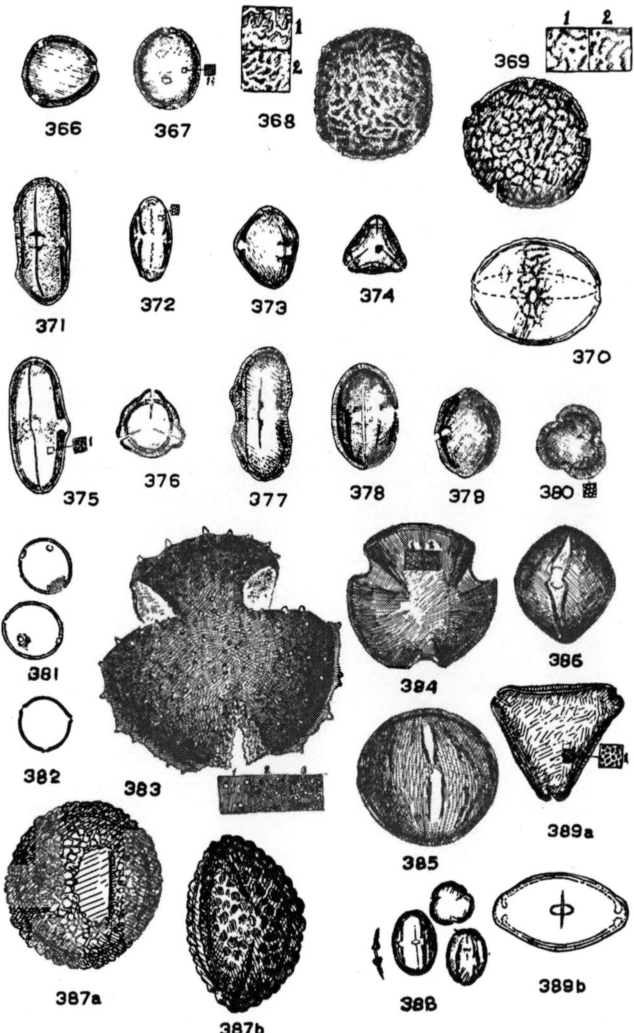

PLATE XX (366–389). — *Ulmaceae* (366–370), *Umbelliferae* (371–380); *Urticaceae* (381–382); *Valerianaceae* (383); *Violaceae* (384–386); *Addenda* (387–389). — 366 and 367, *Celtis aculeata;* 368, *Ulmus laevis;* 369 and 370, *U. scabea;* 371, *Angelica archangelica;* 372, *Azorella peduncularis;* 373 and 374, *Hydrocotyle vulgaris;* 375 and 376, *Oenanthe aquatica;* 377, *Pimpinella saxifraga;* 378–380, *Xanthosia ciliata;* 381, *Urtica dioica;* 382, *U. urens;* 383, *Valeriana excelsa;* 384 and 385, *Viola mauiensis;* 386, *V. palustris;* 387*a* and *b*, *Sciadopitys verticillata;* 388, *Castanopsis chrysophylla;* 389*a* and *b*, *Symplocos crataegoides.*

JENTYS-SZAFER, J., 1928: La structure des membranes du pollen de *Corylus*, de *Myrica* et des espèces européennes de *Betula* et leur détermination à l'état fossile (Bull. Acad. Pol. Sc. Lettr., Cl. Sc. Math. Nat., Sér. B).

KIRCHHEIMER, F., 1938: Bemerkungen über die botanische Zugehörigkeit von Pollenformen aus den Braunkohlenschichten (Planta, vol. 28).

KULCZYŃSKI, S., 1932: Die diluvialen Dryasfloren der Gegend von Przemyśl (Acta Soc. Bot. Pol., vol. IX).

KUMAZAWA, M., 1936: Pollen grain morphology in *Ranunculaceae, Lardizabalaceae* and *Berberidaceae* (Japan. Journ. Bot., vol. VIII).

LEITNER, J., 1942: Ein Beitrag zur Kenntnis der Pollenkörner der *Labiatae* (Österr. Bot. Ztschr. 91:29–40).

MEINKE, H., 1927: Atlas und Bestimmungsschlüssel zur Pollenanalytik (Bot. Archiv).

OVERBECK, F., 1934: Zur Kenntnis der Pollen mittel- und nordeuropäischer *Ericales* (Beih. Bot. Centralbl., vol. LI).

—— und SCHMITZ, H., 1931: Zur Geschichte der Moore, Marschen und Wälder Nordwestdeutschlands, I. Das Gebiet von der Niederweser bis zur unteren Ems (Mitt. Prov.-St. Naturdenkmalpfl. Hannover, Heft 3).

VON POST, L., 1924: Ur de sydsvenska skogarnas regionala historia under postarktisk tid (Geol. Fören. Förhandl., vol. 46).

—— —— 1930: Die Zeichenschrift der Pollenstatistik (*Ibid.*, vol. 51, 1929).

POTONIÉ, H., 1934: Zur Mikrobotanik der Kohlen und ihrer Verwandten, I. Zur Morphologie der fossilen Pollen und Sporen (Arb. Inst. Paläobot. Petrogr. Brennsteine, vol. IV).

RUDOLPH, K., 1935: Mikrofloristische Untersuchung tertiärer Ablagerungen im nördlichen Böhmen (Beih. Bot. Centralbl., vol. LIV, Abt. B).

—— —— und FIRBAS, F., 1924: Die Hochmoore des Erzgebirges. Ein Beitrag zur postglazialen Waldgeschichte Böhmens (*Ibid.*, vol. XLI).

SCHROEDER, D., 1939: Eine bronzezeitliche Wegstrecke in Nordhannover (Darstell. Niedersachs. Urgesch., vol. 4, Hildesheim).

SIMPSON, J. B., 1936: Fossil pollen in Scottish tertiary coals (Proc. Roy. Soc. Edinb., vol. LVI).

SUNDELIN, U., 1919: Über die spätquartäre Geschichte der Küstengegenden Östergötlands und Smålands (Bull. Geol. Inst. Upsala, vol. XVI).

THIERGART, F., 1937: Die pollenflora der Niederlausitzer Braunkohle, besonders im Profil der Grube Marga bei Senftenberg (Jahrb. Preuss. Geol. Landesanst., vol. 58).

TRELA, J., 1928: Zur Morphologie der Pollenkörner der einheimischen *Tilia*-Arten (Bull. Acad. Polon. Sc. Lettr., Cl. Sc. Math. Nat., Sér. B).

WODEHOUSE, R., 1932: Tertiary pollen, I. Pollen from the living representatives of the Green River flora (Bull. Torr. Bot. Club., vol. 59).

—— —— 1935: Pollen grains. Their structure, identification and significance in science and medicine (McGraw-Hill, New York).

—— —— 1936: Pollen grains in the identification and classification of plants, VII. The *Ranunculaceae* (Bull. Torr. Bot. Club, vol. 63).

YAMASAKI, T., 1933: Morphology of pollen grains and spores (Rep. Exper. Forest, Kyoto Imp. Univ., no. 5; Japanese).

ZANDER, E., 1935: Beiträge zur Herkunftsbestimmung bei Honig, I (Berlin).

—— —— 1941: Beiträge zur Herkunftsbestimmung bei Honig, III (Leipzig).

Chapter IX

POLLEN MORPHOLOGY — GYMNOSPERMS

Morphology of Winged Pollen (*cf.* WODEHOUSE 1933 and 1935): — Winged pollen grains consist of body and varying numbers of air-sacs or bladders. The body is spheroidal or slightly flattened (suboblate or oblate), resembling in shape a double lens. The exine is particularly thick in the proximal part of the body — the cap. It consists of ektexine, endexine, and between these, rod-like mesexinous elements in a compact arrangement which gives the exine a complicated, somewhat dotted texture. A distinct boundary frequently occurs between the thick exine of the cap and the thin exine of the distal part of the body. The latter part is occupied mainly by the air-sacs. These are separated from the interior of the grain by endexine; their outer wall consists of ektexine with attached mesexinous elements which protrude into the lumen of the bladders. The mesexinous elements are more widely spaced than those of the body and are much more irregular. Branched or unbranched, single or combined in different ways, they tend to produce an array of different patterns, which, although usually appearing to be more or less reticular, are extremely difficult to draw and to describe. Microtome sections of acetolysed pollen grains embedded in paraffin make these subtle details of pollen construction easier to observe and safer to interpret.

Near the proximal root of the air-sacs there frequently are slight ekt-mesexinous ridges or frill-like projections (marginal crests), varying in appearance in different species. This, however, cannot be used as a reliable guide to identification because of its variations within the same species.

At the distal root of the bladders, where they merge into the distal surface of the body of the grain, the characteristic texture of the bladders comes rather abruptly to an end. The intervening space between the bladders is, morphologically speaking, the furrow (*cf.* also GOEBEL 1933), which, in *Pinus* and similar types, extends from end to end of the grain vertical to the plane which passes through the two bladders. It is covered by an exceedingly thin and flexible membrane, smooth and often devoid of any markings.

Abies (PLATE XXI, FIG. 390). — Grains large, 78 to 111 μ in diameter, mostly over 90 μ. Exine of the cap very thick and of a coarse texture. Marginal crest absent or only faintly suggested by a few slight undulations near the proximal roots of the bladders. Boundary of cap usually sharply defined. Exine of the ventral surface usually smooth. Bladders various, but generally comparatively small in relation to the size of the grain and always forming a sharp re-entrant angle with the cap at their proximal roots (WODEHOUSE, p. 263).

Abies sibirica. — FIG. 390: lateral view, proximal side (cap) down; entire grain 82 by 149 μ; body 82 by 119 μ; Tobolsk.

Cedrus deodara (PLATE XXII). — FIG. 398: lateral view, proximal side down (schematical); outline of the grain in polar view marked by crosses; entire grain 60 by 112 μ; body 60 by 78 μ; Geneva. — Grains generally about 65 μ in diameter (62 to 78 μ). Bladders various, but usually proportionately much smaller than those of the grains of *Pinus* and always more laterally placed, leaving a broader and longer furrow area between them. Cap circular or slightly elongated transversely, its texture merging gradually with that of the bladders (WODEHOUSE). There is usually no re-entrant angle between the cap and the proximal roots of the bladders; this tends to create a certain resemblance between the pollen of *Cedrus* and *Picea.*

Chamaecyparis nootkatensis (PLATE XXII). — FIG. 399: grain from Hort. Bergianus 1935; 27 μ. — Grains acolpate; exine prominently flecked with small granules.

Cryptomeria japonica (PLATE XXII). — FIG. 400; lateral view; distal part with germ pore up; 30 by 34 μ; Visby, Sweden, 1935. — Grains spheroidal, 23.9 to 31.9 μ in diameter (according to JIMBO, 1933, 30 to 35 μ), provided with a single germ pore, consisting of a finger-like projection standing straight up from the surface and slightly bent at the top (WODEHOUSE). Ektexine near the pore thin, smooth, and flecked with small granules; otherwise thicker and, in places, forming small, irregular ridges.

Cunninghamia lanceolata (PLATE XXII). — FIG. 401: oblique polar view (distal part); 31 by 34 μ; Fukien. — Grains approximately spheroidal, ranging from 34.2 to 40 μ in diameter, provided with a single germ pore (WODEHOUSE). Size, according to JIMBO (1933), 30 to 38 μ. Ektexine loosely flecked, presenting a rather rugged surface, just as in the grains of *Cryptomeria japonica.*

Cupressus macnabiana (PLATE XXI). — FIGS. 391, 392: grains from California; 27 μ. — Grains spheroidal, acolpate, psilate.

Dacrydium elatum (PLATE XXII). — FIG. 404: lateral view (distal part upwards); 50 by 70 μ; Tonkin. — Grains with two low, almost rudimentary bladders. The cap, of coarse texture, gradually merges into the uneven, somewhat undulating surface of the bladders. From this surface, long mesexinous elements project into the lumen of the bladders, probably (in part, at least) down to their endexinous floor. The furrow is short, and not surrounded by a thickened rim.

Dacrydium franklinii (PLATE XXII). — FIG. 402: lateral view, distal part up; entire grain 29 by 54 μ; Tasmania. FIG. 403: proximal side of grain; 32 by 49 μ. — Small grains, more flattened on the distal than on the proximal surface, provided with two bladders, inserted on the lateral extremities of the grain. The bladders are proportionately small, have a smooth surface, and form a sharp re-entrant angle with the cap at their proximal roots. The ektexine of the cap presents a particularly solid appearance and has a very faint texture only.

Encephalartos altensteinii (PLATE XXII). — FIG. 405: distal part of grain with open furrow; 32 by 40 μ; Lisbon. — Grains monocolpate; exine of a faint texture. *Encephalartos* belongs to the *Cycadaceae.* Fossil pollen grains referred to this family have been found in Eocene

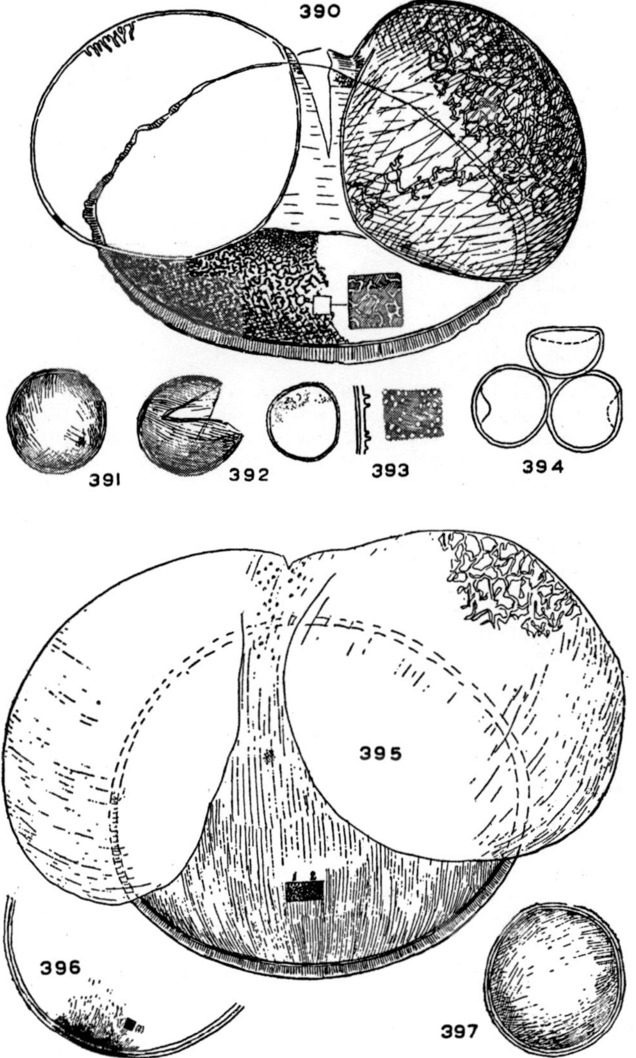

PLATE XXI (390–397) (*Gymnospermae* 1). — 390, *Abies sibirica;* 391 and 392, *Cupressus macnabiana;* 393 and 394, *Juniperus virginiana;* 395, *Keteleeria davidiana;* 396, *Larix decidua;* 397, *Libocedrus decurrens.*

beds in North America and have been described by WODEHOUSE (1933, pp. 483, 484) as *Cycadopites* and *Dioonipites*.

Ephedra antisyphilitica (PLATE XXVI). — FIG. 447: lateral view; 52 by 35 μ; Texas. — Grains prolate to subprolate, provided with approximately thirteen longitudinal ridges separated by well defined grooves. When the pollen grains germinate, the exine dehisces, splitting into two or more parts through the grooves. WODEHOUSE has observed this kind of dehiscence in the grains of *E. intermedia*. The same phenomenon has been recorded by STAPF (1889) for other species: his figures show a grain split that way, with the pollen protoplast emerging from the split end.

Fossil *Ephedra* pollen has been recorded by AUER (1933) from Tierra del Fuego.

Fitzroya cupressoides (PLATE XXII). — FIG. 406: grain from Valdivia; 30 by 25 μ. — Grains acolpate, with somewhat rugged surface and granular texture.

Ginkgo biloba (PLATE XXII). — FIGS. 407, 408: distal part of grains; Kansu; FIG. 407, 34 by 25 μ, FIG. 408, 34 by 25 μ. — Grains monocolpate, elliptical or sometimes, in polar view, nearly circular, 27–32 μ (WODEHOUSE); a grain reproduced by DRAHOWZAL (1936) is about 28 μ long.

Gnetum latifolium (PLATE XXII). — FIGS. 409, 410: grains from Ceram; approximately 19 by 13 and 17 by 13 μ respectively. — Grains subechinate, apparently acolpate. Ruptured grains often present an appearance similar to the pollen of species of *Nymphaea* and many monocotyledons.

Juniperus virginiana (PLATE XXI). — FIG. 393: oblique polar view; pore situated near the upper end; the detail figures show the structure of the exine in optical section and in surface view respectively; 20 μ; Visby 1935. FIG. 394: outline of three young grains, about 21 μ. — Grains spheroidal, exine thin, transparent, easily ruptured; according to WODEHOUSE 21.6 to 25.1 μ in diameter. Exine irregularly granular, provided with a round, faintly marked germ pore. The presence of a germ pore in *Juniperus* pollen was, apparently, first noticed by IVERSEN (1934). The pore is frequently difficult to detect and may possibly be lacking in some species.

Keteleeria davidiana (PLATE XXI). — FIG. 395: lateral view; entire grain 102 to 161 μ; body 90 by 108 μ. — Grains similar to those of *Abies*, provided with two big bladders. Diameter, bladders included, about 140 μ; height of bladders about 72 μ (RUDOLPH 1935). Exine of the cap thinner, its surface smoother, and the texture more delicate (granular, not irregularly reticulate) than in *Abies*.

Larix (PLATES XXI, XXIII). — Grains acolpate, without bladders, spheroidal, psilate, ranging in different species from 62.5 to 102 μ in diameter. Exine comparatively thin, usually rupturing and frequently cast off when the grains are moistened (WODEHOUSE).

Larix decidua (PLATE XXI). — FIG. 396: part of grain; diameter about 70 μ; Västerås 1936. — Ekt-endexine boundary usually poorly defined; exine smooth, with faint traces of an exceedingly fine reticulation. Diameter, according to GERASIMOV (1930), 60 to 92, usually 70 to 80 μ.

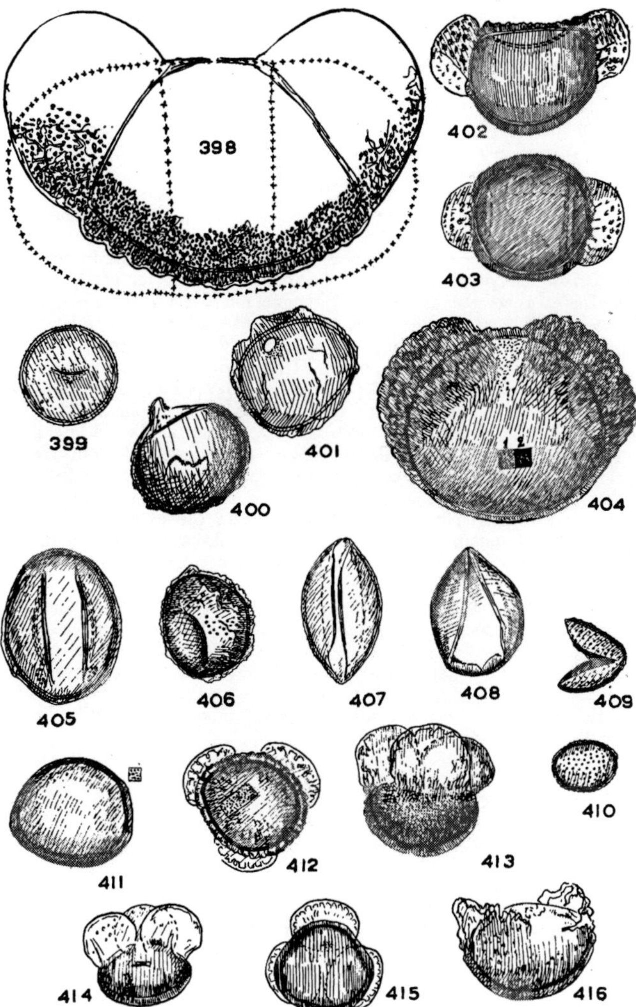

Plate XXII (398–416) (*Gymnospermae* 2). — 398, *Cedrus deodara;* 399, *Chamaecyparis nootkatensis;* 400, *Cryptomeria japonica;* 401, *Cunninghamia lanceolata;* 402 and 403, *Dacrydium franklinii;* 404, *D. elatum;* 405, *Encephalartos altensteinii;* 406, *Fitzroya cupressoides;* 407 and 408, *Ginkgo biloba;* 409 and 410, *Gnetum latifolium;* 411, *Macrozamia spiralis;* 412 and 413, *Microcachrys tetragona;* 414 and 415, *Pherosphaera hookeriana;* 416, *Phyllocladus hypophyllus.*

Larix occidentalis (PLATE XXIII). — FIG. 417: grain from Copenhagen; approximately 79 by 89 μ.

Libocedrus decurrens (PLATE XXI). — FIG. 397: grain from Geisenheim am Rhein; 37 μ. — Grains spheroidal, rather uniform in size, 29.6 to 36.5 μ in diameter, without a pore (WODEHOUSE). Exine surface nearly smooth, provided with small granules.

Macrozamia spiralis (PLATE XXII). — FIG. 411: lateral view; 27 by 33 μ. — Grains rounded, apparently without a furrow; in expanded grains, the proximal side is more flattened than the distal side; exine of a faint texture. Exdexine thin throughout, ektexine thick in what seems to be the proximal part of the grain, otherwise of nearly the same thickness as the endexine. Ektexine not everywhere adhering to the endexine (*cf.* the extreme lateral parts in the illustrations). The thin distal part of the exine probably functions as a furrow.

Microcachrys tetragona (PLATE XXII). — FIG. 412: proximal side of grain; 34 μ (wings included); Tasmania 1937. FIG. 413: lateral view; 33 by 30 μ. — Pollen with several (usually three) air bladders. In polar view, the body of the grain is more or less circular, in lateral view, lens-shaped with the distal surface more flattened than the proximal. The exine is of a coarse texture but its surface is smooth or nearly so, except in the equatorial part of the grain, immediately below the proximal root of the bladders, where the ektexine is frequently slightly undulating. The bladders are directed more upwards than laterally and are comparatively high, forming more than a hemisphere. They are of a faint texture.

Pherosphaera hookeriana (PLATE XXII). — FIG. 414: lateral view; 26 by 31 μ; Tasmania. FIG. 415: proximal part of grain; 26 μ (bladders included). — Grains of about the same type as in *Microcachrys*. Body, in polar view, circular or somewhat triangular with the bladders at the angles. The texture is distinctly granular. Bladders smooth-walled, provided with a few unconnected, speck-like internal thickenings.

Phyllocladus hypophyllus (PLATE XXII). — FIG. 416: lateral view; 25 by 37 μ; Borneo. — Grains with two poorly developed bladders.

Picea (PLATE XXIII). — Grains with two large bladders. Body of the grain biconvex with well rounded corners in lateral view, circular or slightly elliptical in polar view. The bladders are comparatively low, their contours in polar view (FIG. 418), running smoothly into the lateral contours of the body without forming any apparent angles. The thick exine of the cap has a fine granular texture passing gradually into the more or less reticular texture of the bladders (FIGS. 418, 420). There is no real marginal crest. A rim, however, is usually developed as a borderline between the cap and the thin-walled distal furrow area.

The grains are nearly of the same size as those of *Abies*, but they are provided with lower bladders and they show, in lateral view, a smaller re-entrant angle between the proximal root of the bladders and the contour line of the body (this angle, however, varies according to the orientation, eventual compression, etc. of the grains; thus the angles in FIG. 419 are not quite typical). The texture of the cap is finer than in *Abies*, *Cedrus*, and *Pinus*, and of somewhat the same

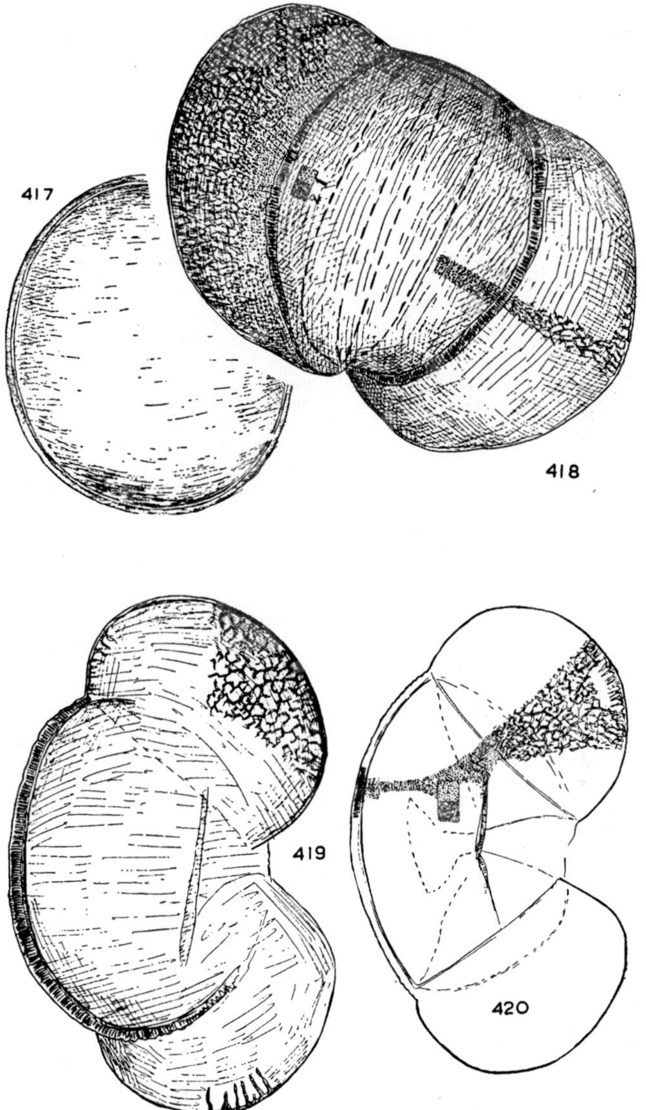

PLATE XXIII (417–420) (*Gymnospermae* 3). — 417, *Larix occidentalis;* 418 and 419, *Picea excelsa* var. *obovata;* 420, *Picea omorika.*

character as in *Keteleeria*. The texture of the bladders is also comparatively delicate.

Picea excelsa var. *obovata*. — FIG. 418: proximal side; 84 by 138 μ (bladders included); Finland. FIG. 419: lateral view; 68 by 130 μ; body 68 by 87 μ; Mustila, Finland, 1936. — The author has not endeavoured to point out any real distinction between pollen grains of this variety of spruce and the pollen grains of *Picea excelsa* itself. The longest axis of the grains in *P. excelsa* is, according to RUDOLPH (1935), 82–131 μ, averaging 110 μ (bladders included), according to ČERNJAVSKI (1935), 75–103 μ (bladders not included). The length of the base of the bladders, measured in lateral view of the grain from the proximal to the distal root of the bladders, is 56–73 μ, averaging about 68 μ (FIRBAS 1925). The corresponding length in FIG. 419 (*P. obovata*) is 59 μ.

Picea omorika (PLATE XXIII). — FIG. 420: lateral view; entire grain 57 by 116 μ; body 57 by 87 μ; length of bladders 51 μ; Serbia. — Texture of the grain usually rather more delicate than in *P. excelsa;* attachment of the bladders to the body more " pinoid " than is usually the case in *Picea*. Length of grain (the distance between the two points where the contour of the cap reaches the proximal roots of the bladders) 60–85 μ, averaging 75 μ (ČERNJAVSKI 1935; pollen grains boiled in KOH-solution — concentration never above 10 per cent — for 1–2 minutes and immediately, without rinsing in water, embedded in glycerine and measured); length of bladders 35–70, generally 45–52 μ (FIRBAS 1927).

Pinus (PLATE XXIV). — Grains with two air-sacs; body round or slightly elliptical when seen from the distal or the proximal side, ranging in diameter from 45 to 65 μ (WODEHOUSE). Total length of the grains, bladders included, ranging from about 61 μ in *P. banksiana* to about 101 μ in *P. pinaster*, *P. ponderosa*, and other species (RUDOLPH 1935). Cap of a well defined granular texture and generally with a conspicuous rim at the point of transition to the distal part of the grain. Marginal crests occur frequently and are sometimes well developed (FIG. 427).

As pointed out by GERASIMOV (1930) there are two general types of bladders. These types were described by RUDOLPH (1935) as the *silvestris*-type and the *haploxylon*-type respectively. In the former, the bladders are more or less contracted at their base and represent more than half of the sphere. In polar view, parts of three intersecting circles form the contour of the grain. *P. banksiana* (FIG. 422), *P. montana* (FIG. 426), and *P. silvestris* (FIG. 428) may be used as examples of this type. In pollen grains of the *haploxylon*-type, the bladders are semicircular, broadly attached to the body. In polar view, the contour of the entire grain is rather elliptical (*cf. P. cembra*, FIG. 425, and *P. peuce*, RUDOLPH *l.c.*, FIG. 1*b*, p. 253).

The grains of *Pseudolarix kaempferi* are of essentially the same type as the grains of *Pinus* (WODEHOUSE) — notably the *silvestris*-type (RUDOLPH *l.c.*) — although the body appears to be more rounded (spherical). The texture of the cap is particularly fine, gradually disappearing towards the distal part of the grain. There is, furthermore, no marginal crest and only a faint limit between the proximal and the distal part of the body.

Pinus banksiana. — FIG. 421: proximal part of grain; 49 by 66 μ; Arnold Arboretum. FIG. 422: oblique lateral view, distal part up; 40 by 65 μ. — Grains small, bladders of the *silvestris*-type; texture of body and bladders well defined; marginal crest frequently well developed. Body 31–41 μ (DEEVEY 1939; grains from Quebec); 47.64 ± 0.34 μ (average of 137 grains; acetolysed material from the New York Botanical Garden, 1939).*

Pinus brutia. — Length of entire grain, according to MARCHETTI (1936), 72–115 μ (grains treated with NaOH), (72–)79–95–108(–115) μ (acetolysed grains), 69–77–92 μ (grains treated according to HÖRMANN 1929).

Pinus cembra. — FIG. 425: proximal part of grain (oblique view); 84 by 103 μ; Stockholm 1935. — Studying the pollen morphology of *P. cembra* and *P. montana*, FURRER (1927) found that the grains of the former were „ im allgemeinen derber gebaut, so z.B. Netzstruktur der Luftsäcke kräftiger und Zellwand wenigstens auf der Rückenseite dicker.... Luftsack mit grosser Ansatzfläche; diese mehr an den gegenüberliegenden Enden der Zelle gelegen, weniger stark als bei *P. montana* gegen die Bauchseite hin, dagegen mehr auf die Rückenseite hinaufreichend ".

Two years later, HÖRMANN (1929) published a paper entitled " Die pollenanalytische Unterscheidung von *Pinus montana*, *Pinus silvestris* und *Pinus cembra* ", and pointed out the possibility of distinguishing between the pollen grains of these species by means of size variation statistics as well as by certain morphological details. HÖRMANN studied pollen grains (from herbarium specimens) boiled in a mixture of alcohol, glycerine, and water, equal parts, and left in this mixture for several days. The average size of the grains, bladders included, was found to be the following: in *P. cembra* 72 μ, in *P. montana* 62–70 μ, in *P. silvestris* 62 μ. The texture of the pollen grains is described as follows (*l.c.*, pp. 226, 227):

„Bei *Pinus montana* erscheint im Mikroskop eine weitmaschige Felderung der Luftsäcke, die sich jedoch bei genauerer Einstellung in zahlreiche kleinere Felder auflöst. Die grossen Felder entstehen dadurch, dass einzelne Leisten des Netzwerkes stärker ausgebildet sind. Der Durchmesser dieser grossen Felder beträgt 3 bis 8 μ, der Durchmesser der kleinen dagegen nur 1 bis 2 μ. Das Netzwerk ist nicht immer geschlossen. Sehr häufig ragen in die Felder Arme, die die gegenüberliegende Querleiste nicht erreichen und so eine unvollständige Teilung des Feldes bewirken.

Die Netzung der Luftsäcke wird gegen das Pollenkorn zu immer enger und geht ganz allmählich in die Struktur des Pollenkorns über. Im Mikroskop erscheinen „grünliche" Rillen, die „rötliche" Grübchen umschliessen. Diese Rillen bilden kein geschlossenes Netz, sondern sind meist unregelmässig hin- und hergebogen und auch oft gabelig geteilt. Nur ganz wenige Rillen werden so kurz, dass sie wie längliche Punkte aussehen. Die Breite dieser Rillen ist im Vergleiche zu *P. silvestris* und *P. cembra* sehr gross; es kommen 8 bis 9 Rillen auf 10 μ.

Der Kamm des Pollens ist ziemlich breit und besonders gegen die Luftsäcke zu oft zackig; er erinnert an den Kamm des *Abies*-Pollens.

Bei *Pinus silvestris* ist im Gegensatz zu *P. montana* nur eine grobe Felderung sichtbar.

* CAIN (1940) was able, with considerable certainty, to determine by means of size-frequency statistics the occurrence of *Pinus banksiana* in buried soils of the Piedmont of South Carolina, an area several hundred miles south of the present limits of the species. Measurements of pollen grain lengths (between the outer points of insertion of the bladders) of 745 fossil grains revealed a strongly trimodal size-frequency distribution. By comparison of the three modes for fossil grains with size-frequency curves for 12 modern species of the eastern United States, it appeared that the small size of the jack pine grains made their determination rather certain. The modes for fossil grains fell at 46.8, 54.6, and 62.4 microns. The means for grains of modern plants, measured in the same manner, were: *Pinus banksiana*, 44.8; *P. glabra*, 53.3; *P. clausa*, 57.4; *P. resinosa*, 58.3; *P. strobus*, 59.1; *P. echinata*, 59.5; *P. rigida*, 61.9; *P. palustris*, 62.3; *P. serotina*, 63.7; *P. taeda*, 66.9; *P. pungens*, 72.1; *P. virginiana*, 72.3 microns.

Diese bildet ein geschlossenes Netz. Die Grösse der Felder schwankt zwischen 2 und 4 µ. Ein feineres Maschenwerk ist zwar hie und da schwach angedeutet, gewöhnlich aber fehlt es vollständig. Eine unvollständige Teilung der Felder durch hervorragende Leisten ist ebenfalls sehr selten.

Auch bei *P. silvestris* geht die Netzung der Blasen allmählich in die Struktur des Pollenkorns über. Diese unterscheidet sich aber von der Struktur des Pollenkorns von *P. montana* ganz wesentlich. Die bei *P. montana* vorhandenen, wurmartig hin- und hergeschlängelten Rillen fehlen. An die Stelle der Rillen treten hier Punkte, die auch hie und da zusammenfliessen und so kurze Striche ergeben. Auf 10 µ kommen ungefähr 10 bis 12 Punkte. Die Struktur ist also auch viel zarter als bei *P. montana*.

Der Kamm des Pollens ist sehr schmal und glatt. Nur ab und zu gegen die Ansatzstellen hin etwas höckerig; er erinnert an den Kamm des *Picea*-Pollens.

Der Pollen von *Pinus cembra* nimmt in seinem Bau eine Mittelstellung zwischen den Pollen von *P. montana* und von *P. silvestris* ein. Es ist auch hier eine weite und eine enge Felderung wie bei *P. montana* vorhanden. Aber nicht alle Felder besitzen diese Unterteilung. Viele grosse Felder sind vollständig ungeteilt. Die Struktur des eigentlichen Pollenkorns setzt sich aus Punkten und auch wurmförmig gekrümmten Rillen zusammen. Auf 10 µ kommen 9 bis 10 Punkte oder Rillen. Der Kamm des Pollens ist schmal und glatt".

Further notes on the morphology of the pollen grains of *P. cembra* have been published by GERASIMOV (1930).

Pinus contorta var. *murrayana.* — FIG. 427: distal side of grain; 56 by 98 µ; Stockholm 1935. — Body spheroidal, of a coarse texture; marginal crest usually well developed.

Pinus excelsa. — Length of body 82.04 ± 0.62 µ (average of 60 acetolysed grains from the New York Botanical Garden 1939).

Pinus halepensis. — Length of entire grains, according to MARCHETTI (1936), 72–*82*–92 µ (acetolysed grains), 64–*74*–90 µ (grains treated according to HÖRMANN).

Pinus laricio. — Length of entire grains, according to MARCHETTI (1936), (72–)77–*82*–102(–113) µ (acetolysed grains), 54–*69*–85 µ (grains treated according to HÖRMANN).

Pinus leucodermis. — FIG. 423: distal part of grain; 63 by 89 µ (Serbia). FIG. 424: proximal part of grain, oblique view; 56 by 94 µ; body 56 by 68 µ. — Size of entire grain 62–*79*–92 µ (MARCHETTI 1936; grains treated according to HÖRMANN); length of body 43–70 µ (ČERNJAVSKI 1935; 2100 grains treated with KOH).

Pinus montana. — FIG. 426: lateral view of grain, distal part (with air-sac) up; Torö, Sweden. — Texture of grains etc., *see* under *P. cembra*. Bladders nearly spherical (GAMS *in* DOKTUROWSKY and KUDRJASCHOW 1924, footnote p. 181). Length of entire grain averages about 68.6 µ (BROCHE 1929), 66–72 µ (VON SARNTHEIN 1936), 55–75 µ (DOKTUROWSKY and KUDRJASCHOW 1924), (54–)67.3(–75) µ (STARK 1927), (50–)60–70(–86) µ (HÖRMANN 1929); length of body averaging 48–57 µ (VON SARNTHEIN *l.c.;* grains treated according to HÖRMANN).

Pinus nigra. — Length of entire grain, according to MARCHETTI (1936), 72–*79*–95 µ (acetolysed grains), or 69–*79*–90 µ (grains treated according to HÖRMANN). Length of body 40–65 µ (ČERNJAVSKI 1935; 1100 grains from 11 trees; KOH).

Pinus peuce. — Length of body 43–70 µ (ČERNJAVSKI 1935; 1400 grains from 8 trees; KOH).

Pinus pinaster. — Length of entire grain, according to MARCHETTI (1936), 85–*105*–118 µ (NaOH); 89–*102*–115 µ (acetolysed grains); 87–*95*–108 µ (grains treated according to HÖRMANN and

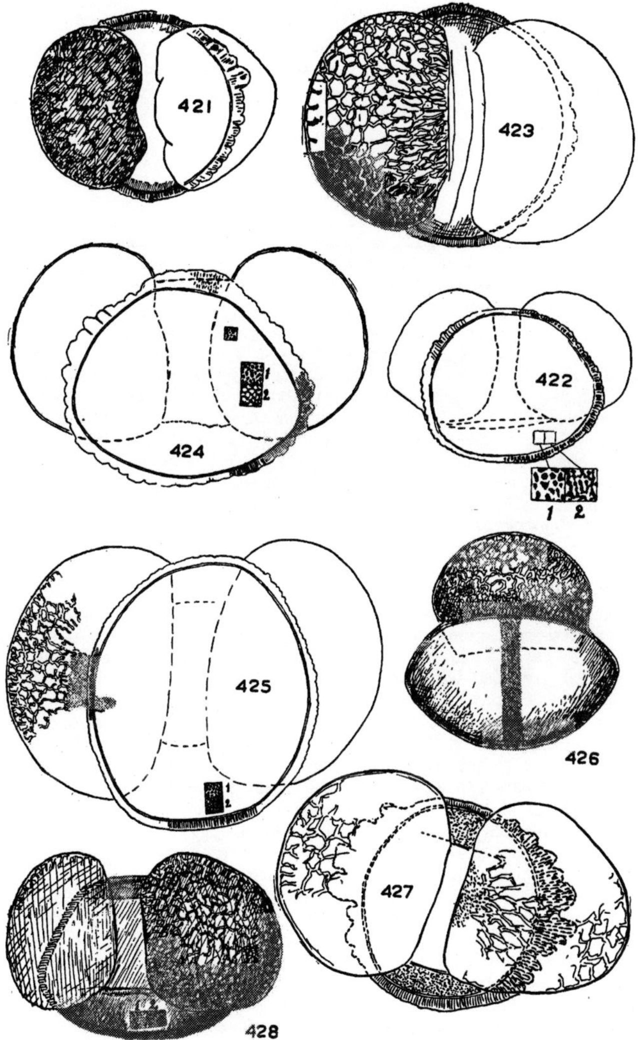

PLATE XXIV (421–428) (*Gymnospermae* 4). — 421 and 422, *Pinus banksiana;* 423 and 424, *P. leucodermis;* 425, *P. cembra;* 426, *P. montana;* 427, *P. contorta* var. *murrayana;* 428, *P. silvestris.*

measured immediately after preparation); 84–*97*–108 μ (same method; after 26 hours); 84–*100*–110 μ (same method; after 72 hours).

Pinus pinea. — Size variation among pollen grains (wings not included) from different localities: 36–51, 39–53, and 37–53 μ (DEEVEY 1939). "Although the number of measurements is small, the figures show clearly that size variability among grains of different trees of the same species can be as great as that found among different species" (DEEVEY *l.c.*).

Pinus rigida. — Size of body: (39–)48(–56) μ (DEEVEY 1939).

Pinus silvestris. — FIG. 428: distal part of grain, oblique view; length of body 55 μ; entire grain 75 μ; Gotland, Sweden. — Length in μ of entire grains (extremes and *average* size): 32.5–*54.7*–68.5 (STARK 1927); about 61.6 (BROCHE 1929); 50–*62*–74 (HÖRMANN 1929; "HÖRMANN method"); 54–*70*–78 (HÖRMANN *l.c.;* grains boiled 13 minutes in H_2O); 54–*70*–82 (HÖRMANN *l.c.;* grains boiled in Eau de Javelle); 46–*54*–62 (HÖRMANN *l.c.;* sample no. 7, p. 220; "HÖRMANN method"); 68–75 (JAESCHKE 1935; fresh pollen in H_2O); 68–80 (JAESCHKE *l.c.;* grains boiled 20 minutes in 10 per cent KOH solution); 65–77 (JAESCHKE *l.c.;* grains boiled one hour in 10 per cent KOH solution); 59–*72*–85 (MARCHETTI 1936; acetolysed grains); 59–*69*–77 (MARCHETTI *l.c.;* "HÖRMANN method"); 54–*66*–78 (VON SARNTHEIN 1936; sample no. 1, p. 551; fresh pollen in H_2O); 51–*66*–78 (VON SARNTHEIN *l.c.;* sample no. 2, p. 551; "HÖRMANN method"); 54–*75*–87 (VON SARNTHEIN *l.c.;* sample no. 3, p. 551; 10 per cent KOH solution, glycerine).

Pinus thunbergii. — Pollen grains small, only slightly larger than the grains of *P. banksiana.* Length of body 52.47 ± 0.36 μ (average of 117 acetolysed grains from the New York Botanical Garden, 1939).

Podocarpus (PLATE XXV, FIGS. 429–431). — Grains more or less spheroidal, provided with two or three well-defined bladders or with bladdery projections. In size, they range from 23 to 39 μ in diameter, except in *P. dacrydioides*, where they may be as large as 45 μ in diameter. Furrow usually long, its boundaries sharply delineated by an abrupt change in texture and a rather pronounced thickening along its rim. Bladders usually large and spreading but tending to be weak and flaccid. They are smooth on their outer surface but conspicuously marked inside by reticulate thickenings. At their proximal roots, their texture merges with that of the cap at its margin, which is sometimes developed as a marginal crest (WODEHOUSE).

Podocarpus dacrydioides. — FIG. 429: part of grain; about 68 μ, body about 50 μ; New Zealand. — Grains, according to WODEHOUSE, about 45 μ in diameter, provided with three bladders. Exine of the cap coarsely granular, especially towards the margin. At the point where it merges into the dorsal roots of the bladders it is coarsely reticulate-granular and thrown into small ridges or folds. Reticular thickenings on the inner surface of the bladders less developed than in *Pinus*.

Podocarpus spicatus. — FIG. 430: distal part of grain; 43 by 78 μ; New Zealand. — Grains resembling the pollen of *Pinus* (the *haploxylon*-type) but smaller; bladders provided with a particularly wide-meshed reticulum.

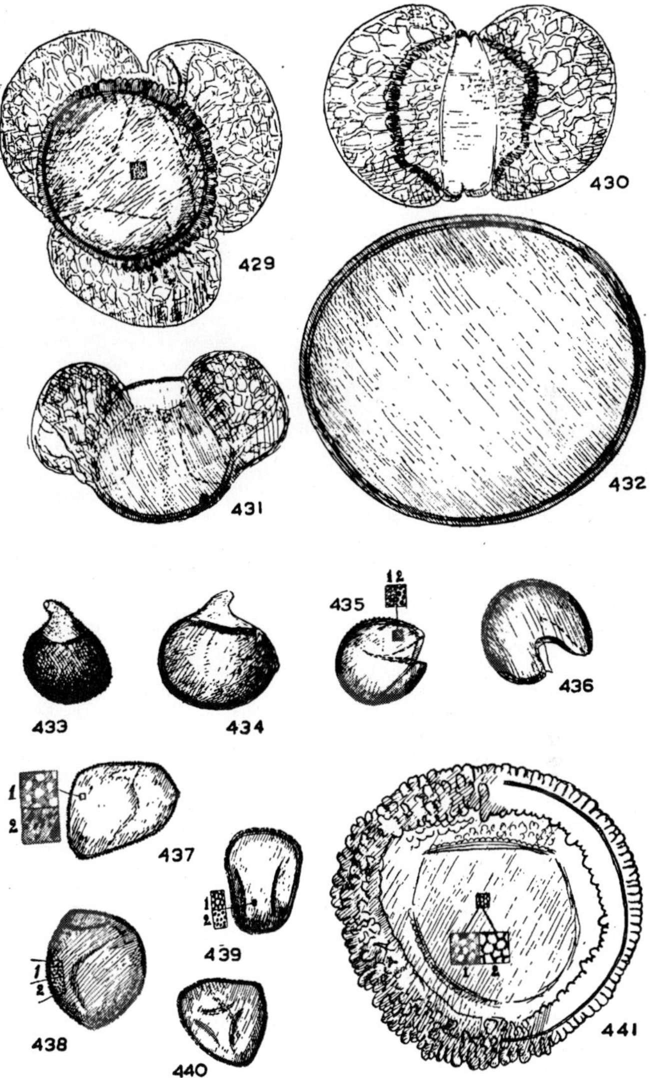

PLATE XXV (429-441) (*Gymnospermae* 5). — 429, *Podocarpus dacrydioides;* 430, *P. spicatus;* 431, *P. spinulosus;* 432, *Pseudotsuga taxifolia;* 433, *Sequoiadendron giganteum;* 434, *Sequoia sempervirens;* 435 and 436, *Taxodium distichum;* 437, *Taxus brevifolia;* 438–440, *T. baccata;* 441, *Tsuga canadensis.*

Podocarpus spinulosus. — Fig. 431: lateral view, distal part up; 37 by 69 μ; N. S. Wales.

Pseudotsuga taxifolia (Plate XXV). — Fig. 432: grain from Västerås 1936; 80 by 93 μ. — Grains spheroidal, without a trace of bladders, pore or furrow, uniform in size, 90–100 μ in diameter, closely resembling those of *Larix.* Exine comparatively thin and smooth (Wodehouse).

Sciadopitys verticillata (Plate XX). — Fig. 387*a*: polar view; distal (?) side up; 42 μ; Japan. Fig. 387*b:* lateral view; 46 by 29 μ. — Grains monocolpate, in polar view more or less circular; the central part of the distal side is often depressed. Ektexine thick, particularly on the proximal (?) side and provided with densely placed rounded warts. Diameter of grain, according to Jimbo (1933), 35–40 μ.

Fossil pollen of *Sciadopitys* has been described as *Sporites serratus* (Potonié and Venitz 1934). Rudolph (1935) has described a pollen type, the " A-type ", which in part at least should be referred to *Sciadopitys* („ Ein sehr steter Bestandteil der untersuchten Pollenspektren, vielleicht aber verschiedener Herkunft. Ähnlicher Pollen findet sich z.B. bei Monokotylen. Der morphologisch gleichfalls ähnliche Pollen von *Magnoliaceen* ist viel grösser. Manche schlüsselförmige Exemplare erinnern auch an *Sciadopitys* "). The true nature of *Sporites serratus* was demonstrated by Thiergart (1937).

Sequoia (Plate XXV). — Grains similar to those of *Cryptomeria,* approximately spheroidal, 29 to 41 μ, provided with a single germ pore, consisting of a conical projection which rises abruptly from the surface and bends sharply to the side, suggesting in appearance the handle of a curling stone (Wodehouse; this description also refers to *Sequoiadendron*).

Sequoia sempervirens. — Fig. 434: lateral view, distal part upwards; about 28 by 32 μ; California. Pollen, when expanded, about 33 μ (Buchholz 1939).

Sequoiadendron giganteum (Plate XXV). — Fig. 433: lateral view, distal part upwards; about 22 by 25 μ; California. — Grains essentially the same as those of *Sequoia;* size of expanded grains about 23 μ (Buchholz 1939).

Taxodium distichum (Plate XXV). — Figs. 435, 436: lateral view of grains; about 24 and 27 μ respectively; Naples. — Grains more or less spheroidal, similar to those of *Juniperus.* They are provided with a single germ pore, consisting of a conical papilla similar to but much less prominent than that of *Cryptomeria* and not bent at the top (Wodehouse). Size 27–31 μ (Wodehouse), 30–37 μ (Potonié and Venitz 1934).

Taxus baccata (Plate XXV). — Fig. 438: grain from Visingsö, Sweden; about 29 μ. Fig. 439: grain from Visby, Sweden; 27 by 21 μ. Fig. 440: grain from Kashmir; about 24 μ. — Grains essentially the same as those of *T. brevifolia.* Ornamentation faint, but easily discernible with the aid of a high resolving power. Diameter about 28 μ (Kirchheimer 1935).

Fossil *Taxus* pollen has been found very seldom and some of the records of it seem to be doubtful. A continuous *Taxus* pollen curve, reaching up to 23 per cent, is shown in the diagram of an interglacial deposit near Posen (Szafer and Trela, Interglaziale Flora von

Szeląg (Schilling) bei Poznań mit besonderer Berücksichtigung der Pollenanalyse. Spraw. Kom. Fizj. Polsk. Akad. Umiej., t. LXIII, 1928).

Taxus brevifolia. — FIG. 437: grain from Oregon; about 22 by 30 μ. — Grains more or less spheroidal or somewhat angular in outline, 24 to 27 μ in diameter. There is no well marked germinal pore or furrow, but usually a slightly bulging area where the exine is visibly thinner than in the rest of the grain (WODEHOUSE). Exine with a faint ornamentation consisting of densely placed, short rods which tend to be longer at the angles of the grain.

Torreya nucifera (PLATE XXVI). — FIG. 446: grain from Japan; about 26 μ. — Grains, according to JIMBO (1933), 24–30 μ in diameter, according to WODEHOUSE, about 29 μ; similar to those of *Taxodium;* somewhat irregular in shape, without bladders but provided with a poorly defined germinal furrow.

Tsuga canadensis (PLATE XXV). — FIG. 441: distal part of grain; about 78 μ; Connecticut. — In polar view, the grains are more or less circular, in lateral view flat or cup-shaped, since the centre of the distal part is frequently somewhat depressed. Diameter in polar view 67–80 μ (BAAS 1932), 62–85, usually about 64 μ (WODEHOUSE). The centre of the distal part of the grain (*cf.* FIG. 441) corresponds to the furrow of monocolpate pollen grains. It is encircled by a well developed marginal fringe (" girdle of air sac ", JIMBO 1933), which, as suggested by RUDOLPH (1935, p. 256), may represent an initial stage in the formation of air-sacs, like those found in *T. pattoniana*, *Microcachrys*, *Pinus*, *Podocarpus*, etc. The fringe is more or less " puffy ", with frequently rather twisted protrusions compactly arranged. The furrow area, which in dry, shrunken grains cannot be protected by this fringe, or " Krause ", is covered by an operculum (homologous to that in *Nymphaea* etc.; *cf.* PLATE XV, FIGS. 260, 261). In the optical section, in the right hand side of FIG. 441, there are four contour lines representing, in order, the outermost contour of the ektexine, the outer and the inner contour of the endexine, and lastly, the inner contour of the ektexine, which often curves inwards, towards the operculum. The intruding projections, set off on the outermost contour, correspond to the internal, " mesexinous ", thickenings of the air-sacs of *Pinus*, etc.

Tsuga diversifolia. — FIG. 442: proximal part; about 96 μ; Hort. Bergianus 1935. FIG. 443: distal part; about 90 μ. — Size generally about 60 to 70 μ (BAAS 1932), 55–100 μ (JIMBO 1933). In FIG. 442, sectors I–III show some details of the exine in the order in which they appear under the microscope (at high, medium, and low adjustment respectively). According to JIMBO, the entire surface of the grain is covered by relatively short, sharp spines.

Tsuga pattoniana. — FIG. 444: proximal part of grain; size of body 47 by 51 μ; the longest axis of the entire grain measures probably about 70 μ. FIG. 445: lateral view; size of body 61 by 68 μ; entire grain 76 by 82 μ. — The pollen grains are either decidedly pinoid (" *silvestris*-type "), with two well defined air-sacs (FIG. 444), or possess a varying number of less well defined bladders clustered on the distal part of the grain (FIG. 445). Sometimes these embryonic bladders merge into a fringe of much the same type as that met with in the pollen grains of *T. canadensis* and *T. diversifolia*. That the fringe — or the bladders — belong to the distal part of the grain may be seen in immature material where the pollen grains are still united in tetrads. The texture of the grains is well defined and granular, and has — unlike the texture of the pollen grains of *Picea*, *Pinus*, etc. — essentially the same pattern in the cap as in the bladders.

Welwitschia mirabilis (PLATE XXVI). — FIG. 448: lateral view, distal part (with furrow) up; 37 by 63 μ; Angola. FIG. 449: distal part; 39 by 58 μ. — Grains ellipsoidal, monocolpate, 51 to 57 by 29 to 32 μ. Exine of smooth texture, marked by 19 or 20 longitudinal grooves and low, rounded ridges (WODEHOUSE). The ektexine is slightly raised above the endexine, particularly at the pointed ends of the grain.

References: —

AUER, V., 1933: Verschiebungen der Wald- und Steppengebiete Feuerlands in postglazialer Zeit (Acta Geographica, 5:2, Helsinki).

BAAS, J., 1932: Eine frühdiluviale Flora im Mainzer Becken (Ztschr. f. Bot., vol. 25).

BROCHE, W., 1929: Pollenanalytische Untersuchungen an Mooren des südlichen Schwarzwaldes und der Baar (Ber. Naturf. Ges. Freib., vol. XXIX).

BUCHHOLZ, J. T., 1939: The generic segregation of the *Sequoias* (Am. Journ. Bot., vol. 26).

CAIN, S. A., 1940: The identification of species in fossil pollen of *Pinus* by size-frequency determinations (Am. J. Bot. 27:301).

CRANWELL, L. M., 1940: Pollen Grains of the New Zealand Conifers (New Zealand Journ. of Science and Technology XXII:1B–17B).

ČERNJAVSKI, P., 1935: Über die rezenten Pollen einiger Waldbäume im Jugoslavien (Beih. Bot. Centralbl., vol. LIV).

DEEVEY, E., 1939: Studies on Connecticut lake sediments (Amer. Journ. Sci., vol. 237).

DOKTUROWSKY, W. S. und KUDRJASCHOW, W., 1924: Schlüssel zur Bestimmung der Baumpollen im Torf (Geol. Arch., vol. 3).

DRAHOWZAL, G., 1935: Beiträge zur Morphologie und Entwicklungsgeschichte der Pollenkörner (Österr. Bot. Ztschr., vol. LXXXV).

FIRBAS, F., 1925: Zur Waldentwicklung im Interglazial von Schladming an der Enns (Beih. Bot. Centralbl., vol. XLI).

—— —— 1927: Beiträge zur Kenntnis der Schieferkohlen des Inntals und der interglazialen Waldgeschichte der Ostalpen (Ztschr. f. Gletscherkunde, vol. XV).

FURRER, E., 1927: Pollenanalytische Studien in der Schweiz (Vierteljahrsschr. Naturf. Ges. Zürich, vol. 72).

GERASIMOV, D. A., 1930: On the distinctive characteristics of the pollen of *Larix* and *Pinus cembra* in peat (Geol. Fören. Förhandl., vol. 52, Stockholm).

GOEBEL, K., 1933: Organographie der Pflanzen, 3. Aufl. (Jena).

HÖRMANN, H., 1929: Die pollenanalytische Unterscheidung von *Pinus montana*, *P. silvestris* und *P. cembra* (Österr. Bot. Zeitschr., vol. LXXVIII).

IVERSEN, J., 1934: Moorgeologische Untersuchungen auf Grönland (Medd. Dansk Geol. Foren., vol. 8).

JAESCHKE, J., 1935: Zur Frage der Artdiagnose der *Pinus silvestris*, *Pinus montana* und *Pinus cembra* durch variationsstatistische Pollenmessungen (Beih. Bot. Centralbl., vol. LII).

JIMBO, T., 1933: The diagnoses of the pollen of forest trees, I (Sc. Rep. Tôhoku Imp. Univ., 4. ser., vol. VIII: 3, Sendai).

KIRCHHEIMER, F., 1935: Die Korrosion des Pollens (Beih. Bot. Centralbl., vol. LIII).

MARCHETTI, M., 1936: Analisi pollinica della torbiera di Campotosto (Appennino Abruzzese) (N. Giorn. Bot. Ital., n.s., vol. XLIII).

POTONIÉ, R. und VENITZ, H., 1934: Zur Mikrobotanik des miocänen Humodils der niederrheinischen Bucht (Arb. Inst. Paläobot. Petrogr. Brennsteine, vol. 5, Berlin).

RUDOLPH, K., 1935: Mikrofloristische Untersuchung tertiärer Ablagerungen im nördlichen Böhmen (Beih. Bot. Centralbl., vol. LIV).

VON SARNTHEIN, R., 1936: Moor- und Seeablagerungen aus den Tiroler Alpen in ihrer waldgeschichtlichen Bedeutung (*Ibid.*, vol. LV).

STAPF, O., 1889: Die Arten der Gattung *Ephedra* (Vienna).

STARK, P., 1927: Über die Zugehörigkeit des Kiefernpollens in den verschiedenen Horizonten der Bodenseemoore (Ber. Deutsch. Bot. Ges., vol. XLV).

THIERGART, F., 1937: Die Pollenflora der Niederlausitzer Braunkohle (Jahrb. Preuss. Geol. Landesanst., vol. 58).

WODEHOUSE, R., 1933: Tertiary pollen, II. The oil shales of the Green River formation (Bull. Torr. Bot. Club, vol. 60).

—— —— 1935: Pollen grains (McGraw-Hill, New York).

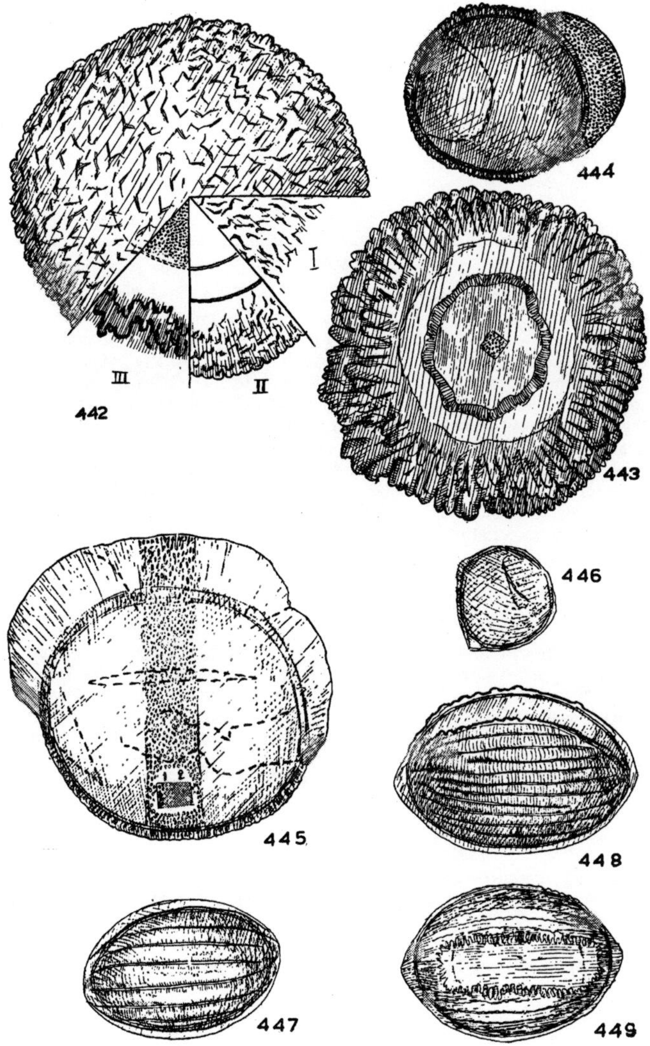

PLATE XXVI (442–449) (*Gymnospermae 6*). — 442 and 443, *Tsuga diversifolia;* 444 and 445, *T. pattoniana;* 446, *Torreya nucifera;* 447, *Ephedra;* 448 and 449, *Welwitschia mirabilis.*

Chapter X

SPORE MORPHOLOGY — PTERIDOPHYTES

Cyathaeaceae: —

Hemitelia grandifolia (PLATE XXVII). — FIG. 450: distal part of spore; about 35 μ; St. Vincent (VON EGGERS no. 6731). FIG. 451: proximal part; about 37 μ. — Spores trilete; in polar view, the contour lines of the endexine form a triangle with rounded corners. The outer part of the spore coat is closely attached to the endexine at these corners, but in other places it is somewhat raised above the surface of the endexine and provided with irregularly located small pits and three large pits, one at the central part of each side of the triangle. The larger pits are situated either near the margin of the distal part of the grain or in the transition between the distal and the proximal part; they cause the spore superficially to resemble *Tilia* pollen. Similar pits are characteristic of the spores of many other species of *Hemitelia*. *Cf.* also KNOX 1938.

Equisetaceae: —

Equisetum arvense (PLATE XXVII). — FIGS. 452, 453: spores from Gotland 1934; diameter about 44 and 40 μ respectively. — Acetolysed spores show no trace of elaters. They are usually cut open in a way which may indicate their representing a transitional type between alete and monolete spores. The wall is sometimes folded; if the folds are few and more or less parallel, a certain similarity to quercoid pollen grains will be observed.

Isoëtaceae: —

Isoëtes echinosporum (PLATE XXVII). — FIGS. 454, 455: lateral view of microspores; proximal part down; 21 by 30 and 23 by 19 μ respectively; Tenhult, Sweden. FIGS. 456, 457: polar view; proximal part up (FIG. 456), down (FIG. 457); FIG. 456, 26 by 17 μ; FIG. 457, 24 by 17 μ. — Length of microspores 22–29 μ, in average about 24 μ (OBERDORFER 1931).

Isoëtes lacustre. — FIG. 458: lateral view; proximal part down; 30 by 44 μ; Tenhult, Sweden. — Length of microspores 31 by 44 μ; in average about 41 μ (OBERDORFER 1931). In polar view, the microspores are elliptical, resembling monocolpate pollen grains, in lateral view and longitudinal orientation, their contour is like that of a cleft orange. The outermost layer of the spore coat hangs down as a thin, double veil from the proximal part of the microspore.

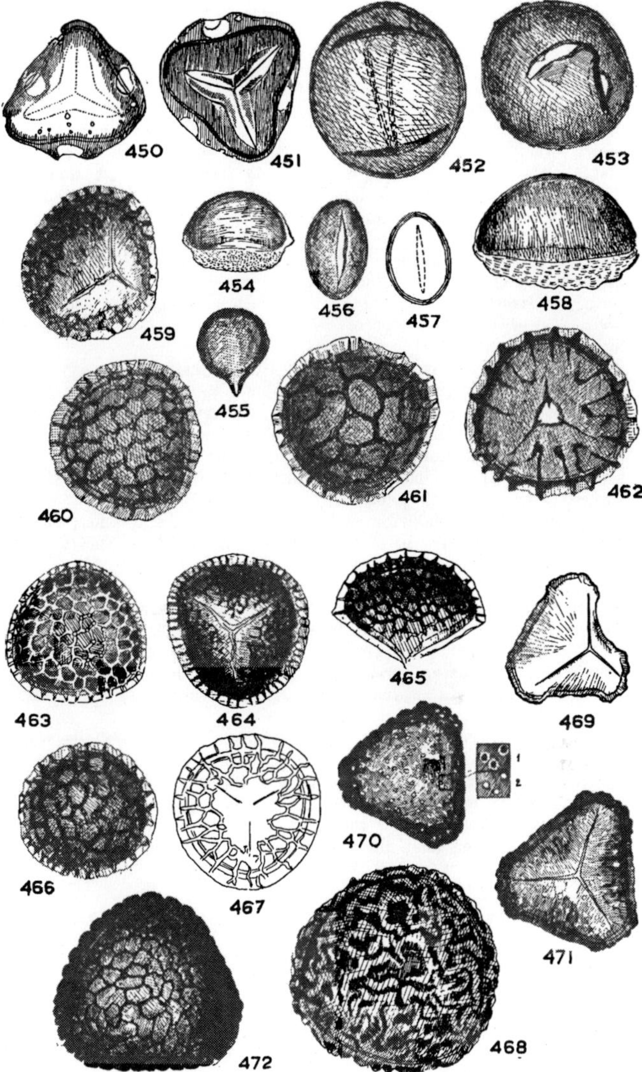

Plate XXVII (450–472) (*Pteridophyta* 1). — 450 and 451, *Hemitelia grandifolia;* 452 and 453, *Equisetum arvense;* 454–457, *Isoëtes echinosporum;* 458, *I. lacustre;* 459 and 460, *Lycopodium alpinum;* 461 and 462, *L. annotinum;* 463–465, *L. clavatum;* 466 and 467, *L. complanatum;* 468, *L. inundatum;* 469–471, *L. selago;* 472, *Botrychium lunaria.*

Lycopodiaceae (Plate XXVII): —

Lycopodium. — Spores trilete, tetrahedral, ranging in size from 23 μ or less in *L. cernuum* (Wilson 1934) to about 45 μ (Knox 1938).

Lycopodium alpinum. — Fig. 459: proximal part of spore, oblique view; about 41 μ in diameter; Snasahögarna, Sweden. Fig. 460: distal surface; 41 μ. — Spores similar to those of *L. complanatum;* distal surface coarsely reticulate; the number of muri appearing on the equator, *i.e.* the spore margin in polar view, is 30 or less (Wilson 1934).

Lycopodium annotinum. — Fig. 461: distal surface; about 44 μ in diameter. Fig. 462: proximal part of spore; 44 μ. — Meshes of the reticulum larger than in any other *Lycopodium* spores mentioned in this book [the greatest diameter of the distal spore surfaces exceeds 4–5 meshes (Rudolph 1935)].

Lycopodium clavatum. — Fig. 463: distal surface; 39 μ; Stockholm. Fig. 464: proximal part of spore; 39 μ. Fig. 465: lateral view; 30 by 42 μ. — Size 26–34 μ, averaging about 28 μ (Kirchheimer 1933). Reticulum finer than that of the spores of *L. alpinum* and *L. complanatum;* the number of muri appearing on the equator 35 or more (Wilson 1934); the greatest diameter of the distal spore surface exceeds 6–10 meshes (Rudolph 1935).

Lycopodium complanatum. — Figs. 466, 467: distal and proximal part of spore; 37 and 40 μ respectively; Stockholm. — Spores essentially similar to those of *L. alpinum.*

Lycopodium inundatum. — Fig. 468: distal surface; 55 μ; Helsinki. — „Sporen auf der kugelschaligen Grundfläche mit kräftigen und zum Teil verbogenen anastomosirenden Leisten unregelmässig und ziemlich dicht besetzt; die durch eine sehr deutliche und fast regelmässige Ringleiste abgegrenzten Pyramidenflächen weniger deutlich kleinmaschig bis fast warzig" (Luerssen 1889). Diameter about 37 μ (Rudolph 1935).

Lycopodium selago. — Fig. 469: proximal part; oblique view. Figs. 470, 471: distal and proximal part of spore; 35 and 36 μ respectively. — In polar view, the outline of the spores is hexagonal and not rounded as in the case of the *Lycopodium*-spores previously mentioned; at the extension of the radii of the triradiate scar there are three short, uneven lines alternating with three longer, nearly smooth ones. The distal surface of the exine is covered with small, rounded pits.

Ophioglossaceae (Plate XXVII): —

Botrychium lunaria. — Fig. 472: distal surface; 48 μ; Norrbotten, Sweden. — Spores trilete, rounded-triangular in polar view and provided with a thick wall. Diameter 37.5–47.5 μ (Karpowicz 1927).

Polypodiaceae (Plate XXVIII): —

For detailed description of the spores mentioned below, the work of Luerssen (1889) and the papers of Erdtman (1923) and Karpowicz (1927) may be consulted.

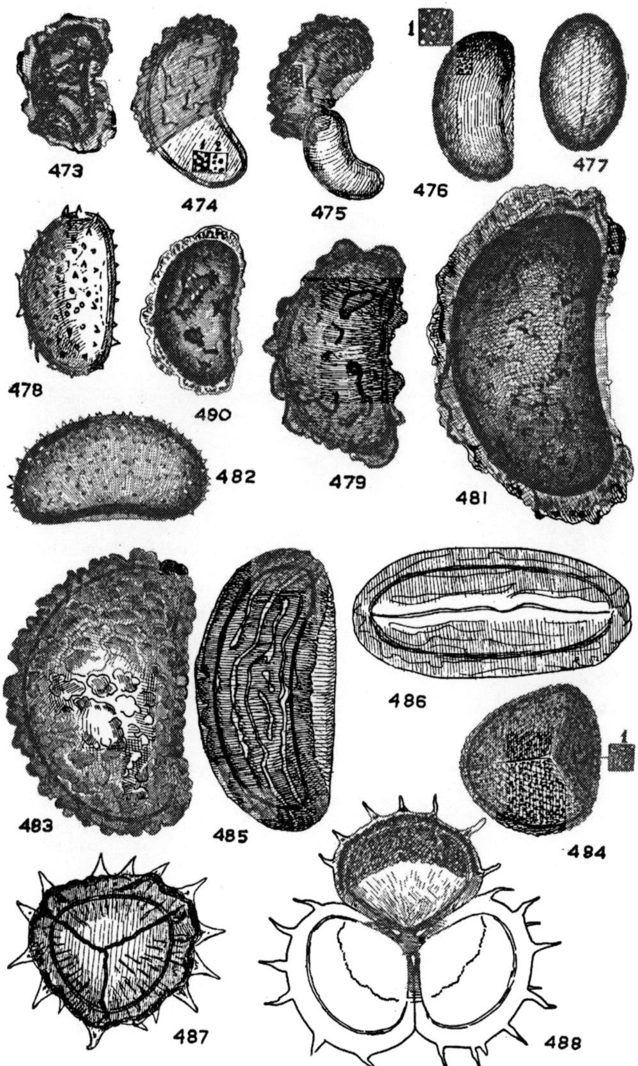

PLATE XXVIII (473–488) (*Pteridophyta* 2). — 473, *Asplenium septentrionale;* 474, *Athyrium alpestre;* 475, *A. crenatum;* 476 and 477, *A. filix-femina;* 478, *Cystopteris fragilis;* 479, *Dryopteris filix-mas;* 480, *D. linnaeana;* 481, *D. phegopteris;* 482, *D. thelypteris;* 483, *Polypodium vulgare;* 484, *Pteridium aquilinum;* 485 and 486, *Schizaea digitata;* 487 and 488, *Selaginella selaginoides.*

Asplenium septentrionale. — FIG. 473: lateral view; 21 by 35 μ; Västerås. — Size 39–45 by 47.5–62.5 μ (KARPOWICZ).

Athyrium a pes.re. — FIG. 474: lateral view; 25 by 48 μ; Storlien, Sweden. — Size 30–37.8 by 42.5–47.5 μ (KARPOWICZ).

Athyrium crenatum. — FIG. 475: lateral view; 19 by 37 μ; Haapaniemi.

Athyrium filix-femina. — FIG. 476: lateral view; 22 by 39 μ; Stockholm. FIG. 477: distal surface; 23 by 36 μ. — Size 32.5–45 by 42.5–62.5 μ (KARPOWICZ).

Cystopteris fragilis. — FIG. 478: lateral view; 22 by 44 μ; Ängelsberg, Sweden. — Size 32.5–45 by 42.5–62.5 μ (KARPOWICZ).

Dryopteris filix-mas. — FIG. 479: lateral view; 34 by 58 μ; Västerås 1937. — Size 30–37.5 by 50–72 μ (KARPOWICZ).

Dryopteris linnaeana. — FIG. 480: lateral view; 26 by 44 μ; Vänersborg, Sweden. — Size 30–37.5 by 47.5–55 μ (KARPOWICZ).

Dryopteris phegopteris. — FIG. 481: lateral view; 48 by 86 μ; Stockholm. — Size 47.5–50 by 55.5–75 μ (KARPOWICZ).

Dryopteris thelypteris. — FIG. 482: lateral view; 29 by 55 μ; Ramlösa, Sweden — Spores somewhat similar to pollen grains of *Nuphar.* Size, according to KARPOWICZ, 30–42.5 by 45–55 μ.

Polypodium vulgare. — FIG. 483: lateral view; 50 by 75 μ. — Size 45–50 by 55–77.5 μ (KARPOWICZ); 70–80 μ (POTONIÉ and VENITZ 1934). Some attention has been paid to the fact that the large and presumably heavy spores of this fern are frequently found in European late-quaternary deposits, in peats and other sediments. *Cf. e.g.* ERDTMAN, Journ. Linn. Soc. (Bot.), vol. XLVI, 1924 (p. 486), and FIRBAS, Planta, vol. 22, 1934 (p. 129: 70 per cent *Polypodium* spores in a surface sample from the isle of Baltrum).

Pteridium aquilinum. — FIG. 484: proximal part; about 38 μ. — Spores tetrahedral, trilete; diameter 30–40 μ (KARPOWICZ).

Schizaeaceae (PLATE XXVIII): —

Schizaea digitata. — FIG. 485: lateral view; 37 by 74 μ. FIG. 486: proximal view; 34 by 74 μ. — Spores bilateral, monolete, provided with longitudinal, sometimes bifurcated and anastomosing ridges.

Selaginellaceae (PLATE XXVIII): —

Selaginella selaginoides. — FIG. 487: proximal part of microspore; about 59 μ (spines included). FIG. 488: tetrahedral tetrad: one of the microspores taken away. — Microspores trilete, more or less rounded; distal surface with long, pointed, sometimes slightly bent spines [according to REEVE (1935) their number is 25–32] and of a distinct granular texture; proximal surfaces smooth, sometimes provided with small radiating folds; germinating slits nearly as long as the longest radii of the microspores.

Microspores of *Selaginella* are frequently found in late glacial deposits, the maximum frequency hitherto recorded being probably 1100 per cent (FIRBAS, Beih. Bot. Centralbl., vol. LII, 1934, p. 127). Of course their presence may not always be taken as an indication of late glacial conditions; for example the author has found up to 120 per cent *Selaginella* microspores in peats of varying age in the Shetland Islands.

References: —

ERDTMAN, G., 1923: Beitrag zur Kenntnis der Mikrofossilien in Torf und Sedimenten (Ark. f. Bot., vol. 17:10, Stockholm).

KARPOWICZ, W., 1927: Studien über die Entwicklung der Prothallien und der ersten Sporophytblätter der einheimischen Farnkräuter (*Polypodiaceae*) (Bull. Acad. Pol. Sc. Lettr., Cl. Sci. Mat. Nat., sér. B).

KIRCHHEIMER, F., 1933: Die Erhaltung der Sporen und Pollenkörner in den Kohlen sowie ihre Veränderungen durch die Aufbereitung (Bot. Arch., vol. 35).

KNOX, E., 1938: The spores of Pteridophyta, with observations on microspores in coals of Carboniferous age (Trans. Proc. Bot. Soc. Edinb., vol. XXXII).

—— 1939: The spores of Bryophyta compared with those of Carboniferous age (*Ibid.*).

LUERSSEN, C., 1889: Die Farnpflanzen (Rabenhorst's Kryptogamen-Flora, Ed. 2, vol. 3).

OBERDORFER, E., 1931: Die postglaziale Klima- und Vegetationsgeschichte des Schluchsees (Schwarzwald) (Ber. Naturf. Ges. Freiburg, vol. XXXI).

POTONIÉ, R. und VENITZ, H., 1934: Zur Mikrobotanik des miocänen Humodils der niederrheinischen Bucht (Arb. Inst. Paläobot. Petrogr. Brennsteine, vol. 5).

REEVE, R. M., 1935: The spores of the genus *Selaginella* in north central and north eastern United States (Rhodora, vol. 37).

RUDOLPH, K., 1935: Mikrofloristische Untersuchung tertiärer Ablagerungen im nördlichen Böhmen (Beih. Bot. Centralbl., vol. LV).

WILSON, L. R., 1934: The spores of the genus *Lycopodium* in the United States and Canada (Rhodora, vol. 36).

Chapter XI

POLLEN ANALYSES AND THE GRAPHIC PRESENTATION OF THEIR RESULTS

Tabulation and Calculation of Percentages: — After chemical treatment of the pollen-bearing material (*cf.* Chapter IV) the pollen grains are identified under the microscope and counted by the use of the micrometer stage. As a rule the grains can be identified at a magnification of about 200 diameters, although a higher magnification — about 400 diameters — is sometimes preferable. This is particularly true when the grains are very concentrated: *e.g.* when they average more than ten in a single visual field, or when all pollen types, not only the familiar grains of forest trees, are to be recorded. Two microscopes must be at hand: one of them for the analysis; the other for identification of pollen grains by comparison with slides from a reference set of pollen preparations.

Important analyses may be checked by permanent slides. Such slides must also be made whenever immersion power is used. There is, however, a decided disadvantage with these slides inasmuch as the pollen grains are fixed, or nearly so, and cannot be turned from one side to another, which may be essential to ensure a clear conception of the shape of the grains.

According to common practice, initiated by VON POST (1918), about 150 tree pollen grains from each sample are counted. The relative frequencies of the different tree pollen species are calculated on this number. If more than 150 grains are counted, the percentages, as a rule, will not change; if they do change, the modification will be very slight. This holds good in countries where, in conformity with most parts of extra-mediterranean Europe, the number of tree pollen types in the peat usually does not exceed a dozen. If there is a greater number of tree pollen types present, the percentages should be based on a correspondingly greater number of pollen grains. In quoting percentages, the number of pollen grains on which they are founded should always be reported [*cf.* diagram, TEXTFIG. 11, and the attached tree pollen table (p. 160). This is a simplified diagram, exhibiting the main features of the pollen table. The decline (during the katathermic period) of the oak pollen frequency is most distinct. In order to accentuate this fact, the oak pollen curve in this special case has been laid out by means of a hatched line]. In this connection, it may be added that pollen analysts have as yet not devoted much attention to the purely statistical aspects of quantitative pollen analysis or pollen statistics as it is also, and quite correctly, called. Only a few papers have been published on this subject, *e.g.* by BARKLEY (1934) and ORDING (1934); *cf.* also LINDER (*in* MAURIZIO 1939).

Complete tallies of the fossil forms of each kind are presented in tabular form; in complicated analyses, it is convenient to make the tabulations on special blanks. Textfig. 8 shows a type of blank used by Faegri in recording pollen grains of late-quaternary deposits in Norway. Inserted is an imaginary record of 372 pollen grains, each one marked by a separate line (tetrads are also marked in the same way).

The contents of the record are as follows:

Picea: no grains.

Pinus: 27 grains. Some lines are made up by two halves, as wings, half grains, etc., are tabulated separately.

Betula: 104 grains. Columns 6–12 show the distribution of the grains into size classes.

" QM ", or " *Quercetum mixtum* ": 18 grains (14 *Quercus*, 4 *Ulmus;*

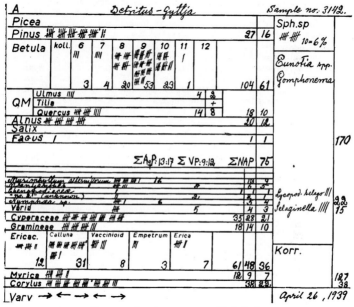

A Detritus-Gyttja Sample no. 3142.
Picea
Pinus 27 16
Betula koll. 6 7 8 9 10 11 12
3 4 20 53 23 1 104 61
QM Ulmus 4 2
Tilia +
Quercus 14 8 18 10
Alnus 20 12
Salix
Fagus 1 1 1
ΣAqP: 13:17 Σ VP: 9:12 ΣNAP 75
Myriophyllum alterniflorum 16 13 9
Menyanthes 8 6 5
Chenopodiaceae 1 1 1
"no. 21" (unknown) 2 2 1
Nymphaea sp. 6 5 4
Varia 5 4 3
Cyperaceae 35 28 21
Gramineae 18 14 10
Ericac. 12 Calluna 31 Vaccinioid 8 Empetrum 3 Erica 7 61 48 36
Myrica 12 9 7
Corylus 38 22
Varv → ← → ← →
Sph.sp 10=6%
Eunotia spp.
Gomphonema
170
Lycopod. Selago
Selaginella
22
15
Korr.
127
38
April 26, 1939

Textfigure 8. — Analysis record (K. Faegri).

pollen of *Tilia* was also present but not recorded during the analysis proper). L. von Post unites *Quercus, Tilia,* and *Ulmus* under the heading " mixed oak-forest ", or *Quercetum mixtum,* and most authors have followed this example. In this way, however, quite heterogeneous elements may be included in the " *Quercetum mixtum* ": in their composition and postglacial history, the uniform *Quercus sessiliflora*-forests of southwestern Sweden are thus very different from the " mixed oak-forests " of central Europe. Furthermore, in the countries along the southeastern shores of the Baltic, *Ulmus effusa* is to be considered rather as a constituent of " Auwälder " than of mixed oak-forests.

Alnus: 20 grains; *Fagus:* one grain.

The number of tree pollen grains (*Alnus, Betula, Fagus, Pinus, Quercus,* and *Ulmus*) totals 170. This is the $\sum$ AP, or total of tree (lat. arbor) pollen grains. The number of pollen grains of aquatic plants (AqP) totals 22 (16 grains of *Myriophyllum alterniflorum*, 6 grains of *Nymphaea* sp.). The frequencies of the AqP-constituents may be based on the $\sum$ AP or on the $\sum$ NAP (non-arboreal pollen). In the first case their frequencies will be 9 and 4 per cent respectively, in the last case 13 and 5 per cent.

In the centre of the blank, above figures 16 and 6, it may be seen that the $\sum$ AqP amounts to 13 per cent of the $\sum$ AP and to 17 per cent of the $\sum$ NAP. The pollen grains of *Menyanthes* and those of an unknown type (" no. 2 ") and of " varia " (" varia s. str. "; according to Faegri unidentified pollen grains of insignificant or sporadic appearance only) are included under the heading " Varia " (*s. lat.*). Their total, the $\sum$ VP, and their individual frequencies are expressed in the same way as in the AqP.

Chenopodiaceae, Cyperaceae, Ericaceae (including *Empetrum*), *Gramineae*, and *Myricaceae* pollen form the $\sum$ NAP, or the total of non-arboreal pollen grains. Three figures are seen on the line for each member of this group; first, the number of tabulated grains; second, the relative frequency calculated on the $\sum$ NAP; and, third, the relative frequency expressed as a percentage of the $\sum$ AP.

Finally 38 pollen grains of *Corylus* are recorded. The frequency, 22 per cent, is expressed as a percentage of the $\sum$ AP.

During analysis the preparation is moved alternately left and right by means of the micrometer stage. The direction is indicated by arrows in the lowest horizontal row of the blank. If the analysis be suddenly interrupted, it may thus be resumed without confusing the direction. After the analysis proper the PF (pollen frequency or number of pollen grains per square centimeter of the preparation) may be written in the same row.

In the vertical column at the extreme right are found, from top to bottom, the total number of tabulated AP, AqP, VP, NAP, and *Corylus* pollen. The next column has three sections: in the upper, the number of *Sphagnum* spores is recorded (their frequency is expressed as a percentage of the $\sum$ AP); the central section is reserved for notes of different kinds, such as the quotation of microfossils which are not counted (above) and records of spores (below); in the lowest section notes on corroded pollen grains, etc., are entered.

Although the basic principle of computing the pollen percentages was set forth in a paper by Holst (1909), it was first clearly expounded by L. von Post in 1916 and more thoroughly defined in 1929. It implies that pollen grains of vegetation units representing fixed types, *e.g.* the forests, must form the basis for calculating the frequencies of pollen grains from plants which do not represent fixed types in the same sense, *e.g.* the scrubby or herbaceous undergrowth of a forest. L. von Post's clear argument for so doing is quoted here in detail (von Post 1929, pp. 549–551, 557, 558): —

„Sobald ich mir darüber klar war, dass die absoluten Pollenfrequenzen mich nicht weiterbringen würden, galt es, eine anwendbare Berechnungsbasis für die relativen Frequenzzahlen zu finden. Am nächsten lag die Pollensumme. Aber welche Pollensumme? Die Ericazeen, die Zyperazeen, die Gramineen und die dann und wann angetroffenen Kräuter mussten natürlich ausserhalb der Pollensumme gelassen werden. Denn die Frequenzvariationen dieser Pollenarten waren in den vorliegenden Fällen offenbar fast gänzlich durch extrem lokale Umstände bedingt. Nur die Waldbaumsumme, deren Hauptkonstituenten meistens ihren Standort ausserhalb der Moorareale gehabt hatten, konnte in Betracht kommen. Wie sollte man aber diese Waldbaumpollensumme abgrenzen? Wie verhielten sich jene Bäume, die, wie die Erle, die Birke, die Fichte und die Kiefer, den Mutterformationen etlicher Torfarten angehörten? Und welche von den Bäumen der trockneren Böden waren in derartiger Weise in der Pollenflora repräsentiert, dass sie als Konstituenten der Waldbaumpollensumme anerkannt werden sollten? Diese Fragen konnte nur, nachdem eine gewisse Erfahrung gewonnen war, und durch Prüfung verschiedener Alternativen beantwortet werden.

Von Beginn an war es erwünscht, die areellen Verschiebungen der Waldtypen so weit wie möglich aus den Pollendiagrammen herauslesen zu können. Dies lag nahe, weil die meisten der in der Pollenflora vertretenen Bäume ja mehr oder weniger reine Waldbestände bilden, und weil man eben in den zeitlichen Arealschwankungen dieser Bestände einen unmittelbaren Ausdruck der Klimavariationen erwarten konnte. In Schweden müssen die Kiefer, die Fichte, die Birken, die Rotbuche und, in gewissem Grade, die Weissbuche im grossen und ganzen als gute Vertreter fixierter Waldtypen betrachtet werden. Mischwälder von diesen Elementen bezeichnen meistenteils nur Zwischenstufen einer fortlaufenden Entwicklung, die durch Störungen des Gleichgewichts hervorgerufen worden ist, und die auf einen neuen Stabilitätszustand abzielt. Anders verhält es sich mit den Eichenarten, der Linde, der Ulme, den Erlen, der Hasel, den Salweiden und den Weiden. Die edlen Laubbäume, die Erle und die Hasel erwiesen sich als die Hauptvertreter der wärmezeitlichen Bewaldung. Keine dieser Baumarten vermag aber an und für sich einen gewissen Waldtypus zu repräsentieren. Sämtliche kommen gegenwärtig als Glieder einer Reihe von waldlichen Verbänden vor, und ganz entsprechende Verhältnisse gingen sehr bald aus den Frequenzrelationen der betreffenden Pollenarten in den Lagerfolgen deutlich hervor. Der Eichenmischwald mit *Quercus, Tilia* and *Ulmus* in wechselnden Mengen und der Auwald mit der Erle als Vertreter waren die wärmezeitlichen Waldtypen, deren Dasein der fossile Pollenregen der südschwedischen Moore bezeugte. Ebenso unverkennbar war die normale Rolle der Hasel nur die eines regelmässig vorhandenen Unterholzes innerhalb dieser beiden Baumverbände gewesen. Der *Salix*-Pollen erschien meistens nur gelegentlich und in durchaus zu vernachlässigenden Mengen. Allein in subarktischen Schichten war seine Frequenz eine derartige, dass er den Habitus des Pollenspektrums beeinflusste. Hier mussten aber die Weiden die Vertreter einer um das Bodenareal konkurrierenden Pflanzensoziologischen Einheit — des Tundragebüsches — gewesen sein und folglich als den Waldbäumen gleichstehend betrachtet werden.

Die logische Konsequenz der Verhältnisse, die mir meine ersten pollenstatistischen Versuche enthüllten, war die, dass jede einzelne der Pollenmengen der Birke, der Erle, der Fichte, der Kiefer, der Rotbuche, der Weissbuche und der Weiden als der Vertreter eines Stückes Bodenareal in die Waldbaumpollensumme eingereiht werden musste, die Pollenmenge der Eiche, der Linde und der Ulme aber nur als Gesamtvertreter des Eichenmischwaldes und die der Hasel, als die eines Nebenbestandteiles dieses Waldtypus und des Auwaldes, überhaupt nicht. . . .

. . . Die Grundregel, dass die 100%-Summe aus jenen Elementen der fossilen Pollenflora, welche Bodenareale vertreten, gebildet werden soll, ist auch in solchen Fällen zu befolgen, wo eine derartige Pollensumme ganz anders als in den Waldgebieten Europas aufgebaut werden muss, z.B. wenn sich eine Möglichkeit eröffnet, die regionalen Verschiebungen zwischen Wald und waldlosen Arealen zu ermitteln. Infolge der Wertlosigkeit der absoluten Pollenzahlen wird dies nur unter der Voraussetzung gelingen, dass die waldlose Area irgendwie in der Pollenflora positiv vertreten wird. Es dürfte kaum völlig ausgeschlossen sein, dass Getreidepollenfrequenzen die sukzessive Erweiterung des Ackerbodens zu registrieren vermöchten. Ob die westeuropäische Heide in entsprechender Weise durch Ericazeen-Pollen vertreten sein möchte, ist zwar noch mehr fraglich. Es hat sich aber bereits herausgestellt, dass man mit Hilfe des Pollens der Gräser und gewisser anderen Steppenpflanzen die Steppe entwicklungsgeschichtlich dem Wald gegenüberstellen kann. In solchen Fällen muss natürlich nicht nur das Waldbaumpollen, sondern, mit gleichem Recht, auch das Pollen der Steppenvertreter in die Prozentsumme aufgenommen werden".

With the exception of the principles set forth by von Post there are no rules for computing the frequencies of different pollen categories. But, as pollen analysis still is in its early stages of evolution, conditions can hardly be expected to be otherwise. A discussion of the *pros* and *cons* of different suggestions relative to the calculation of pollen frequencies lies beyond the scope of this book. For this reason, the detailed analysis by Faegri already quoted (p. 153) must suffice. It refers to conditions in northern Europe and demonstrates, in a representative way, both how the AP (tree pollen), NAP (non-tree pollen), AqP (aquatic plant pollen) and VP (various pollen) frequencies may be calculated.

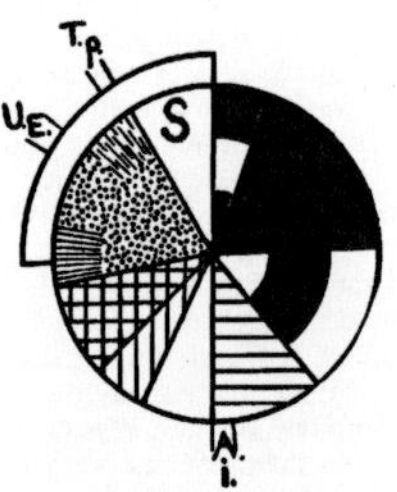

Textfigure 9. — Graphical representation of pollen spectrum. — In the spectrum proper concentric circles are reserved for gymnosperms, vertical and horizontal lines for angiosperms. The nine main sectors are as follows (enumerated clockwise, beginning at the north radius; *cf.* also the text, p. 156):

1. Ab (*Abies*): 5 per cent = 18 degrees; black, white, black;
2. P (*Pinus*): 20 per cent; black;
3. Pc (*Picea*): 15 per cent; white, black, white;
4. A (*Alnus*): 11 per cent, 2 of which as marked outside of the sector, referring to A.i. (*Alnus incana*); horizontal lines (distance between the lines equalling a tenth of the radius);
5. B (*Betula*): 7 per cent; no pattern;
6. C (*Carpinus*): 5.5 per cent; vertical lines (distance between the lines equalling a tenth of the radius);
7. F (*Fagus*): 9 per cent; meshes (distance between the lines the same as in 4 and 6);
8. QM ("Quercetum mixtum"): 20 per cent; dot pattern; in this special case also the pollen frequencies of the individual genera of the QM are shown by special patterns (*cf.* the text), *viz. Quercus* (6 per cent), *Ulmus* [8 per cent, 2 of which, as marked outside of the sector, refer to U.e. (*Ulmus effusa*)], and *Tilia* [6 per cent; 1 per cent refers to T.p. (*Tilia platyphyllos*)]. The keystone of the spectrum is:
9. (*Salix*): 7.5 per cent.

The circle outside of the spectrum proper indicates the *Corylus* pollen frequency (26 per cent).

Pollen Spectra: — The relative frequency numbers of the different pollen species in a sample constitute the pollen spectrum (a term first used by Jessen, 1917, p. 24) of the sample. Visual representations of pollen spectra can be effected in many ways. For example, a circular area, representing the $\sum$ AP, may be divided into sectors with areas proportionate to the percentages of the spectrum (von Post 1924). The sectors are lettered or marked with special symbols (*cf.* Trela 1934, Gams 1937, Schwickerath 1937, p. 27). The first sector begins at the northern radius, the other sectors follow clockwise (textfig. 9). Direct comparability is attained by marking off the pollen frequencies in a fixed order. With particular reference to conditions in

Europe, the gymnosperm pollen frequencies may be marked off first, in alphabetical order of the abbreviations of the generic and specific names of the pollen producers:

A : *Alnus*	GS : *Gramineae spontaneae*
Ab : *Abies*	Hi : *Hippophaë*
Ac : *Acer*	Ix : *Ilex*
B : *Betula*	Lx : *Larix*
C : *Carpinus*	Malt: *Myriophyllum alterniflorum*
Co : *Corylus*	P : *Pinus*
Cs : *Castanea*	Pc : *Picea*
Cyp : *Cyperaceae*	Q : *Quercus*
Emp: *Empetrum*	QM : *Quercetum mixtum*
Er : *Ericaceae*	S : *Salix*
F : *Fagus*	SS : *Sphagnum* spores
Fx : *Fraxinus*	T : *Tilia*
G : *Gramineae*	U : *Ulmus*
GC : *Gramineae cultae*	

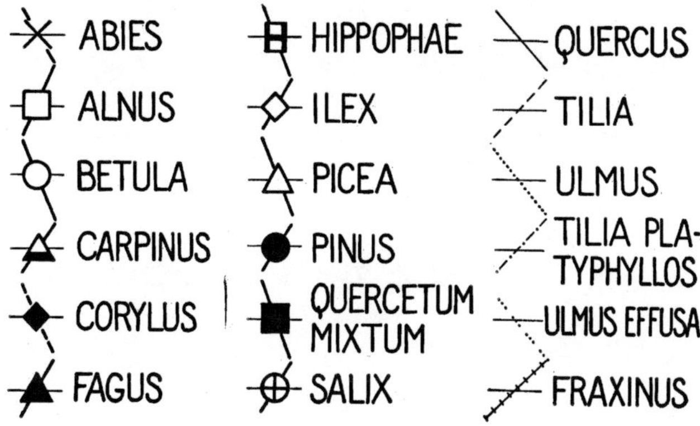

TEXTFIGURE 10a. — SYMBOLS USED IN POLLEN DIAGRAMS.

The *Pinus* sector (P) precedes the *Picea* sector (Pc), a fact which is not without significance, because by so doing the *Abies* and *Picea* sectors, which otherwise may easily be confused, are nearly always separated by the black *Pinus* sector. Then follow, also in alphabetical order, the pollen frequencies of the deciduous trees. The pollen frequencies of the mixed-oak-forest constituents, however, are as a rule brought together in a special QM-sector (dotted pattern). The outer part of this sector (*i.e.* the area between the periphery and a circle drawn through two thirds of the radius) may be reserved for marking the share of the different QM-genera, *viz. Quercus* by radiating lines in full; *Ulmus* by dots (same pattern as in the inner part of the sector); *Tilia* by radiating broken lines; *Fraxinus* by small open circles.

Pollen frequencies of different species of the same genus can be shown outside of the generic sector by extending the radii encompassing each specific subsector and inserting proper initials or abbreviations between the extensions. The pollen frequency of *Corylus* is denoted by an outer circle or a part of it (100 per cent = a complete circle; if the frequency is " 250 per cent " there must be two and a half circles; the periphery of the (first) *Corylus* circle should be drawn at a distance from the original circle equalling one sixth of the diameter of the latter, etc. The *Corylus* circle starts at the northern radius and proceeds counter-clockwise; it will, therefore, usually border on that part of the spectrum to which it phytogeographically is most closely related — *viz.* the QM-sector). Overcrowding reduces the clarity of the spectra and ought to be avoided. The NAP-frequencies (*Corylus* excepted), AqP- and VP-frequencies require special spectra drawn along lines which have to be adapted according to the individual cases.

Graphical representations of pollen spectra in maps may convey

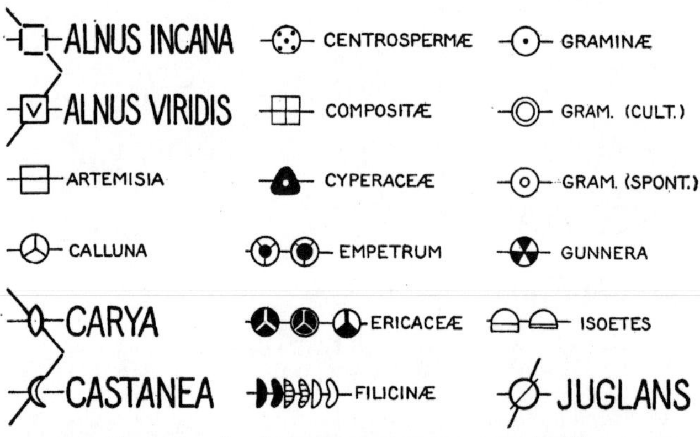

TEXTFIGURE 10*b*. — SYMBOLS USED IN POLLEN DIAGRAMS (*contd.*).

some idea of the composition of forests at different periods (*cf.* VON POST 1924, FIRBAS 1927, RUDOLPH 1928 and 1930; further references in GAMS 1937). Pollen frequencies of single species may be indicated by filled circles, proportionate in size to the frequency numbers. Total absence of pollen is indicated by a dot surrounded by a small circle (VON POST 1924, plates 3 and 4; LUNDQVIST 1928).

Pollen Diagrams: — On the basis of a series of pollen spectra from a sectional boring in a bog, a pollen diagram may be constructed. In a pollen diagram curves for the single species or for a group of species offer both a visual representation of the composition of the pollen flora, and the oscillations in frequency which have taken place reciprocally between the pollen-curves during the formation of the bog.

The principles of pollen diagrams have been fully discussed by VON POST (1916, 1929). In a typical pollen diagram, the pollen frequencies are marked off on ordinates, while the depths under the surface are indicated by abscissae. Parallel with the diagram there should be a column representing the stratigraphy by the use of suitable symbols for the various peat and sediment types encountered (*cf.* FAEGRI *in* FAEGRI and GAMS 1937).

Pollen and spore symbols are tabulated in TEXTFIG. 10. Several of these have been used by L. VON POST since the beginning of modern pollen analysis. Others are younger and a few are proposed here for the first time. With the exception of the signs for *Nuphar*, *Nymphaea*, and " aquatic herbs ", all symbols are of equal height. Unfilled tree pollen symbols have thick contours; herb pollen symbols less prominent contours. Tree pollen symbols are connected by thick lines; symbols of *Corylus*, *Myrica*, and other species (the pollen frequencies of which are expressed as percentages of the tree pollen total) by thick, broken lines. Pollen frequencies of *Quercus*, *Tilia*, *Ulmus*, etc. (*i.e.*

TEXTFIGURE 10c. — SYMBOLS USED IN POLLEN DIAGRAMS (*contd.*).

the constituents of the mixed oak-forests) are indicated by thin lines of different design (*cf.* right column of TEXTFIG. 10*a*). In order to prevent the diagrams from defeating their purpose by overcrowding, the curves of these trees are often shown in accessory diagrams.

The symbol of *Alnus incana* refers to the comparatively large aspides of the pollen grains, while that of *Myriophyllum alterniflorum* recalls the big collars of the pores. The number of dots and rings in the signs of *Centrospermae* and *Plantago* may be varied in order to indicate different pollen types or different pollen species. Similarly, the sign which is used for the *Compositae* may be varied, *e.g.* by making two opposite squares black. The symbol of the *Cyperaceae*, proposed by VON POST, is very appropriate, although a little too intricate to expect unanimous acceptance. It may be exchanged for a black triangle provided with a central white dot. Two *Empetrum* signs are proposed: the figure on the left may be retained for *Empetrum nigrum* (diameter of outer circle: diameter of inner circle = 10:4). The ratio between the diameters of the circles of the symbol for cultivated grasses is 10:7; that of wild grasses, 10:3.

POLLEN PERCENTAGES. — Muskeg near Twin Lakes, Itasca State Park, Minnesota (*cf.* diagram, p. 161). In addition to the pollen grains quoted below, there were found stray pollen grains of *Carpinus* and *Juglans* and, in sample 29, one grain of *Carya*. — tr, trace.

Sample	ABIES	ALNUS	BETULA	LARIX	OSTRYA	PICEA ALBA	PICEA MARIANA	PINUS	QUERCUS	TILIA	ULMUS	CORYLUS	Number of pollen grains counted
1	1.4	5.7	8.3	1.4	—	0.6	4.3	77.1	0.3	—	0.9	—	352
2	2	0.5	3	2	—	tr	2	90	0.5	—	—	—	175
3	tr	1.5	2.5	1.5	—	—	2	92	—	—	0.5	0.5	154
4	tr	1	4	tr	—	—	3	91	1	—	—	tr	166
5	0.5	0.5	5	1	—	—	2	89	2	—	—	1	175
6	2.5	3	4	0.5	0.5	0.5	3.5	83.5	1	—	1	—	200
7	—	2	9	2.5	—	—	3	79	4	—	—	—	162
8	tr	4.5	11	1.5	—	—	tr	79	2	0.5	1.5	1.5	160
9	1.5	5.5	26	tr	—	—	1	62	2.5	1	—	2.5	166
10	tr	5	19	—	tr	—	—	71	1	—	4	2	188
11	—	6	25	0.5	3	—	—	55	9	0.5	1	1	167
12	—	5	32	—	3	—	—	49	7	tr	4	3	218
13	0.5	9	30	tr	tr	—	1	45	10	1.5	3	2	152
14	—	7	39	—	7.5	—	—	19	21	0.5	6	0.5	187
15	0.5	9	32.5	—	12	—	—	20	20	1	5	0.5	214
16	—	11	45	—	6	—	—	19	17	tr	2	11	178
17	—	6	56	—	tr	—	tr	10	26	—	2	10	155
18	—	8.5	34	—	2	—	0.5	34	19	—	2	2	169
19	—	6	45	—	2	tr	—	15	30	tr	2	10	164
20	—	9	44	—	1.5	tr	—	18	26	—	1.5	5	151
21	—	9	25	—	1	—	—	34	28	0.5	3	1	151
22	—	7	30.5	—	—	tr	—	27	35	—	0.5	3	177
23	—	8	31	—	4	—	—	30	25	—	2	2	150
24	—	14	32.5	—	—	—	—	26	25	2	0.5	2.5	160
25	—	8.5	17	—	—	—	—	31	42	0.5	1	2	155
26	—	9	16	—	—	tr	—	32	41	—	2	8	150
27	—	5	17	—	—	—	—	26	50	1	1	7	150
28	—	6	16	—	3	—	—	22	50	—	2	3	122
29	—	6	12	—	1	tr	—	21	58	—	2	14	209

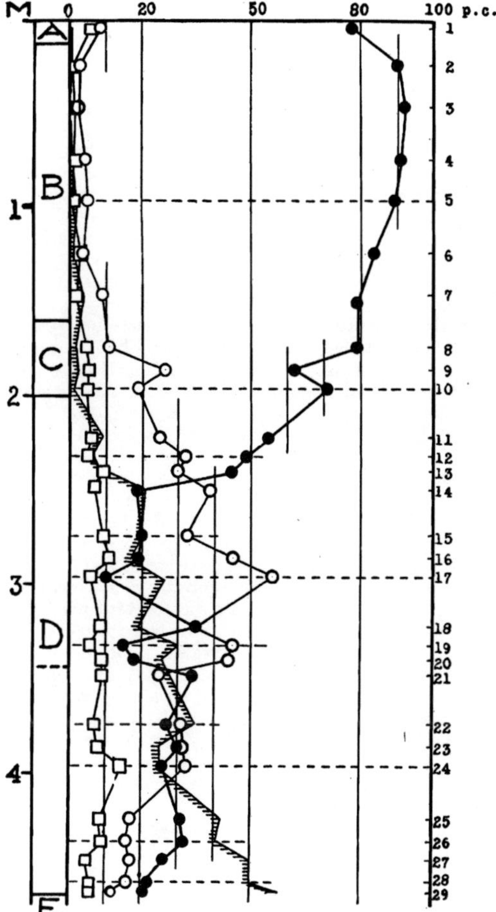

TEXTFIGURE 11. — POLLEN DIAGRAM. — Muskeg near Twin Lakes, Itasca State Park, Minnesota (G. ERDTMAN and C. O. ROSENDAHL, June 1931). Analyses by G. ERDTMAN. Squares, *Alnus;* open circles, *Betula;* filled circles, *Pinus;* hatched line, *Quercus;* A, living *Sphagnum;* B, swamp peat, raw; C, swamp mud ("dy"); D, marl; E, rock or stones. The pollen percentages are given in the table on p. 160.

In conformity with the pollen frequencies of the constituents of the mixed oak forests, the NAP and AqP frequencies are often indicated with special diagrams. The illustration of frequency values in tables will render the results particularly easy to analyze.

Very clear diagrams are obtained if the pollen curves are drawn in different colours. The following colours have been proposed by von Post: *Alnus*, red; *Betula*, bright green; *Carpinus*, violet; *Fagus*, yellow; *Picea*, dark green; *Pinus*, dark blue; *Salix*, light blue. *Corylus* and *Quercetum mixtum* are indicated by sepia; broken lines for the former, full for the latter (*cf.* von Post in von Post, von Walterstorff, and Lindqvist 1925). Diagrams in black and red have been published by Thomaschewski (1930). Important shifts in the composition of the spectra can also be indicated at a glance by shading the area of the diagram represented by the selected genus.

Pollen diagrams formerly exhibited only the frequencies of forest trees and hazel (*cf.* textfig. 11). During the last few years, however, diagrams of a more detailed type have been published in ever increasing number. Textfigs. 12 and 13 (reproduced from Faegri 1941) may serve as examples of such diagrams. Textfig. 12 is a diagram from a Norwegian lake. It should particularly be noted that the samples analyzed come from a uniform sequence of limnic strata. The oldest samples exhibit high *Salix* and NAP frequencies. The marked transition at 3.90 m lends support to the suggestion that continuous forests were established. From this level to 1.6 m the NAP percentage is usually less than 25. This indicates that the forest maintained its position during the corresponding period. At 1.5 m and higher, however, increasing NAP values register the decrease and final disappearance of the forests, as a consequence probably of a climatic deterioration favouring the establishment of wet heaths with much *Sphagnum*. The diagram is incomplete, since it does not show the pollen flora of the whole sequence of layers. It will, however, serve as an example of the characteristic behaviour of the QMP-curve, first emphasized by von Post, *viz.* increasing, maximum, and decreasing values during the three main climatic subdivisions of postglacial times. These subdivisions have been termed by Chiarugi (1936) as follows:

1. The anathermic period, or the stage of climatic amelioration, characterised by the appearance and the first increase of relatively thermophilous trees of different kinds;

2. The hypsothermic period, or the stage of climatic optimum, characterised by the culmination of these forest elements;

3. The catathermic period, or the stage of climatic deterioration, characterised by the decrease in characteristic trees of the hypsothermic period and the appearance or the return of the predominant forest constituents of the present day.

The diagram, textfig. 13, is also from a Norwegian deposit. Stratigraphically very varied, this deposit is in striking contrast to that shown in textfig. 12. A layer of coarse, muddy detritus is covered by a thin layer of *Sphagnum* peat. This peat, again, is overlain by shore peat with remnants of grasses, then by marine sand, and, finally, by alder brushwood peat. In the main diagram, the NAP frequencies — based on the AP total — vary considerably, obviously quite irregularly. To the right of the main diagram, however, is a special NAP diagram with the NAP frequencies based on the NAP

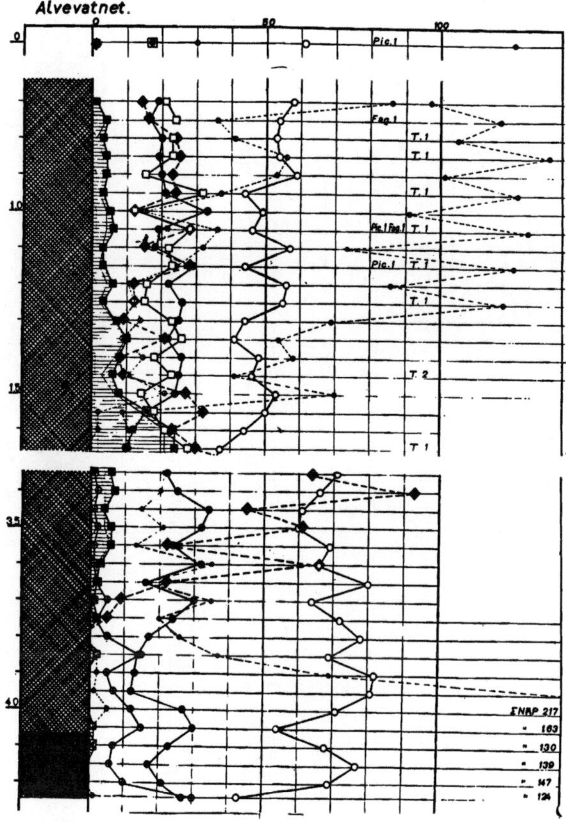

TEXTFIGURE 12. — POLLEN DIAGRAM FROM LAKE ALVEVATNET, NORWAY (K. FAEGRI). — The QM-frequencies are particularly emphasized by thin equidistant lines drawn through the area between the zero-line and the QM-pollen-curve at right angles to the former. Pollen symbols explained in TEXTFIG. 10*a*, *b*, *c*. Supplement and exception: small filled circles denote NAP, small circles with three radii *Sphagnum* spores.

total. This is most instructive and shows three main subdivisions. The first of these is characterised by AqP, the second by dominating grass pollen, the third and last by high *Chenopodiaceae* pollen frequencies. As emphasised by FAEGRI, this suggests a transition from open

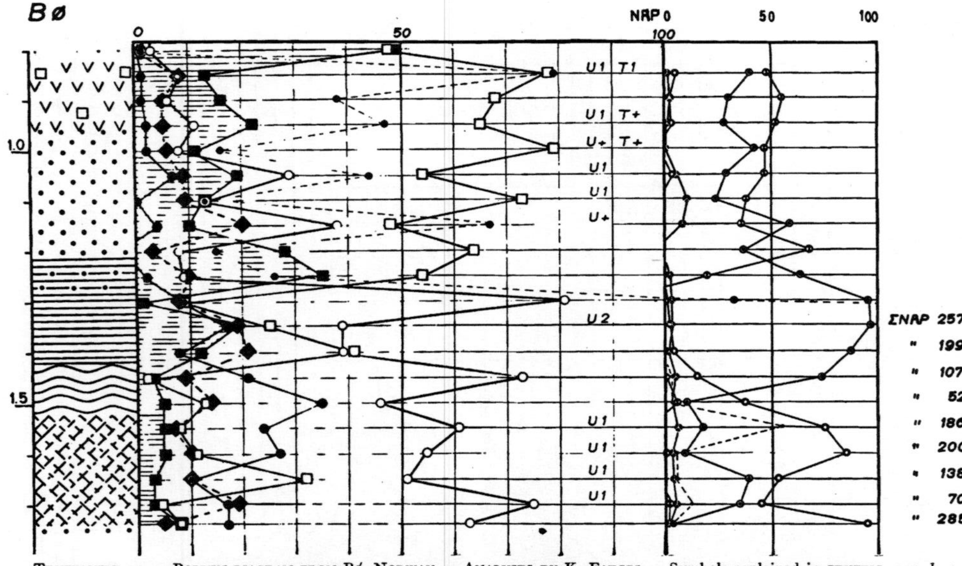

TEXTFIGURE 13. — POLLEN DIAGRAM FROM Bø, NORWAY. — ANALYSES BY K. FAEGRI. — Symbols explained in TEXTFIG. 10*a*, *b*, *c*. — Supplement and exceptions: small filled circles denote NAP, small circles with thick contour *Cyperaceae* pollen, small circles with thin contour and black center *Gramineae* pollen, small circles with two filled sectors *Chenopodiaceae* pollen, small black points AqP. Of these the symbols for NAP and AqP are connected by broken lines, the other symbols by full lines.

water to a grass-moor, and then to a sandy deposit, the marine character of which is testified by the dominance of pollen from salt-marsh plants. Since no diatoms were found in the sand, there was no other means of establishing its marine character.

Diagrams either are open or continuous. Diagrams of the last type are based on analyses of continuous series of samples taken, for example, at intervals not exceeding one centimeter. Diagrams of the first type offer a less complete record of the fossil pollen flora because of the greater intervals at which the samples are taken. The " Lupendiagramme " by SCHROEDER (1930, pp. 26, 27) may be mentioned as examples of continuous diagrams. It takes considerable time to elaborate such diagrams and mere irregularities, such as a local " over-representation " of certain pollen types, may now and then influence the general course of the pollen curves. Some of these drawbacks, however, may be reduced or eliminated. It is thus sometimes possible to procure a continuous peat pillar which is cut into pieces of suitable size. These pieces are dried and reduced to a powder. In this way, the pollen grains are evenly apportioned, and the influence of local " over-representation " lessened. The ensuing analysis will thus provide an average in the pollen flora, undisturbed by any irregularities.

It is not necessary to consider here the abundance of suggestions and modifications which have appeared from time to time concerning pollen diagrams. Most of them have been considered in a paper by GAMS (1929). It will suffice to mention that, following a proposal by VON POST (1924), the diagrams may be disintegrated — *i.e.*, divided into as many parts as there are pollen curves. (Several modifications are of course possible). The ensuing detail diagrams consist of the diagram area of each species — *i.e.*, the area enclosed between the zero-line and the pollen curve of the species in question. These areas appear black in the illustrations and are arranged in a horizontal row, appropriately in the same order as the sectors of a visual representation of a pollen spectrum (TEXTFIG. 9). Diagrams of this type are not very useful when making detailed comparisons, but they may be profitably used in wider regional reviews of different diagram types, etc. (examples in VON POST 1929). A modification has been proposed by SEARS (1933); more recently this has been particularly clearly presented by RICHARDS (1938, p. 132). It implies the replacement of the homogeneous diagram areas by horizontal lines, proportionate in length to the percentages and drawn at those very levels where the pollen species in question has been found. Spore diagrams of the same type have been drawn by THIERGART (1937). In such diagrams (drawn on paper with mm squares) both the number of samples analysed and the exact percentages may be seen at a glance. TEXTFIGS. 14*a* and *b*, finally, may serve as examples of good pollen diagrams illustrating the pollen flora at different parts of a peat deposit. In the upper part of TEXTFIG. 14*a* is a section showing the stratigraphy of a bog in northern Sweden; borings were taken as close as every fifth meter. In the lower part of the figure is a diagram showing the composition of the fossil pollen flora of peat samples from just above the subsoil. TEXTFIG. 14*b* is an ordinary diagram from the top to the bottom of the peat at point 20 in the section TEXTFIG. 14*a*. Diagrams of this kind are

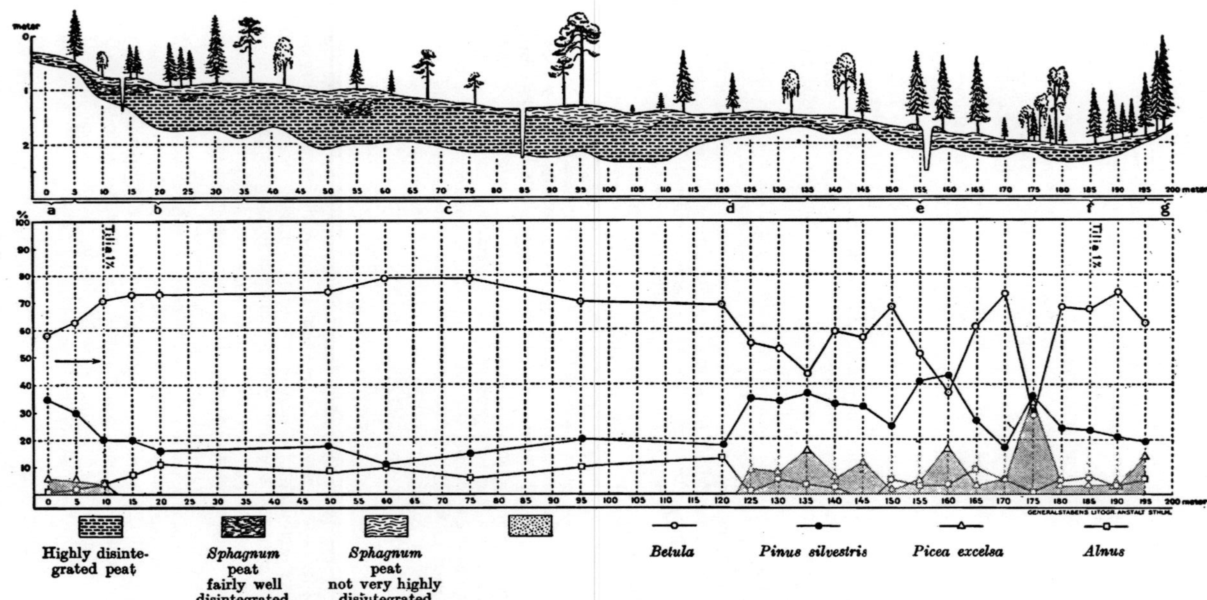

TEXTFIGURE 14*a*. — SECTION THROUGH THE MIDDLE PART OF THE EXPERIMENTAL FIELD, Kulbäcksliden Experimental Forest, Västerbotten (from MALMSTRÖM). — The diagram below the picture of the profile shows the relative frequency of tree pollen in peat samples collected at intervals of 5 m. along the line close to the moraine bed below the peat. At point 50 are thus found 74% birch pollen, 18% pine pollen and 8% alder pollen. At point 135: 44% birch pollen, 37% pine pollen, 16% spruce pollen and 3% alder pollen.

very helpful in a study of the evolution and growth of peat bogs. Thus it is evident from TEXTFIGS. 14*a* and *b* that the formation of the peat did not start contemporaneously in the different parts of the bog; in some parts peat began to accumulate previous to the immigration of the spruce, in other parts just at or even after that time. To make the diagram, TEXTFIG. 14*b*, more comprehensible a shaded tone has been applied to those parts of the diagram which are enclosed between the ordinate and the spruce pollen curve.

The evidence furnished by pollen diagrams may often be substantially corroborated and complemented by supplementary information,

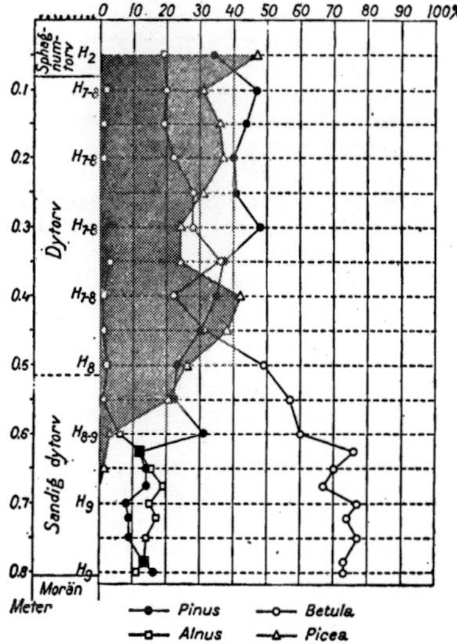

TEXTFIGURE 14*b*. — POLLEN DIAGRAM from point 20 in the section FIG. 14*a* (from MALMSTRÖM).

sometimes in the form of special diagrams, relative to the occurrence in the peat of wood samples (*cf. e.g.* BERTSCH 1926), fruits and seeds (*cf. e.g.* HESMER 1935, LUNDBLAD 1936, SELLING 1938; illustrations of fruits and seeds are found in BERTSCH 1941), bud scales etc. (HESMER *l.c.*), and diatoms (papers by AARIO, FROMM, HALDEN, HYYPPÄ, LUNDQVIST, SUNDELIN, THOMASSON, and others, cited in chapter XVI; *cf.* also the diagrams by M. FJAESTAD-FLORIN published in SELLING *l.c.*).

Isohylochrones and Isopollen Lines (Isopolls): — With a great number of pollen diagrams from a certain area at hand, it is sometimes profitable to transcribe and transfer important details to special maps showing the regional distribution of the phenomena to which these details may be due. AUER (1933) has published a map of Tierra del Fuego with a number of "isohylochrones", or lines indicating the position of the forest frontiers at different periods. Easily identifiable layers of volcanic ash traceable from bog to bog all over the district served as datum lines. Peat-lands with intercalated strata of volcanic ash also occur in northwestern U. S. A. (Washington), Alaska, Kamtchatka, Japan, Iceland, etc. In these countries little or nothing has been done so far along the lines suggested by AUER.

About fourteen years ago (when it was more customary than now to adopt BLYTT's and SERNANDER's division of postglacial times into the climatic phases pre-boreal, boreal, atlantic, sub-boreal, and sub-atlantic — using them also to indicate time) the author constructed a series of maps indicating, in terms of these periods, the immigration and presumed maximum occurrence of the main forest trees in extra-Mediterranean Europe. In one map, *e.g.*, data referring to the distributional movements of hornbeam were gathered. By inserting lines separating areas invaded during the boreal period from those invaded during the atlantic period, etc., an idea of the main distributional tendencies was obtained. Evidently an advance from southeastern Europe had occurred, following the Carpathians and continuing west-, north-, and eastwards, even to places beyond the present area of the hornbeam. These maps were compiled from data supplied by a great number of pollen analysts, but they were never published since it was felt that the age-determinations were too often uncertain.

From maps of this kind to maps with isopollen lines, or isopolls, as drawn by SZAFER (1935), is but a short step. Isopolls are lines referring to a certain period and drawn through places with equal averages of pollen percentages. Thus the "sub-boreal" five per cent *Fagus* isopoll separates areas with sub-boreal average beech-pollen frequencies exceeding five per cent from areas where the corresponding frequencies are less than five per cent. Among others, SZAFER has drawn a series of maps showing the relative amounts of spruce pollen in the post-glacial deposits of Poland. Subarctic layers yielding more than two per cent spruce pollen have only been found in the northeastern and the southwestern corners of the country. In the former, there is one isopoll, the 2 per cent line; in the later four, *viz.* the 2, 5, 10, and 20 per cent lines. Thus, the main spruce centre of that time was probably situated in the mountains of the southeast. The later, post-subarctic, development involves a gradual increase of the spruce pollen frequencies and a gradual advance, both from north and south, of the isopoll systems towards the Bug-Wistula line.

Uncritical readers may overrate the value of isopoll maps. It must always be kept in mind that such maps indicate nothing but figures and often not the realities behind them. Even if the figures be averages, under certain conditions, they may mean something quite different than under other conditions. For example, five per cent of a certain pollen type may indicate either dense or thin forests, depending on climate, the direction of winds and a number of other factors. There might even have been no forests at all at the corresponding isopoll, since the tree pollen grains, particularly in countries with large open areas, may have been produced in distant forests.

References: —

AUER, V., 1933: Verschiebungen der Wald- und Steppengebiete Feuerlands in postglazialer Zeit (Acta Geographica, 5:2).

BARKLEY, F. A., 1934: The statistical theory of pollen analysis (Ecology, vol. XV).

BERTSCH, K., 1926: Die Pflanzenreste aus der Kulturschichte der neolithischen Siedlung Riedschachen bei Schussenried (54. H. Schr. Ver. Gesch. Bodensees u. seiner Umgebung).

—— —— 1941: Früchte und Samen. Bestimmungsbuch zur Pflanzenkunde der vorgeschichtlichen Zeit (Stuttgart).

CHIARUGI, A., 1936: Ricerche sulla vegetazione dell'Etruria Marittima, I (N. Giorn. Bot. Ital., N. S., vol. XLIII).

FAEGRI, K., 1941: Quartärgeologische Untersuchungen im westlichen Norwegen, II. Zur spätquartären Geschichte Jaerens (Bergens Mus. Årbok 1939–40, Naturvitensk. rekke, no. 7).

FAEGRI, K. und GAMS, H., 1937: Entwicklung und Vereinheitlichung der Signaturen für Sediment- und Torfarten (Geol. Fören. Förhandl., vol. 69).

FIRBAS, F., 1927: Die Geschichte der nordböhmischen Wälder seit der letzten Eiszeit (Beih. Bot. Centralbl., vol. 43).

GAMS, H., 1929: Bemerkungen über Vorschläge zur Abänderung der Pollendiagramme (Geol. Fören. Förhandl., vol. 51).

—— —— 1937: Darstellungsweise und Zeichenwahl für waldgeschichtliche Karten (*Ibid.*, vol. 59).

HESMER, H., 1935: Samen- und Knospenschuppenanalysen in Mooren (Ztschr. Forst- u. Jagdwesen).

HOLST, N. O., 1909: Postglaciala tidsbestämningar (Sveriges Geol. Unders., ser. C, no. 216).

IVERSEN, J., 1941: Land Occupation in Denmark's Stone Age (Danm. Geol. Unders., II. R., No. 66).

JESSEN, K., 1917: Bidrag til vegetationens historie i Randers fjord-dal (pp. 21–43 *in* A. C. JOHANSEN, Randers Fjords Naturhistorie, Copenhagen).

LÜDI, W., 1939: Die Signaturen für Sedimente und Torfe (Ber. Geobot. Forsch.-Inst. Rübel f. 1938:87–91).

LUNDBLAD, K., 1936: Svartökärr. En torvgeologisk och utvecklingshistorisk studie (Svenska Mosskulturfören. Tidskr., vol. 50).

LUNDQVIST, G., 1926: Studier i Ölands myrmarker (Sveriges Geol. Unders., ser. C, no. 353).

MALMSTRÖM, C. and TAMM, O., 1927: Försöksparken Kulbäcksliden (Program Svenska Skogsvårdsfören. exk. Västerbotten 19–21 juni 1927).

MAURIZIO, A., 1939: Untersuchungen zur quantitativen Pollenanalysen des Honigs (Mitt. Gebiet Lebensmittelunters. u. Hyg., Bern, vol. XXX).

ORDING, A., 1934: On new methods and facilities concerning pollenanalytical investigations; Norwegian with English summary (Meddel. norske skogsforsøksvesen, vol. V, no. 18).

VON POST, L., 1918: Skogsträdspollen i sydsvenska torvmosselagerföljder (Forhandl. 16. skand. naturforskermøte, Kristiania 1916).

—— —— 1924: Some features of the regional history of the forests of southern Sweden in post-arctic time (Geol. Fören. Förhandl., vol. 46).

—— —— 1929: Die Zeichenschrift der Pollenstatistik (*Ibid.*, vol. 51).

—— —— 1931: Problems and working-lines in the postarctic forest history of Europe (Proceed. 5th Int. Bot. Congr., Cambridge 1930).

VON POST, L., VON WALTERSTORFF, LINDQVIST, S., 1925: Bronsåldersmanteln från Gerumsberget i Västergötland (Monografiserie utg. av K. Vitt. Hist. Antikv. Akad. Stockholm, no. 15).

POTZGER, J. E., 1941: Pollen spectra as time markers (Amer. Midland Nat. 25:224–227).

RICHARDS, R., 1938: A pollen profile of Otterbein bog, Warren county, Indiana (Butler Univ. Bot. Studies, vol. IV).

RUDOLPH, K., 1928: Die bisherigen Ergebnisse der botanischen Mooruntersuchungen in Böhmen (Beih. Bot. Centralbl., vol. 45).

—— —— 1932: Die natürliche Holzartenverbreitung in Deutschland nach den bisherigen Ergebnissen der Pollenanalyse (Forstarchiv).

SCHROEDER, D., 1930: Pollenanalytische Untersuchungen in den Worpsweder Mooren (Abh. Nat. Ver. Brem., vol. XXVIII).

SCHWICKERATH, M., 1937: Die nacheiszeitliche Waldgeschichte des hohen Venns und ihre Beziehung zur heutigen Vennvegetation (Abh. Preuss. Geol. Landesanst., N. F., H. 184).

SEARS, P., 1933: Climatic change as a factor in forest succession (Journ. Forestry, vol. XXXI).

SELLING, O., 1938: Entwicklungsgeschichtliche Studien im Molken-See mit besonderer

Rücksicht der Frequenzwechsel der Makrofossilien (Geol. Fören. Förhandl., vol. 60).
SZAFER, W., 1935: The significance of isopollen lines for the investigation of the geographical distribution of trees in the post-glacial period (Bull. Acad. Polon., Cl. Sc. Math., Sér. B).
THIERGART, F., 1937: Die Pollenflora der Niederlausitzer Braunkohle (Jahrb. Preuss. Geol. Landesanst., vol. 58).
THOMASCHEWSKI, M., 1930: Pollenanalytische Untersuchung der Moore Stangenwalde und Sakoschin im Gebiet der Freien Stadt Danzig (Bull. Acad. Pol. Sc. Lettr., Cl. Sc. Math. Nat., Sér. B, 1929).
TRELA, J., 1934: Die postglaziale Waldentwicklung des süd-östlichen Teiles des Sandomieren-Urwaldes auf Grund der pollenanalytischen Untersuchung (Acta. Soc. Bot. Pol., vol. XI).

Chapter XII

CORRELATIONS

Modern pollen analysis was primarily intended to make stratigraphical correlations possible when other methods had failed. The pollen grains would be used as guide fossils: *i.e.*, not the single grain, but the pollen spectra, or groups of spectra, exhibited by the curves in the pollen diagrams.

No investigator has done more along this line than has L. VON POST. VON POST started from the assumption that two isochronous beds ought to contain about the same pollen flora if they were not situated too far apart from each other and if they were formed under fairly similar conditions. In 1916, he made a correlation of two profiles of slightly humified *Sphagnum* peat, both in the same bog, but situated about 250 m apart (*cf.* VON POST 1918). This correlation was founded on an almost identical course of the pollen curves of *Fagus*, *Quercetum mixtum*, *Pinus*, and *Betula*. He also endeavoured (VON POST 1916, p. 262) to correlate, with pollen statistics, part of an Auwald-deposit with a stratigraphically undifferentiated raised-bog peat about 730 m away. LUNDQVIST (1920) correlated deposits situated 13 km from each other, and ERDTMAN (1921) showed that correlations could readily be extended for more than 100 km. In so doing, a central bog was selected to serve as a prototype and its pollen spectra, from surface to bottom, designed by numbers, *e.g.* 3, 6, 9, 12, etc. A level in neighbouring bogs was designed as level no. 2 if its spectrum appeared to be somewhat younger than that of the prototype level no. 3, respectively as level no. 4 if it was considered to be slightly older, etc. In this way, a system of about 100 isochronous levels was established. In later investigations it has been customary to refer the levels to certain pollen-statistical periods or time-zones: in the Swedish late-quaternary deposits eleven, designated as I–XI, and, if needed, these zones are subdivided and referred to as I*a*, I*b*, etc. (*cf.* VON POST 1925, NILSSON 1935).

Correlations of pollen diagrams may sometimes advantageously be based on " pollen limits ". While some pollen curves run through the whole diagrams, others have a lower or an upper limit or both. VON POST (1918) has studied several lower limits: *viz.*, the combined *Alnus-Ulmus-Tilia-Corylus*-pollen limit, a palaeofloristic guide level coinciding with the period immediately preceding the maximum extension of the Ancylus Lake; the *Quercus*-pollen limit, which is less sharply defined and corresponding approximately to the time immediately following the Ancylus maximum; and, finally, the *Fagus-Picea*-pollen limit. This limit is discussed more thoroughly and is partly recognized as an *empirical limit*, or the level at which the said pollen species begin to occur in a frequency amounting to at least 1 per cent, partly as a *rational limit*, or the level where a steady increase of the *Fagus*-

and *Picea*-pollen frequencies begins. An exposition of the mathematical aspect of the calculation of empirical pollen limits has been given by ORDING (1934*a*).

As described above, correlations are exclusively pollen-statistical and should, if possible, be supported and supplemented by evidence derived from other branches of science such as palaeoclimatology, geochronology and other geological disciplines, archaeology, etc.

Palaeoclimatology (recurrency levels): — In bog investigations, geological and climatological points of view are often inseparable. Thus, the pioneer work of BLYTT (*cf. e.g.* BLYTT 1876) and SERNANDER (*cf. e.g.* SERNANDER 1910) was a result of combined stratigraphical and palaeo-climatological investigations. Since their work, it has been customary, at least so far as extra-mediterranean Europe is concerned, to adopt the division of the post-glacial period into the climatic phases: Pre-Boreal, Boreal, Atlantic, Sub-Boreal, and Sub-Atlantic, using them also to indicate periods of time. While the retention of these terms to denote the type of climate prevailing at different times is convenient, too rigid a use is not desirable, since the increasing volume of evidence shows more and more plainly that the story is in some respects simpler, and in other respects more complex than this fivefold division would suggest. Taking VON POST's threefold scheme of amelioration, optimum, and deterioration as a basis, GRANLUND's (1932) work in Sweden has demonstrated the possibility of superimposing upon this groundwork a number of smaller climatic cycles. He has discovered and dated in the raised bogs of south and central Sweden five so-called recurrence levels (Swedish: rekurrens-ytor, abbreviated RY), each marking a stage at which the moss has again started to grow luxuriantly after a period of retardation due probably to dryness. Of these five, RY III, the junction between Sub-Boreal and Sub-Atlantic times, has usually been considered the most important one, and that which could be seen so strikingly in many raised-bogs of northwestern Europe. Similarly, in the bogs of Norwegian coastal districts ORDING (1934*b*) has found evidence of three dry phases since the transition between the Sub-Boreal and Sub-Atlantic. In bogs in Shropshire, England, there are also signs of comparable phenomena (HARDY 1939). As the recurrence levels are held to be contemporary in wide areas, they may be readily used for correlating purposes, thus checking any correlations made in other ways.

Besides the recurrency levels, there are several other palaeoclimatological features, which, although usually of a more local bearing, also may afford important evidence in connection with pollen analysis and help us penetrate to the very core of the problems.

Geology and Archaeology: — In formerly glaciated districts, correlations may be made with interglacial epochs and their substages as well as with the consecutive stages of the postglacial period. Laminated clays may sometimes yield pollen, primary as well as secondary (*cf.* p. 202), and provide in this way the means of introducing an exact geochronological time-scale into vegetation history (*cf.* SANDEGREN 1924, FROMM 1938).

Correlations may be further established with eustatic changes of sea-level, as shown by intramarine deposits, submerged peats, etc.; with isostatic changes of level; and with those intricate situations which may result as a consequence of eustatic and isostatic movements taking place contemporaneously.

In countries which have been endowed by a generous nature with peat bogs or other polleniferous deposits, pollenanalytical notices sometimes slip over the confines of the scientific periodicals and appear in the daily newspapers. As a rule the object of such notices is to inform the public about the dating of prehistoric objects. It is only natural, therefore, that pollen analysis, in the layman's opinion, seems to be more or less an aide-de-camp to archaeology, to be called upon whenever a dating of some article found in peat is required.

Even though a pollen analyst considers himself capable of attributing a certain age to an object found in peat, the determination is almost never a direct one. On the contrary, the analyst has to rely on the chronology — relative or absolute — of archaeology itself, and to correlate this chronology with certain spectra or phases in the pollen diagrams. An object which cannot be dated by the archaeological experts but which occurs at a level with a characteristic pollen spectrum may thus nevertheless be dated. It is not always necessary to collect peat samples at the very place where the find was made, since crevices and fissures in the object itself may yield sufficient peat for an ordinary pollen analysis. Thus, the cleaning of archaeological objects may sometimes destroy their only chance of ever being dated.

Although technically accurate, pollen-analytical evidence in datings may be misleading. They provide more or less exaggerated maximum figures in cases where the object has sunk through the surface of the preserving deposit into older layers (*cf.* Odum 1920). A wooden anchor loaded with a heavy stone, as used by the Vikings, may have ultimately come to rest in clay deposited thousands of years previous to the Viking age. The object may even have been forced down into older layers. The pollen flora in peat adhering to a wooden shoe of comparatively modern pattern, encountered in a bog in southern Sweden, thus turned out to be approximately 8000 years old. As for objects not directly deposited in bogs but, in the course of time, buried under more recent peat, only the minimum age can be ascertained by means of pollen analysis.

The literature on pollen-analytical datings of archaeological finds is very varied. For northwestern Europe, the following papers may be consulted: Jessen 1920, 1935, Sundelin 1922, von Post 1925, Thomson 1928, Bertsch 1928, 1931, Isberg 1930, Thomasson 1934, Clark 1936, Schütrumpf 1935, Gross 1937, 1938, 1939, etc.

References: —

Bertsch, K., 1928: Klima, Pflanzendecke und Besiedlung Mitteleuropas in vor- und frühgeschichtlicher Zeit nach den Ergebnissen der pollenanalytischen Forschung, XVIII (Ber. röm.-germ. Kommission).

—— —— 1931: Paläobotanische Monographie des Federseerieds (Bibl. Bot. H. 103).

—— —— 1940: Geschichte des deutschen Waldes (Jena, Fischer).

Blytt, A., 1876: Essay on the immigration of the Norwegian flora during alternating rainy and dry periods (Kristiania).

Clark, J. G. D., 1936: The Mesolithic settlement of Northern Europe. A study of the food-gathering peoples of Northern Europe during the early post-glacial period (Cambridge, University Press).

ERDTMAN, G., 1921: Pollenanalytische Untersuchungen von Torfmooren und marinen Sedimenten in Südwest-Schweden (Ark. f. Bot., vol. 17:10).

FLORSCHÜTZ, F., 1938: Archaeologie en Palaeobotanie (Berichten der Nederlandsche Anthropologische Vereeniging over 1938).

FROMM, E., 1938: Geokronologisch datierte Pollendiagramme und Diatomeenanalysen aus Ångermanland (Geol. Fören. Förhandl., vol. 60).

GRANLUND, E., 1932: Geologie der schwedischen Hochmoore, ihre Bildungsbedingungen, Entwickelungsgeschichte und Verbreitung, sowie der Zusammenhang von Hochmoorbildung und Versumpfung (Sveriges Geol. Unders., ser. C, no. 373).

GROSS, H., 1937: Pollenanalytische Altersbestimmung einer ostpreussischen Lyngbyhacke und das absolute Alter der Lyngbykultur (Mannus, vol. 29).

—— —— 1938: Auf den ältesten Spuren des Menschen in Altpreussen (Prussia, vol. 32).

—— —— 1939: Moorgeologische Untersuchung der vorgeschichtlichen Dörfer im Zedmar-Bruch (*Ibid.*, vol. 33).

HARDY, E. M., 1939: Studies of the postglacial history of British Vegetation, V. The Shropshire and Flint Maelor Mosses (New Phytologist, vol. XXXVIII).

ISBERG, O., 1930: Das Vorkommen des Renntiers (*Rangifer tarandus* L.) in Schweden während der postarktischen Zeit nebst einem Beitrag zur Kenntnis über das dortige erste Auftreten des Menschen (Ark. f. zool., vol. *21a*, no. 12, Upsala).

JESSEN, K., 1920: Bog-investigations in North East Sjaelland (Danmarks Geol. Unders., II. R., no. 34).

—— —— 1935: Archaeological dating in the history of North Jutland's vegetation (Acta Archaeologica, vol. V, fasc. 3).

LUNDQVIST, G., 1920: Pollenanalytiska åldersbestämningar av flygsandsfält i Västergötland (Svensk Bot. Tidskr., vol. 14).

NILSSON, T., 1935: Die pollenanalytische Zonengliederung der spät- und postglazialen Bildungen Schonens (Geol. Fören. Förhandl., vol. 57).

ORDING, A., 1934*a*: On new methods and facilities concerning pollenanalytical investigations (Medd. Norske Skogforsøksvesen).

—— —— 1934*b*: Orienterende pollenanalyser fra norske kystdistrikter (Pollenanalyses of peat samples from Norwegian coastal territories) (*Ibid.*).

ØDUM, H., 1920: Et Elsdyrsfund fra Taaderup paa Falster (Danm. Geol. Unders., IV. Raekke, vol. 1, no. 11).

VON POST, L., 1916: Einige südschwedische Quellmoore (Bull. Geol. Inst. Upsala, vol. XV).

—— —— 1918: Skogsträdspollen i sydsvenska torvmosselagerföljder (Förhandl. 16. skand. naturforskermøte 1916, Kristiania).

—— —— 1925 *in* VON POST, L., VON WALTERSTORFF, E., LINDQVIST, S.: Die bronzezeitliche Mantel von Gerumsberget in Västergötland (Monografiserie utg. av Vitterhets-, Historie- och Antikvitetsakad., no. 15, Stockholm).

—— —— 1928: Pollenanalyse (Reallexikon der Vorgeschichte, vol. X, Berlin).

RYTZ, W., 1930: Neue Wege in der prähistorischen Forschung mit besonderer Berücksichtigung der Pollenanalyse (Mitt. Antiquar. Ges. Zürich, vol. XXX).

SANDEGREN, R., 1924: Ragundatraktens postglaciala utvecklingshistoria enligt den subfossila florans vittnesbörd (Sveriges Geol. Unders., ser. Ca, no. 12).

SCHÜTRUMPF, R., 1935: Pollenanalytische Untersuchungen der Magdalenien- und Lyngby-Kulturschichten der Grabung Stellmoor (Nachrichtenbl. Deutsche Vorzeit, 11. Jahrg.).

SEARS, P. B. and JANSON, E., 1933: The rate of peat growth in the Erie Basin (Ecology 14:348).

SERNANDER, R., 1910: Die schwedischen Torfmoore als Zeugen postglazialer Klimaschwankungen (Postglaz. Klimaänderungen, Stockholm).

SUNDELIN, U., 1922: Råbelövssjöns och Nosabykärrets senkvartära historia och de där gjorda stenåldersfynden (Geol. Fören. Förhandl., vol. 44).

THOMASSON, H., 1934 *in* ALIN, NIKLASSON och THOMASSON: Stenåldersboplatsen på Sandarna vid Göteborg (Göteborgs Vet. Vitt. Samh. Handl., Femte följden, Ser. A, vol. 3, no. 6).

THOMSON, P., 1928: Das geologische Alter der Kunda- und Pernaufunde (Beitr. zur Kunde Estlands, vol. IV).

Chapter XIII

OUTPUT AND DISSEMINATION OF POLLEN

The previous chapters have dealt chiefly with practical aspects of pollen analysis, such as the collecting of peat and pollen samples, identification of pollen grains, etc. In this and the two following chapters some of the theoretical fundamentals of pollen analysis will be discussed: output and dissemination of pollen (the present chapter); what may happen to the grains after settling (differential buoyancy and surface receptivity; pollen flora of surface samples, etc.). Chapter XV presents some features of the background against which fossil pollen grains must be considered in order that they may be correctly rated and interpreted.

Pollen Production: — Recently it has been the ambition of several scientists, particularly POHL (1933*a*, *b;* 1937*a*, *b*), to replace subjective estimations of the pollen output from different plants by objective calculations. Although as yet no formulae are able to indicate accurately the relation between a pollen spectrum and the plants that produced the pollen grains, such calculations are of great interest in the theory of pollen statistics.

The pollen output is computed by emptying the thecae, suspending the grains in water and counting all grains of a fixed portion of the suspension (POHL 1937*a*). Alternatively, flowers, inflorescences, etc., collected immediately before the opening of the anthers, may be macerated by means of suitable chemicals, dissolving all but the exines of the pollen grains (*cf.* ERDTMAN 1938). The exines are then counted by aid of a counting chamber.

Here follow some of the figures — slightly rounded off — given by POHL:

Pollen output per stamen: *Rumex acetosa* 30.000, *Secale cereale* 19.000, *Fraxinus excelsior* 12.500, *Acer platanoides* 1.000, *Calluna vulgaris* 500 (tetrads).

Pollen output per flower. Gymnosperms (male strobili): *Pinus nigra* 1.480.000, *Picea excelsa* 590.000 (the output may vary widely according to the size of the strobilus; thus in one strobilus ERDTMAN found 605.000 pollen grains, in another 1.800.000), *Juniperus communis* 400.000 (ERDTMAN), *Pinus montana* 300.000, *Pinus silvestris* 158.000; in *Abies alba* the output per strobilus may be much less than in *Picea excelsa* (a single observation — from a cultivated tree from near Stockholm — gave a number of 150.000 grains). Angiosperms: *Rumex acetosa* 180.000, *Secale cereale* 57.000, *Tilia cordata* 43.500, *Quercus sessiliflora* 41.200, *Fagus silvatica* 12.000, *Acer platanoides* 8.000, *Calluna vulgaris* 4.400 (tetrads).

Pollen output per inflorescence: *Pinus nigra* 22.500.000, *Pinus montana* 7.500.000, *Pinus silvestris* 5.775.000; *Acer pseudoplatanus*

25.000.000, *Betula pubescens* 6.000.000 (ERDTMAN), *Betula verrucosa* 5.450.000, *Alnus glutinosa* 4.445.000, *Corylus avellana* 3.930.000, *Quercus robur* 1.250.000 (ERDTMAN), *Quercus sessiliflora* 555.000, *Fagus silvatica* 175.000; herbs: *Rumex acetosa* 393.000.000, *Typha angustifolia* 174.000.000, *Secale cereale* 4.250.000.

Pollen output of ten year old branch systems (millions): *Pinus silvestris* 346, *Acer pseudoplatanus* 336, *Alnus sibirica* 302, *Corylus avellana* 244, *Pinus nigra* 120, *Betula verrucosa* 118, *Quercus sessiliflora* 111, *Picea excelsa* 107, *Carpinus betulus* 95, *Tilia cordata* 89, *Pinus montana* 52, *Fagus silvatica* 28.

Total pollen output in millions of grains (relative values equal to the product of the average pollen output of ten year old branch systems and the number of cubic metres of branches and stems, less than 7 cm in diameter, from pure and high grade stands, 100 by 100 m): *Pinus silvestris* 12.500, *Picea excelsa* 11.000, *Acer pseudoplatanus* 10.415, *Alnus* 6.950 (calculation based on *A. glutinosa* and *A. sibirica; cf.* POHL, *l.c.*, footnote p. 417), *Carpinus betulus* 5.900, *Corylus avellana* 5.600, *Tilia cordata* 5.600, *Betula verrucosa* 5.570, *Quercus sessiliflora* 3.500, *Fagus silvatica* 2.050.

Total pollen output, expressed in millions of grains, for a period of fifty years [calculation as above under "Total pollen output" and with due consideration to any intervals between the years of blossom; figures in brackets give the output in terms of the pollen output of *Fagus silvatica* (assumed to be equal to 1)]: *Alnus* 362.720 (17.7), *Pinus silvestris* 322.750 (15.8), *Tilia cordata* 280.490 (13.7), *Corylus avellana* 280.450 (13.7), *Betula verrucosa* 278.480 (13.6), *Picea excelsa* 274.750 (13.4), *Carpinus betulus* 144.700 (7.7), *Pinus montana* 59.260 (2.9), *Quercus sessiliflora* 34.410 (1.6), *Fagus silvatica* 20.450 (1.0).

The pollen output may also be expressed as the total number of grains to the square meter of soil. This involves calculation of the area of the projection of the tree or shrub on the ground, followed by a determination of the actual number of flowers or inflorescences. Calculated in this way, the number of millions of pollen grains per square meter is, according to POHL, in *Calluna vulgaris* 4060, *Alnus glutinosa* 2160, *Secale cereale* 1270, *Corylus avellana* 965, *Carpinus betulus* 710, *Fraxinus excelsior* 118.

If these figures and those previously quoted are to be correlated with the actual occurrence of the pollen-producing species, several circumstances which cannot be dealt with here must be considered. They include, among others, the proportion of male and female individuals in dioecious plants and the prolonged intervals between the years of flowering encountered in marginal districts of the potential area of distribution for any given species. To these considerations may be added such extremely complex factors as the connection between the pollen production and the status of the forests, their age, density, exposure, etc.

In a wider sense than has hitherto been the case, pollen statistics ought to be carried out parallel with quantitative seed and fruit analyses. In this connection, figures illustrating the ratio between the number of pollen grains and the number of ovules, capable of development, would be of a certain interest. To each ovule in *Corylus avellana* there are some 2.550.000 pollen grains, in *Fagus silvatica* 637.000, in *Acer pseudoplatanus* 94.000, in *Secale cereale* 57.000, in *Tilia cordata* 43.500, in *Betula verrucosa* 6.700 (POHL *l.c.*).

Weight of Pollen Grains: — Pollen grains, as a rule, are very hygroscopic and their absolute weight therefore varies with the amount of moisture contained. TABLE 10, first column, shows, according to POHL (1937*b*), the average absolute weight per single pollen grain (dried over 90 per cent sulphuric acid) for a dozen species.

By computing the average volumes of the grains, POHL has also been able to supply figures illustrating the specific gravity of air-dried pollen (TAB. 10, column 2). As a rule the volumes were ascertained from measurements of grains suspended in 50 per cent alcohol. The volumes, obtained in this way, are considered to be almost identical with those of air-dried grains (TAB. 10, column 3). On absorbing water, the specific gravity changes slightly, decreasing in grains with a specific gravity in air-dried condition exceeding 1, increasing in the others.

Rate of Sinking: — The fourth column, TAB. 10, gives the theoretical rate of sinking of the pollen grains in air. These values have been calculated by POHL, using the figures in the second and third columns. These values, as a rule, are somewhat less than the empirical figures obtained from experiments conducted by BODMER (1922), KNOLL (1932, and in REMPE, 1937), FIRBAS–REMPE (1936), and DYAKOWSKA (1937). The figures within brackets in the fourth column are the extreme values found by these authors. In addition the following empirical values in cm/sec. may be mentioned: *Abies pectinata* 38.71 (DYAKOWSKA), *Larix polonica* 12.29 (*id.*), *L. decidua* 9.9 (KNOLL), 12.5–22.0 (BODMER), *Secale cereale* 6.0–8.8 (*id.*), *Carpinus betulus* 4.5–6.79 (DYAKOWSKA, KNOLL), 2.2–2.9 (BODMER), *Pinus cembra* 4.46 (DYAKOWSKA), *Ulmus glabra* and *Tilia cordata* 3.24 (*id.*), *T. platyphyllus* 3.2 (KNOLL), *Salix caprea* 2.16 (DYAKOWSKA), *Alnus viridis* 1.7 (KNOLL).

As to the theory of pollen analysis, too much importance should not be attached to these figures. This has been stressed particularly by FIRBAS and REMPE (1936). A vertical air current, rising at a speed of 2 m per second, would, in fifteen minutes, reach an altitude of 1800 m, carrying *Corylus* pollen 1782 m (rate of fall about 2 cm per second) and *Larix* pollen 1638 m if we assume that the average rate of fall is 16 cm per second.

The rate at which the grains fall is, of course, closely dependent on how much water they contain. The hygroscopicity may vary from species to species. Detailed investigations do not seem to have been made yet in which these factors have received due consideration.

The Extent of Dissemination by Wind: — Although exact information concerning the dissemination of pollen grains and spores is difficult to obtain, there is available considerable evidence showing that they can be transported for very great distances. HESSELMAN (1919) exposed plates to trap pollen on lightships in the Gulf of Bothnia. The experiments were carried on from 24th May to 26th June 1918.

Table 10: ABSOLUTE WEIGHT, SPECIFIC GRAVITY, VOLUME, AND RATE OF FALL IN DIFFERENT POLLEN SPECIES. From POHL (1937*b*); the figures within brackets in the fourth column are the extreme values found by BODMER (1922), KNOLL (1932, and in REMPE, 1937), FIRBAS and REMPE (1936), and DYAKOWSKA (1937): —

ABSOLUTE WEIGHT ($g \times 10^{-9}$)		SPECIFIC GRAVITY		VOLUME (μ^3)		RATE OF FALL (cm/sec)		
Picea excelsa	72.8 ± 2.3	*Typha lat.*	1.161	*Picea*	132200	*Picea*	5.96	(8.7)
Fagus silvatica	37.0 ± 1.8	*Corylus*	1.008	*Fagus*	51770	*Fagus*	4.88	(5.5–6.0)
Pinus silvestris	18.4 ± 0.6	*Dactylis*	0.981	*Pinus silv.*	47030	*Pinus silv.*	2.93	(1.59–4.4)
Pinus montana	15.7 ± 0.1	*Betula*	0.808	*Pinus mont.*	37070	*Dactylis*	2.84	(3.1)
Dactylis glomerata	14.3 ± 0.4	*Alnus*	0.752	*Dactylis*	14600	*Pinus mont.*	2.63	(3.21)
Corylus avellana	10.2 ± 0.6	*Typha ang.*	0.747	*Corylus*	10150	*Corylus*	2.34	(1.7–2.5)
Alnus glutinosa	6.8 ± 0.2	*Fagus*	0.713	*Juniperus*	9460	*Alnus*	1.57	(2.77)
Betula verrucosa	6.1 ± 0.2	*Taxus*	0.579	*Alnus*	9070	*Typha lat.*	1.52	(2.9–4.4)
Typha latifolia (tetrads)	5.4 ± 0.3	*Picea*	0.550	*Betula*	7537	*Betula*	1.52	(1.3–2.9)
Taxus baccata	4.1 ± 0.1	*Pinus mont.*	0.496	*Taxus*	7131	*Typha ang.*	1.26	
Typha angustifolia	4.0 ± 0.1	*Juniperus*	0.405	*Typha ang.*	6776	*Taxus*	1.02	(1.1–2.3)
Juniperus communis	3.8 ± 0.2	*Pinus silv.*	0.391	*Typha lat.*	4669	*Juniperus*	0.89	

In one of the ships, 30 km off the coast, 103,037 pollen grains were trapped. This makes an average per square millimeter of about 16.2 grains of which 7 were *Picea*, 6.8 *Betula*, and 2.4 *Pinus*. In the other, 55 km off the coast, 56,075 grains were trapped, averaging 8.8 per square millimeter (4.1 *Picea*, 3.6 *Betula*, 1.1 *Pinus*).

In 1919, identical experiments were performed by MALMSTRÖM (*cf.* MALMSTRÖM 1923) at Degerö Stormyr in a densely wooded part of northern Sweden. *Picea* pollen comprised a fairly great share of the grains trapped during the week preceding the first opening of the male strobili of the spruce. MALMSTRÖM inferred that this pollen had been carried from southern Sweden, a distance of approximately 700–1000 km.

Evidence of long-distance dissemination has also been produced by peat investigations. In seven peat samples from the Faroë Islands JESSEN and RASMUSSEN (1922) found 17 pollen grains of *Alnus*, *Betula*, *Corylus*, *Pinus*, and *Tilia*, *i.e.* trees which do not grow in the islands at the present day and which presumably never have done so during postglacial times. It is suggested, therefore, that the pollen grains have been carried by the wind from Scotland (about 420 km), Norway (585 km) or, as far as the birch pollen is concerned, from Iceland (430 km).

Considerable interest has been aroused by the fact that pollen grains of *Tilia*, an entomophilous tree, have been found not only in the Faroës but also in other places far outside of its actual area of distribution, such as the Shetland Islands (ERDTMAN), Novaya Zemlya (KUDRJASCHOW), and Petsamo (AARIO). Suspicion has been aroused, although, as it seems, without any reason, that these grains have been produced by a species having pollen grains essentially like those of *Tilia*.

In peat collected in the Chatham Islands, more than 700 km east of New Zealand, pollen grains of *Podocarpus* and *Dacrydium* formed about 1–5 per cent of all pollen grains counted, although nowadays conifers do not occur in these islands (ERDTMAN 1924). A peat sample from southwestern Greenland contained a total of about 28,000 tree pollen grains per gram (calculated on the dry-weight basis) of which 355 were conifer pollen. Most of these were referred to *Picea mariana*, said to be the dominant tree in Labrador. There were also many grains of *Pinus banksiana*. None of these trees is native to Greenland, and it may be safely assumed that these pollen grains have been carried by the wind from the continent of America: *i.e.*, for a distance of 1000 km or more (ERDTMAN 1936).

This does not, however, constitute a proof of the idea which formerly was widely held, that the winged conifer pollen travels further than that of the angiospermous trees. On the contrary, conclusive evidence has been produced which discounts the assumption that the winged conifer pollen is particularly liable to over-representation by long-distance wind transport (*cf.* TAB. 10, p. 178). From measurements of the rate of fall of the pollen grains, it appears that although the air sacs do have a large effect, it is offset by the large size of the conifer grains, so that the conifers show somewhat higher rates of sinking than some angiospermous species, such as birch and alder. The known cases of very high values of conifer pollen in unwooded regions may thus be attributed to other causes such as great local preponderance of conifer woods in adjacent regions.

In order to obtain an idea of the absolute amount of pollen grains disseminated by the winds across the seas, pollen grains were trapped by means of vacuum cleaners during a voyage from Gothenburg to New York 29th May to 7th June 1937 (ERDTMAN 1937). The chief results are shown in the map, TEXTFIG. 15, and in the table underneath. The map shows the route followed, the dates, the direction and the velocity of the wind at every sixth hour, the distance from the nearest land, and the numbers of the filter bags exposed (Roman figures; the pollen grains were caught in bags of filter paper which later were chemically dissolved and the pollen grains isolated by centrifugation). Filled track indicates when the vacuum cleaners were operated, and open track when they were not in operation, owing to bad weather.

The absolute number of pollen grains and spores (spores of mosses and fungi excepted) per 100 cubic metres of air declines from 18.0 in section I (the North Sea) to 0.7 in section III (mid-ocean), rises to 6.0 off Newfoundland, falls to 3.5 south of Nova Scotia, and finally rises to 15.0 off the coast of New England. These figures may be contrasted with the very rough average of 18,000 pollen grains per 100 cubic metres of air obtained from experiments conducted from the 1st April to 1st June 1937 on the top of the water tower in Västerås, about 110 km west of Stockholm and 70 m above sea-level.

Fairly well defined pollen rains were encountered on three occasions (heavy figures in the table): in the North Sea on the 30th May (*Pinus*), in section V 250–660 km off Newfoundland 4th June (*Alnus viridis* and *Cyperaceae*), and in the section VII 220–300 km off the coasts of Nova Scotia and Massachusetts 6th June (*Gramineae, Plantago, Rumex*).

Stray pollen grains of *Alnus viridis*, forerunners of the alder pollen rain mentioned, were found in section IV, more than 650 km from the nearest land. Other pollen grains of undoubted American origin include *Carya, Juglans,* and *Fraxinus.* They were caught in section VI, about 300 km south of Nova Scotia. In section VII pollen grains of *Carya, Ilex* (or *Nemopanthes*), *Juglans*, and spores of *Osmunda* were gathered. Remarkable is the presence in section II — roughly midway between Iceland and Ireland — during strong western and northwestern winds not only of pollen grains of trees and shrubs (*Alnus, Betula, Corylus, Juniperus, Myrica, Picea, Pinus, Populus, Quercus, Salix, Tilia, Ulmus*) but also of pollen grains of herbs (*Chenopodiaceae, Cruciferae, Cyperaceae, Ericaceae, Gramineae, Plantago, Umbelliferae, Urtica*) and of spores (*Dryopteris* and *Lycopodium clavatum*). However, some of these may be due to contamination. It is extremely improbable that the amount of pollen grains and spores due to contamination should exceed the figures supplied in the lowest row of the table, *viz.* the maximum number of pollen grains and spores found in any filter during the return journey from New York to Gothenburg. Thus it is extremely probable that pollen grains of *Betula* and *Pinus, Quercus* and *Salix,* as well as of certain sedges and grasses are carried in great quantities by the wind for more than 1000 kilometers into the middle of the ocean. It is but natural that during a couple of hours the

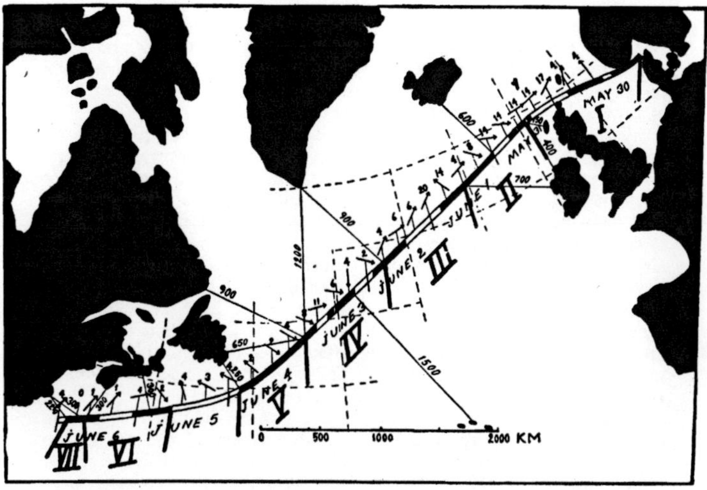

TEXTFIGURE 15. — MAP OF THE ROUTE FOLLOWED. — With the dates, the direction and velocity of the wind at every sixth hour, the numbers of the filterbags exposed (Roman figures) and the distance from nearest land (in kilometers). Filled track, vacuum cleaners operated; open track, vacuum cleaners not in operation. — The table below gives the pollen grains and spores found in the seven sections: —

Besides the pollen grains and spores enumerated in this table there were found a few pollen grains and spores of a number of other species, e. g. Carpinus (1 pollen in section II), Juniperus, Picea, Populus, Caryophyllaceae, Compositae, Cruciferae, Umbelliferae, Urtica, Dryopteris, cf. Equisetum, Lycopodium clavatum and selago, mosses and Uredinae.

Filter	Number of pollen grains and spores pro 100 m³ air	Number of pollen grains and spores counted²)	Number of tree pollen³)	Number of herb pollen and spores³)	Alnus	Betula	Carya	Fagus	Fraxinus	Ilex	Juglans	Myrica	Pinus	Quercus	Rhus	Salix	Chenopodiaceae	Cyperaceae	Ericaceae	Gramineae	Plantago	Rumex	Osmunda
I	18.0	213	179 (84 p. c.)	34 (16 p. c.)	1	10	—	—	—	—	—	—	**157**	8	—	—	1	6	1	4	5	5	—
I a¹) ...	7.0																						
II	1.4	168	134	34	8	35	—	—	—	—	—	1	64	7	—	3	2	2	2	12	4	—	—
III	0.7	59	40	19	—	9							21	3	—	2	1	1	—	4	—	1	—
IV	1.4	75	53	22	6	10	—	(1)⁴)	—	—	—	—	23	5	—	—	—	1	—	8	—	—	—
V	6.0	434	384 (88 p. c.)	50 (16 p. c.)	**328**	11	—	—	—	—	—	—	28	9	—	—	—	**34**	—	—	—	2	—
VI	3.5	190	143	47	20	39	2	—	1	—	2	—	66	5	—	2	1	1	—	11	1	30	—
VII	15.0	235	38 (16 p. c.)	197 (84 p. c.)	4	5	5	—	—	1	1	—	14	4	1	—	—	2	1	**52**	**22**	**85**	3
Total......	(average 6.6)	1374	971 (71 p. c.)	403 (29 p. c.)	367	119	7	(1)⁴)	1	1	3	1	273	41	1	7	5	47	4	91	32	123	3
Maximum number found in any filter during the return voyage (9.—18. June).					2	7	—	1	—	—	—	—	12	2	—	—	2	—	1	6	1	8	—

¹) this filter was exposed on both sides of the Greenwich meridian for six hours in all.

²) spores of mosses and fungi are not included.

³) pollen grains of some shrubs are, resp. may be, included.

⁴) found on the return journey in the same section.

vacuum cleaners could catch only a few such grains. A dispersal of this kind over wide areas, however, seems, as mentioned above, to account fully for the presence of pollen grains of conifers in Greenland for example and has to be considered when studying the pollen flora of tundra deposits and peats of distant lands.

A great number of experiments and observations besides those referred to above may easily be quoted. As a rule, however, they are not primarily concerned with pollen statistics and, in any case, do not afford new problems of any principal bearing. Nor is it necessary to deal more exhaustively with investigations of the vertical distribution of pollen grains and spores in the air by trapping on plates exposed in airplanes at different altitudes. Such investigations, incidentally, have revealed the attack of rust on Canadian crops due to windblown spore infection from the south, and rust spores have been found as high as 5000 m (STAKMAN 1923). In the free atmosphere — at least up to 2000 m — no evident segregation takes place in spite of wide range in size, weight, etc., among pollen grains (FIRBAS and REMPE 1936). This seems to be a particularly interesting fact tending to show that calculations of the transportability, rate of sinking, etc., among pollen grains, although of theoretical interest, are as yet of less importance as far as practical pollen analysis is concerned, than the study, season after season, of the pollen grains settling on the surface of bogs and of how they are successively incorporated in the peat.

Several suggestions have been made as to the classification of the various components of the pollen rains on a basis of transport distance. ERDTMAN (1921) defined the long distance component as grains carried by the wind at least 150–200 km. RUDOLPH and FIRBAS (1927) considered the same component as consisting of grains carried more than 20 km; whilst the distant component was comprised by grains transported 10–20 km; and the local component any grains carried less than 10 km. A third classification has been proposed by HESMER (1933). Distinction is here made between the bog tree component from trees growing on the bog, the marginal component from trees on mineral soils around the bog, the regional component from the region up to 500 m from the bog, but excluding the margin, the distant component from 500 m to 10 km and, finally, the long distance component from > 10 km. In most cases, it will be a difficult task to estimate or determine which components are present in any particular sample. Evidence of this may come not only from the pollen data themselves but also from megascopic fossils, the stratification of the deposit, and knowledge of its development, topography, etc.

Absolute Pollen Frequency of the Air: — The absolute pollen frequency of the air may be defined as the total number of pollen grains in one cubic metre of air. This number can be ascertained in several ways, *e.g.* by means of vacuum cleaners, trapping the atmospherical dust, including the pollen grains, in bags of filter paper. Such experiments have recently been performed in Västerås. The amount of air passing through the vacuum cleaner per minute was calculated to average 0.93 cubic metre by means of a measuring tube provided with a Pitot tube, placed on the exhaust side of the cleaner. The dynamic pressure in the measuring tube was taken with a differential manometer. The measuring arrangement had been calibrated with the assistance of a normal meter whereby the air volume flowing through was taken as the function of the dynamic pressure in the measuring tube. With the help of the calibration graph thus obtained and readings taken from time to time of the manometer, the total volume of the air-flow could be easily ascertained. After filtering the air (usually for 24 hours constantly), the filter bag was taken out of the machine, folded, and put in a glass jar with ground stopper. Twenty grams of dried and powdered calcium chloride tied up in filter paper was put into each jar to absorb the moisture.

In the laboratory each filter was transferred to a glass beaker and soaked with a mixture of 15 cc glacial acetic acid and 15 cc acetic anhydride. In a smaller beaker were mixed 45 cc acetic anhydride, 15 cc glacial acetic acid, and 7 cc concentrated sulphuric acid (slowly added in a narrow jet). After stirring with a glass rod, the reaction mixture was poured on the filter which had just been soaked in the manner described; the filter was then dissolved within five minutes (emergent parts were pressed down into the fluid; when some part of the filter heated too rapidly during the dissolution — indicated by the appearance of a brownish colour — the part in question was swiftly broken up with the aid of the glass rod and the pieces were equally distributed in the fluid).

After the dissolution of the filter paper, the pollen grains, spores, etc., were separated from the yellow-reddish fluid by centrifuging. After decanting, a second acetolysis, as described on p. 27, was usually carried out in order to attain total dissolution of any remaining cellulose fibers. After centrifuging and decanting the sediment was suspended in acetone, transferred to a copper centrifuge tube, and treated with hydrofluoric acid. Finally, the sediment was suspended in dilute glycerine (a fixed quantity, *e.g.* 3 cc) and 0.1 cc of the suspension transferred to a counting chamber. The transfer was made by means of a Hensen " Stempel pipette " (from Schweder, Kiel). The counting chamber was specially made by Leitz, Wetzlar. It takes even the biggest pollen grains and at the same time allows the use of high powers. It is 0.075 mm deep and is provided with four grooves. The pollen-bearing drop is deposited in the area between the two central grooves and a cover glass is gently laid over the drop. When the clamping screws have been applied, the fluid fills the central area, the central grooves, and the area between the central and the marginal grooves. Only the pollen grains in the large central area are counted. The method described is undoubtedly rather intricate as so many different steps are involved. It should be mentioned, however, that other procedures have been considered and tried.

The simplest and at the same time most inexpensive way would be to hydrolyse the filters with sulphuric acid. The time required for transferring the cellulose into water-soluble products is, however, rather long. During this time, the pollen grains are altered by swelling and by partial or even more radical destruction. This method is, therefore, impractical.

The pollen grains in the air can also be caught using an alternative method. The filter bag is exchanged for a series of perforated metal plates smeared with glycerine. When the air is drawn through this apparatus, the particles stick to the plates. The apparatus is washed and the pollen-bearing dust concentrated by centrifuging. In this case the chemical treatment is very easy, but the cleaning of the apparatus offers considerable difficulty, and it is difficult to avoid contaminations. On the other hand a filter bag is exchanged for another in a few seconds.

The modern conimeter constructed by the Zeiss company (*cf.* Löbner 1937) works according to similar principles, but does not seem to be of much use in connection with pollen investigations. It has many forerunners, *e.g.* the aeroscope invented by Pouchet and the aeroconiscope of Maddox. A modification of the former was used by Miquel in his calculations of the amount of dust, pollen, etc., in the air over different parts of Paris (*cf.* Pouchet 1859, Maddox 1870, Miquel 1883).

Further notes on aerobiological techniques are found in the recent paper " Techniques for appraising air-borne populations of microorganisms, pollen, and insects " (Phytopathology, vol. XXXI, 1941, 201–225), edited by the Committee on Apparatus in Aerobiology, National (U.S.) Research Council. *Cf.* also Wodehouse 1942.

References: —

Bodmer, H., 1922: Über den Windpollen (Natur u. Technik, 3).

Dyakowska, J., 1937: Researches on the rapidity of the falling down of pollen of some trees (Bull. Acad. Pol. Sc. Lettr., sér. B).

Erdtman, G., 1921: Pollenanalytische Untersuchungen von Torfmooren und marinen Sedimenten in Südwest-Schweden (Ark. f. Bot., vol. 17).

—— —— 1924: Studies in Micro-Palaeontology, IV. Peat from the Chatham Islands and the Otago District, New Zealand (Geol. Fören. Förhandl., vol. 46).

—— —— 1936: New methods in pollen analysis (Svensk Bot. Tidskr., vol. 20).

—— —— 1937: Pollen grains recovered from the atmosphere over the Atlantic (Meddel. Göteborgs Bot. Trädg., XII).

—— —— 1938: Pollenanalys och pollenmorfologi. Nya metoder och undersökningar (Svensk Bot. Tidskr., vol. 32).

Firbas, F. und Rempe, H., 1936: Über die Bedeutung der Sinkgeschwindigkeit für die Verbreitung des Blütenstaubes durch den Wind (Bioklimat. Beiblätter, H. 2, Braunschweig).

Hesmer, H., 1933: Die natürliche Bestockung und die Waldentwicklung auf verschiedenartigen märkischen Standorten (Ztschr. Forst- u. Jagdwesen).

Hesselman, H., 1919: Iakttagelser över skogsträdspollens spridningsförmåga (Meddel. Statens Skogsförsöksanst., h. 16).

Jessen, K. og Rasmussen, R., 1922: Et profil gennem en Tørvemose paa Faerøerne (Danm. Geol. Unders., IV. R., vol. 1, no. 13).

Knoll, F., 1932: Über die Fernverbreitung des Blütenstaubes durch den Wind (Forsch. u. Fortschr., Jahrg. 23–24).

Löbner, A., 1937: Methodik und Ergebnisse von Staubmessungen im Freien mit dem Zeiss'schen Freiluft-Konimeter (Gesundheits-Ingenieur, 60. Jahrg.).

Maddox, R. L., 1870: On an apparatus for collecting atmospheric particles (Monthly Microscop. Journ., vol. III).

Malmström, C., 1923: Degerö Stormyr (Meddel. Statens Skogsförsöksanstalt, h. 20).

Miquel, P., 1883: Les organismes vivants de l'atmosphère (Paris).

Pohl, F., 1933*a*: Freilandversuche und Bestäubungsökologie der Stieleiche (Beih. Bot. Centralbl., Abt. 1, vol. LI).

—— —— 1933*b*: Untersuchungen über die Bestäubungsverhältnisse der Traubeneiche (*Ibid.*, vol. LI).

—— —— 1937*a*: Die Pollenerzeugung der Windblütler. Eine vergleichende Untersuchung mit Ausblicken auf den Bestäubungshaushalt tierblütiger Gewächse und die pollenanalytische Waldgeschichtsforschung (*Ibid.*, Abt. A, vol. LVI).

—— —— 1937*b*: Die Pollenkorngewichte einiger windblütiger Pflanzen und ihre ökologische Bedeutung (*Ibid.*, Abt. A, vol. LVII).

POUCHET, F., 1859: Traité de la génération spontanée.
REMPE, H., 1937: Untersuchungen über die Verbreitung des Blütenstaubes durch die Luftströmungen (Planta, vol. 27).
RUDOLPH, K. and FIRBAS, F., 1927: Die Moore des Riesengebirges (Beih. Bot. Centralbl. vol. XLIII).
STAKMAN, E., HENRY, A., CURRAN, G., CHRISTOPHER, W., 1923: Spores in the upper air (Journ. Agric. Res., vol. XXXIV).
WODEHOUSE, R. P., 1942: Air-borne Pollens as Allergens (Aerobiology, Publ. Am. Ass. Adv. Sci. No. 17, pp. 8–31).

Chapter XIV

SURFACE SAMPLES

Composition of Pollen Rains: — Although principally of biological interest, investigations like those by POHL, described in the previous chapter, are of value also for the theory of pollen statistics. Yet, it seems still more important to obtain some idea of the total amount of pollen grains, not of the tree pollen only, settling per square unit in the periods of one summer, one year, or a series of years. Recent investigations with this aim have been carried out in the U. S. A. by DUKE and DURHAM (1928), WODEHOUSE (1935), ROSENDAHL, ELLIS, and DAHL (1940) and in Switzerland by LÜDI and VARESCHI (1936). FEINBERG and BERNSTEIN (1940) have reviewed the literature in this field.

The investigations by DUKE and DURHAM were carried out in Kansas City, Missouri, during 1927. Pollen plates were made each day by smearing a thin layer of white petrolatum on an ordinary glass slide and exposing it to the air for 24 hours. Three inches above the slide was a cover, large enough to protect it from the rain. A total of 41,643 pollen grains were counted. This means a pollen sedimentation of about 11,755 grains per year per sq. cm (32 grains per day per sq. cm). Out of the 11,755 grains, more than half, or 6300, come from ragweeds (all species); 1100 from *Quercus;* 970 from *Ulmus;* 490 from *Fraxinus;* 330 from timothy and redtop; 325 from bluegrass and orchard; 305 from *Juglans;* 105 from box elder; while the rest (1830) were classed as " miscellaneous ".

WODEHOUSE trapped grains by exposing microscope slides coated with glycerine jelly. (A similar technique, incidentally, was tried in America as early as in the sixties; *cf.* SALISBURY, Amer. Journ. Sci. Arts, Sept. 1867). The experiments were carried out in the city of Yonkers, N. Y., from the middle of March to the middle of October 1932. The region about Yonkers is comparatively well wooded, with oak as the predominant tree. The results obtained are expressed in a graph, somewhat rounded off to eliminate diurnal fluctuations. Among other things the graph shows that grains of *Juniperus* were the first to be caught on the slides in the springtime. Pollen grains of *Ulmus americana*, which is abundant in the region, were represented throughout the month of April, the maximum frequency (number of grains caught per 1 sq. cm during 24 hours) being 20. The birch pollen season lasted from May 1 to about June 10 (maximum frequency 70); the oak pollen season from the middle of May to the middle of June; the maximum frequency of the oak pollen, 370, was greater than that of any other plant. Many pollen grains of herbs were trapped, particularly pollen of grasses (May–September; maximum number 45) and *Ambrosia* (season two months, beginning about August 10; maximum number 85).

The experiments by LÜDI and VARESCHI were conducted at different altitudes near Davos. The grains were trapped in dishes filled with dilute glycerine. After treatment with 10 per cent KOH, centrifuging, etc., the approximate number of grains per day per square unit was calculated. At Davos Platz, approximately 1500 m above sea level and surrounded by spruce forests attaining their upper limit at about 1950 m, a few grains (less than 0.01 per sq. cm and day) were recovered during the winter. They may have been carried there by the wind from the trees, houses, rocks, or any other place not covered by snow. The highest figures for the period May–August 1935 (averages for periods of ten days) were considerably lower than those obtained by WODEHOUSE: *viz.* 2 tree pollen and 10 grains of grasses and sedges per day per sq. cm.

Experiments have also been performed in Germany by F. BERTSCH (1935). The grains were trapped on microscope slides, 1 sq. cm of which was smeared with glycerine. During 61 days between 23rd March and 15th July 1933, 1396 grains, *i.e.*, about 23 grains per sq. cm per day, were trapped at Ravensburg, Württemberg. Sixty-nine per cent of these came from *Picea*, which makes up about 63 per cent of the forests in the district, and 9 per cent from *Fagus*, supposed to form about 12 per cent of the forests. The following figures were obtained for other trees (areal index within brackets): *Pinus* 8 (7), *Betula* 4 (1), *Abies* 3 (12), *Carpinus* 2 (1), *Ulmus* 2 (1), *Alnus* 1 (2), and *Quercus* 0.5 (2). Stray grains caught a considerable time before or after the proper flowering time are regarded as proofs of long distance pollen transportation. However, we should hardly be justified in endorsing this opinion, since these grains may be local — once deposited, then whirled up, and, finally, redeposited.

A close study of the results of these and other experiments (*cf. e.g.* MÄDE and STROHMEYER 1937) tends to show that pollen sedimentation in areas as small as one sq. cm, or one sq. dm, may be very varying. This has also been shown by OVERBECK (1928) in a study of the pollen flora in the tips of growing *Sphagna*. The water was simply squeezed out from the mosses and the pollen grains condensed with the centrifuge. Some regularity could be traced, although, as a whole, the percentages were rather irregular. Thus — owing presumably to prevailing westerly winds — birch pollen was more frequent in samples collected eastward of a birch grove than in samples gathered to the westward.

In order to obtain a tolerably objective idea of the composition of the pollen rain settling, *e.g.* on the surface of a bog, it would be necessary to investigate the pollen sedimentation of a large number of sample areas. Another method, tried tentatively by the author, is based on the assumption that the composition and amount of pollen sediment in a certain area is proportionate to the composition and amount of pollen grains in the air above the area in question. By trapping the pollen grains by means of vacuum cleaners, it would be possible to record continuously the pollen flora per cubic meter of the air. Uninterrupted drift is secured by enclosing the vacuum cleaners in special cases made from iron plate to prevent any interference from rain and larger objects such as leaves, insects, etc. The vacuum cleaner is placed vertically within the casing with the inlet upwards and the outlet downwards. The amount of air sucked through per minute is calculated by means of a manometer installed in a box attached to the casing. The whole apparatus, or " aerosole collector ", is about 150 cm high, its weight about 70 kg. It is fastened to a heavy concrete square holding it in position even in high winds. If the motor in a vacuum cleaner is worked continuously for weeks it must be overhauled, or even exchanged, once a month. The atmospheric dust is trapped in bags of filter paper, which are later chemically dissolved and treated as described on p. 183.

Each bag is exposed for 24 hours, or for a somewhat longer or shorter period, according to the circumstances. The apparatus does not record any dust carried directly down by rain drops, hailstones, or snowflakes. The approximate amount of such dust could, however, be easily ascertained from a complementary investigation. Microscope slides may be exposed in order to find out the relation between their catch and that of the aerosole collector.

The first experiments with vacuum cleaners were made between Christmas 1936 and New Year's Day 1937. Practically all tree pollen types — even *Tilia* — of Swedish peats, were obtained, some of them in considerable numbers. There were also several grains of corn pollen, easily recognized by their large size (*cf.* PLATE II, FIG. 18). Their presence was correlated with the fact that several corn plants had flowered in the autumn in a garden about 100 m away. Incidentally there was no snow during the time the experiment was on. It seems probable, therefore, that the major part of the grains had been whirled up together with dust from roofs, streets, bare fields, etc., and then trapped when they were settling again.

The knowledge of the composition of the pollen rains would be substantially promoted by pollen-analytical investigations of seasonally deposited sediments such as certain banded clays or different kinds of gyttja (ooze) with a microzonal structure (*cf.* papers by B. W. PERFILIEV in Ber. Biol. Borodin-Station, vol. V, 1927, Verhandl. Int. Ver. Limnologie, vol. IV, 1929, p. 107, and vol. V, 1931, p. 298). Such investigations would make it possible to trace pollen rains, preserved in the sediments, from decade to decade, even from year to year. So far, almost nothing has been accomplished along this extremely important and interesting line.

Flotation and Water Transport: — The winged conifer pollen differs considerably from angiosperm pollen in its powers of flotation, and, where present in considerable amounts, forms more or less conspicuous " Seeblüten ". Winds and currents carrying the surface water to the shore would tend to raise the conifer pollen percentage in the shore deposits compared with those which form in the centre of the lake. This tendency, however, is not a regular one, since pollen washed ashore is often destroyed within a short time. As a rule, flotation does not constitute any serious source of error in pollen analysis. This may be indicated by comparing pollen spectra from sediments with those of contemporaneously formed peats (*cf. e.g.* ERDTMAN 1921, MALMSTRÖM 1923, LUNDQVIST 1924, STARK 1927, SCHMITZ 1930, WASMUND

1931). LÜDI (1939) points out that this may hold good even if the sediments come from deep lakes, as demonstrated by a study of the pollen flora of postglacial marls in Lake Geneva (160–1000 m off the shore; depths ranging from 6 to 39 m).

In order to ascertain the buoyancy of pine pollen, POHL (1933) carried out the following experiment. He kept pine pollen grains for four years in water to which some formaline had been added. After that time, almost all grains were still floating in spite of the fact that the suspension had been thoroughly shaken from time to time. On the other hand much pine pollen has been recovered from water samples from different depths in the North Sea in connection with plankton investigations (GRAN 1912). These facts seem to be somewhat conflicting. Critical investigations, including direct comparison between subaerial and subaquatic pollen sedimentation, seem to be necessary in order to reconcile the somewhat contradictory evidence hitherto reported.

Experiments should also be made in order to calculate the exact amounts of pollen carried by rivers into lake- and flood-plain deposits. It has been stated that pollen grains do travel in this way but as yet there are available no figures illustrating the intensity of this avenue of travel.

Surface Receptivity: — *See* p. 78.

Pollen Flora of Surface Samples: — The present provides the key to the past. Therefore, it is only natural that even at an early stage in the development of pollen analysis, special attention was devoted to the tree pollen contents of the superficial layers of living (*i.e.*, growing, peat-forming) bogs, and to the relation between this pollen flora and the composition and distribution of present day forests.

L. VON POST (1924) has published surface maps showing the occurrence of beech, oak, and spruce pollen in superficial samples, usually unconsolidated débris from growing bogs in southern Sweden. The results indicate that the recent pollen flora is closely tied up with the actual forests: just as the proportion of different pollen types varies, so does that of the pollen-shedding trees. However, there is some evidence to show that beech and oak pollen may tend to be under-represented. Spruce pollen, on the other hand, may be well represented, even outside spruce-bearing areas. This occurs, for example, along the sparsely forested southwestern coast of Sweden and in the barren southern extremity of the island of Gotland. According to VON POST, this indicates that such grains, which are particularly easily carried by the winds, are apt to be over-represented in areas with poor forests or with no forests at all. This important fact has later been corroborated with ample evidence. While the rule itself cannot be doubted, VON POST's assumption that spruce pollen grains are carried by the winds with particular ease is not quite correct (*cf.* TAB. 10; p. 178).

ERDTMAN (1912) has also made a systematic survey of the tree pollen flora of surface samples from southern Sweden: *viz.*, from an area comprising about 1000 sq. km not far from Gothenburg. The coastal part of this area is occupied by heaths interspersed with oak woods (*Quercus sessiliflora*); the inland by mixed coniferous forest (pine and spruce). The demarcation line between these vegetational types is very distinct. Western winds prevail, tending to lessen the admixture of pollen grains produced outside of the area. The surface samples can be classed in two groups: those with dominating conifer pollen; and those with the bulk of pollen made up of grains from deciduous trees. As a rule, samples of the first group were taken from tussocks covered with living *Sphagnum*, whereas those of the second group came from muddy pools surrounding these tussocks. It was inferred that the pool samples were older than the *Sphagnum* samples, since their pollen content was much the same as that in samples taken from the interior of the tussocks at about the same level as the pool samples.

In the *Sphagnum* samples, the amount of alder and birch pollen seemed to bear a fairly close relationship to the actual distribution and abundance of these trees. Pollen grains of hornbeam were absent. It is 150 km in a straight line to the nearest hornbeam groves. The pine pollen frequency varied from 50 to 66 per cent which means that the pine is over-represented. This is probably due to the pine forests in the region at large since local pine stands, which now and then occur in the bogs, do not seem to affect the pollen spectra to the same degree as stands of birch and alder.

The willow pollen frequency, as expected, was very low. The mean oak pollen frequency attained 3 per cent, in striking contrast to the extensive oak forests in the district. The oak, therefore, is in most cases decidedly under-represented when compared both with the conifers and with some of the deciduous trees, *e.g.*, the beech. Elms and lime trees, chiefly single individuals, are found in different parts of the area. Their pollen was rarely met with. The representation of *Corylus* pollen seemed to be consistent with the actual distribution of the hazel.

Similar investigations have been undertaken by LUNDQVIST (1928) in the isle of Oeland, in eastern Sweden. Parts of this island are devoid of forest and apparently receive much pollen from the woodlands of the neighbouring headland. Practically all evidence of the tree pollen flora of surface samples was obtained from the uppermost parts of young sediments still *in statu nascendi* and presumably only a few years old. The results indicate that the amount of *Alnus*, *Betula*, and *Corylus* pollen is in accord with the actual distribution of these trees and shrubs. *Fraxinus* is obviously under-represented, while high conifer pollen frequencies indicate either absence of forests or dominance of *Pinus* or *Picea* in the vicinity of the spot where the surface samples were collected. Pollen of *Carpinus* and *Fagus* did not seem to be carried very far by the winds.

Further investigations along this line have been made in Alberta, Canada, where, in contrast to the Swedish districts, trees, the pollen of which is not preserved in peat, frequently are dominant (ERDTMAN 1935). The area investigated is situated chiefly in the parkland or the wide transition zone between the northern forest and the prairie that stretches across central Alberta. Originally this country was rather continuously covered by various species of poplar and scrub thickets. The dominant tree is the aspen and, although much has been cleared for agricultural land, probably the major part of the area is still tree-clad. The balsam poplar is locally abundant. Thickets of willows are frequent, and throughout the country may be found *Picea albertiana,* the western form of *Picea canadensis,* occurring either singly or in groups. The coniferous covering is greatly extended by the numerous muskegs in certain regions. The chief tree here is the black spruce, *Picea mariana,* although the larch frequently accompanies it. Birch is not abundant and occurs chiefly on the sides of some of the valleys and on muskegs. Groves of jack-pine, *Pinus banksiana,* are found in some sand-hill areas, whereas the lodge-pole-pine, *Pinus murrayana,* is confined to the western extremity of the district, close to the foothills of the Rockies.

Pollen grains of the most common tree, *Populus,* do not occur in the surface samples. On the other hand the jack-pine contributes more than any other tree, or with about 50 per cent, to the average pollen spectrum of the surface samples, although it has such a local distribution that a botanist roaming about in this vast district might not notice it for weeks. Tamarack pollen has been found, but it is rare in spite of the fact that there are hardly any muskegs without tamarack. The birch pollen sometimes attains a frequency of 25 per cent or thereabouts, due to local over-representation caused by the pollen of *Betula glandulosa.* The pollen of the climax tree of the country, the white spruce, has a low frequency, up to 14 per cent. The representation of the pollen of the black spruce is very variable; the highest percentages, about 50, were found in muskegs with a dense growth of black spruce. The alders are decidedly less common than the willows, but their pollen nevertheless has nearly the same frequency as the willow pollen, about 6 per cent.

From all of this we can infer that if conclusions as to the composition of the present and the former conditions of forests in Alberta should be drawn in the customary way on the basis of pollen statistics only, the resulting conclusions would turn out to be entirely misleading. There is no pine-time in Alberta at present, although the average pollen spectrum of the surface samples is that of a "pine-time"; and there is no record of the dominant trees, aspen and poplar, and there is also equally none or only a scanty one of such widely distributed trees and shrubs as *Larix* and *Corylus.*

A particularly interesting illustration of the results of the comparison of pollen deposition with contemporary forest composition is seen in a work published by RUDOLPH and FIRBAS in 1926. Superficial peat samples were collected from different altitudes in the Riesengebirge. Seven samples were obtained between 740 and 950 m, in the forest belt with dominant spruce (*Picea*), abundant fir (*Abies*) and beech (*Fagus*); three samples between 1040 and 1220 m in the pure spruce belt; six samples at about 1240 m in the margin between the spruce and mountain peat (*Pinus montana*) belts; three samples from mountain tops in the dwarf pine belt (*Pinus pumilio*); and various samples above the upper limit of the dwarf pines.

In samples from the lowest forest belt, the spruce pollen was dominant. *Fagus,* though abundant as a tree, gave only 12 per cent of the pollen, whereas the pine, quite absent from the woods, was represented to the extent of 20 per cent, its pollen having been carried there by the wind. The fir was even more scantily represented than the beech. Throughout the spruce belt, the pine pollen was by far the most abundant and remained so through the dwarf pine belt. On the treeless mountain tops, however, there was a striking increase in the pollen from the lower mountain zones due to the weakening of local forest influences, allowing the effects of long distance wind transport to appear. Pollen such as that of oak, lime, and elm must have travelled at least 10 km in these cases.

AARIO (1932) studied superficial peat samples from southern Finland collected along a line from the coast 100 km landwards. In accordance with the composition of the present forests, the samples from the coastal district (a belt about 10 km wide) were very rich in *Alnus* and *Betula* pollen, while pollen of *Pinus* was by far the most dominant in the other samples.

In northern Finland, H. PREUSS collected a series of superficial samples of raw humus, which was later analysed by FIRBAS (1934). In samples from the pine forest region, pollen of *Pinus* dominated, followed by pollen of *Betula* and *Picea.* The non-tree pollen frequency (expressed as percentages of the total tree pollen) was less than 25 per cent. In samples from the birch region, pollen of *Betula* as a rule dominated. The pine pollen frequency reached 10–40, occasionally up to 54 per cent, while pollen grains of *Picea* were few or absent. The non-tree pollen frequency varied from 20 to 50 (–70) per cent. As in the pine region, the *Salix* pollen frequency was low or none, even in places with dense willow shrubs. In samples from the barren tundra region, usually *Betula*, sometimes *Pinus* and, in one case, *Salix* pollen dominated. The non-tree pollen frequencies were usually considerable, but very variable (25–1000 per cent). This may, according to FIRBAS, be due chiefly to the irregular occurrence of the dwarf birch (*Betula nana;* the pollen grains of this species were not distinguished from those of other species and were, as a consequence, recorded as tree pollen instead of non-tree pollen).

Many other investigations have been pursued along similar lines. In this short review, however, the examples mentioned must suffice. They have been chosen from investigations in well-wooded countries as well as in the borderlands between forest and unforested areas of varying character and may convey some idea of the problems which must be faced and taken into account in this particular field of pollen analysis.

In so far as the main problem — " Does the present provide the key to the past? " — is concerned, there is clear evidence that this virtually is the case. But unfortunately there is no universal key. On the contrary: every pollen analyst must make his own key, or, better, a set of keys, since a key which fits in one place probably will be of no use in another.

This involves studies of actual pollen rains; of the surface receptivity of the bogs; of actuopalaeontological processes or the gradual transformation of recent material into fossil substance (RICHTER 1928); and, further, of the ultimate result of these processes, *i.e.*, the youngest fossil microflora. Wide fields of scientific activity are here still awaiting the attention of competent investigators. Future investigations of superficial samples should include all kinds of pollen grains and spores in order to disentangle, as far as possible, the complex interrelationship between tree and non-tree pollen, etc., which so intimately is connected with advanced pollen analysis.

Since the above was written the author has devoted special attention to one of the problems indicated, *viz.* the relation between a plant community and its pollen flora — or "pollen picture" (*German:* Pollenbild) — as revealed by an investigation of the pollen grains contained in the uppermost layers of peat in live bogs. As peat may not occur in all plant communities it should be noted that — in accordance with the experience hitherto gathered — mosses and lichens on old stumps, stem bases, etc. exhibit much the same pollen flora as do the bog samples (from each locality about a dozen samples of mosses and lichens are taken at random, dried and powdered; part of the powder is then bleached and subjected to acetolysis). In the same way as a plant geographer identifies the plant communities in his path a pollen analyst ought to be familiar with their pollen (and spore) pictures. Like an expert scout he should also do his best to trace the tracks of man as produced by fire, cereals, and weeds, etc. in the pollen pictures. Following are examples of such pollen pictures from three localities in south Sweden.

Locality 1. Supposedly oldest beech stand in Sweden (at Bjurkärr, Skatelöv parish). Area about 0.1 sq. km; beeches approximately 225 years old, their average height about 33 m. Cubic mass per hectare 596 m^3 [beech 551.3 m^3 (92.7 per cent), oak 28.3 m^3 (4.5 per cent), remainder 16.5 m^3 (2.8 per cent)]; *cf.* RUDBECK in "Bygd och Natur", Stockholm 1941 (pp. 139, 140).

Pollen flora (analysis of powdered mosses from 20 stumps, stem bases, boles, and fallen trunks of *Fagus silvatica* from the interior of the stand):

Alnus	1.5	*Corylus*	0.5
Betula	20.5	*Artemisia*	0.5
Fagus	52.0	cultivated grasses	1.0
Picea	2.5	*Plantago*	0.5%
Pinus	18.5	*Rumex*	trace
Quercus	5.0		
	100.0%		
	Σ AP:372		

As expected the beech pollen frequency is very high (52 per cent); however, the frequency is by no means high when compared *e.g.* with the cubic mass of beech wood (92.7 per cent of the total mass). *Pinus* and *Betula* could not be seen from the spot where the specimens were collected; in spite of this their pollen frequencies are as high as 20.5 and 18.5 per cent respectively. Although *Fraxinus, Tilia, Ulmus*, and *Myrica*, etc. occur in the neighbourhood, pollen grains of these trees and shrubs are totally lacking.

Locality 2. Largest hornbeam forest in Sweden (Högsrum parish, Oeland; 2 analyses — A and B — from different parts of the forest; analysis C made on living *Sphagna* from under a row of birches along the margin of the forest, near the sea).

	A (%)	B (%)	C (%)
Acer	0.5	—	—
Alnus	2	1	trace
Betula	22.5	21.5	86
Carpinus	25.5	56	2.5
Fraxinus	1	1	trace
Picea	3	1.5	1.5
Pinus	21	12	8
Quercus	15	6.5	2
Salix	0.5	—	—
Tilia	3	trace	—
Ulmus	6.5	0.5	—
Σ AP	200	200	200
Corylus	8.5	2.5	—

Locality 3. Cultivated land at Lida, Åker parish (with wheat, rye, oats, etc.) surrounded by coniferous forest (spruce, pine) fringed by a scattered growth of birch, aspen, oak, hazel, etc. Analysis A: lichens on a row of planted maples surrounded by open fields. Analysis B: lichens from the posts of a local electric power line about 100–200 m from the place where the samples in analysis A were taken. $\sum$ AP in both cases 150.

	A (%)	B (%)
Acer	3.5	—
Alnus	6	5
Betula	49	43.5
Fraxinus	0.5	2
Picea	3.5	5
Pinus	37	39
Quercus	0.5	5
Salix	3	0.5
Tilia	—	0.5
Corylus	1	1
Artemisia	0.5	0.5
Centaurea cyanus	—	3
cultivated grasses	16	17
Rumex	3	4
Umbelliferae	0.5	0.5

The study of " pollen pictures " has provided some hints towards the solution of some interesting problems. Of these three will be considered here. In forest-clad regions in Sweden pollen grains of *Tilia cordata* are almost entirely lacking in samples collected beyond the present range of this tree. Fossil grains of *Tilia* are regularly encountered in the peats and the raw humus cover of the mountainous parts of southern Lapland. This would no doubt imply that *Tilia* was formerly an inhabitant of these mountains, 600 m or more above sea-level, while in our days it has its northernmost outpost by the shores of the Gulf of Bothnia southeast of the mountains mentioned.

The study of the pollen pictures has also revealed the antecedents of the venerable virgin spruce forests in the same mountains. The basal parts of the raw humus characteristic of these monotonous forests contain much the same pollen flora as the pollen pictures of the diversified *Betuleta geraniosa*, so rich in herbs and ferns.

As to *Corylus*, finally, no counterpart of the famous " hazel forests " of ancient times seems to exist in Sweden today, at any rate judging from the pollen pictures presented by hazel groves in southern Sweden. In parts of Manitoba and Minnesota the author traversed aspen and poplar forests with a compact undergrowth of hazel. Seeing that, as a rule, peat does not preserve *Populus* pollen, such forests should consequently offer pollen pictures more or less similar to the pollen spectra of the hazel forest period in southernmost Sweden. This, however, is not the case.*

References: —

AARIO, L., 1932: Pflanzentopographische und paläographische Mooruntersuchungen in N-Satakunta (Comm. Inst. Forest. Fenn. 17:1).

BERTSCH, F., 1935: Das Pfrunger Ried und seine Bedeutung für die Florengeschichte Südwestdeutschlands (Beih. Bot. Centralbl., vol. LIV).

DUKE, W. and DURHAM, O. C., 1928: Pollen content of the air. Relationship of the symptoms and treatment of hay fever, asthma and eczema (Journ. Amer. Med. Assoc., vol. 90).

ERDTMAN, G., 1921: Pollenanalytische Untersuchungen von Torfmooren und marinen Sedimenten in Südwest-Schweden (Ark. f. Bot., vol. 17:10).

—— —— 1935: Pollen statistics (pp. 110–125 *in* WODEHOUSE, R. P.: Pollen grains. New York, McGraw-Hill).

FEINBERG, S. M. and BERNSTEIN, T. B., 1940: Asthma and hay fever contributions during 1939 (Jour. Allergy 11:281).

FIRBAS, F., 1934: Über die Bestimmung der Walddichte und der Vegetation waldloser Gebiete mit Hilfe der Pollenanalyse (Planta, vol. 22).

GRAN, H., 1912: The plancton production in the north European waters in the spring of 1912 (Bull. plankt. pour l'année 1912, Copenhague).

LÜDI, W., 1939: Analyse pollinique des sédiments du lac de Genève (Mém. Soc. Phys. Genève, vol. 41:5).

—— —— and VARESCHI, V., 1936: Die Verbreitung, das Blühen und der Pollenniederschlag der Heufieberpflanzen im Hochtale von Davos (Ber. Geobot. Inst. Rübel f. 1935).

LUNDQVIST, G., 1927: Methoden zur Untersuchung der Entwicklungsgeschichte der Seen (Abderhalden, Handb. biol. Arb.-Meth., vol. IX:2).

—— —— 1928: Studier i Ölands myrmarker (Sveriges Geol. Unders., Ser. C, no. 353).

MÄDE, A. und STROHMEYER, G., 1937: Zur Methodik von Pollenflugversuchen (Der Züchter, vol. 9, Berlin).

MALMSTRÖM, C., 1923: Degerö Stormyr (Meddel. Statens Skogsförsöksanst., h. 20).

OVERBECK, F., 1928: Studien zur postglazialen Waldgeschichte der Rhön (Ztschr. Bot., vol. 20).

POHL, F., 1937: Die Pollenkorngewichte einiger windblütiger Pflanzen und ihre ökologische Bedeutung (Beih. Bot. Centralbl., vol. LVII).

VON POST, L., 1924: Ur de sydsvenska skogarnas regionala historia under postarktisk tid (Geol. Fören. Förhandl., vol. 46).

RICHTER, R., 1928: Aktuopaläontologie und Paläobiologie, eine Abgrenzung (Senckenbergiana, vol. 10).

ROSENDAHL, C. O., ELLIS, R. V., and DAHL, A. O., 1940: Air-borne pollen in the Twin Cities area with reference to hay fever (Minn. Med. 23:619).

RUDOLPH, K. und FIRBAS, F., 1926: Pollenanalytische Untersuchung subalpiner Moore des Riesengebirges (Ber. Deutsch. Bot. Ges., vol. XLVI).

SCHMITZ, H., 1930: Pollenregen-Seeblüte und Pollenanalyse (Paläont. Ztschr., vol. 12).

STARK, P., 1927: Die Moore des Badischen Bodenseegebiets, II (Schr. Naturf. Ges. Freiburg Br., vol. 28).

WASMUND, E., 1931: Pollenregen auf ostholsteinischen Seen und seine Bedeutung für die Pollenanalyse (Centralbl. f. Min. etc., Abt. B).

WODEHOUSE, R. P., 1935: Pollen Grains (New York, McGraw-Hill).

* G. CARROLL (Am. J. Bot. 30:361, 1943) has compared the recent pollen deposition in bryophytic polsters with the adjacent forest composition of the primeval spruce-fir forest of the Great Smoky Mountains. Pollen from forest species foreign to the spruce-fir formation presented an average over-representation or "contamination" of 23% (including *Tsuga*, *Pinus*, *Carya*, *Quercus*, *Tilia*, and *Liquidambar*). Of the normal codominants of the forest, *Picea* was under-represented by 15 per cent and *Abies* and *Betula* were over-represented by 5 and 10 per cent, respectively. Her results indicate that valuable information on wind-transport and over- and under-representation can be obtained by such investigations. With a considerable number of such studies, it may be possible to establish correction factors for fossil pollen spectra which will allow a more accurate interpretation of past forest composition.

Chapter XV

POLLEN FLORA OF PEAT SAMPLES

Differential Resistance of Pollen: Specific Resistance to Decay: — In LINDLEY's and HUTTON's "Fossil Flora of Great Britain" (vol. III, London 1873), there is a description of an experiment which was made in order to ascertain the differential resistance of plant remains to decay. This experiment is described as follows: "On the 21st of March, 1833, I filled a large iron tank with water, and immersed in it 177 specimens of various plants, belonging to all the more remarkable natural orders; taking care in particular to include representatives of all these which are either constantly present in the Coal Measures, or are universally absent. The vessel was placed in the open air, left uncovered, and left untouched, with the exception of filling up the water as it evaporated, till the 22nd of April 1835". Some seventy-five years later, similar experiments were conducted by LAGERHEIM in order to find out the resistance of different pollen types to processes going on in putrescent water. With pollen analysis later in full swing, experimental points of view were usually neglected. At the same time, presumed irregularities in the fossil pollen flora were often correlated with supposed weak powers of resistance to decay in certain pollen species. In the literature, we may sometimes even discern faint suggestions of a classification of the grains into different classes of resistance: from particularly resistant grains (among which *Picea*, *Pinus*, *Tilia*, and *Ericaceae* are as a rule mentioned) to grains such as those of *Acer*, *Fraxinus*, *Juniperus*, *Larix*, *Lauraceae*, *Populus*, *Rosaceae*, *Taxus*, etc., which are often erroneously supposed to be absent from fossil sites.

Arguments against the differential preservation of pollen have been presented by ERDTMAN (1921) and BRINCKMANN (1934). By means of a great number of pollen diagrams from neighbouring bogs, ERDTMAN showed that the characteristic features of the fossil pollen flora could be traced in detail from one deposit to another irrespective of widely varying types of stratification. BRINCKMANN made the same observations in studying the pollen flora of soils formed in the tidal flats, marshes, and bogs in the Jade territory.

However, these observations do not exclude the possibility of differential pollen preservation. Investigations by ZETZSCHE and others have taught us that the chemical properties of pollen exines vary somewhat. But it should be emphasized that the destructive forces, concerning which practically nothing definite is known, are probably equally varying. These destructive forces may sometimes even attack robust armours, such as those of *Tilia* pollen, more severely than exines which are supposed to be of a weaker constitution (*cf.* C. A. WEBER, Engler's Bot. Jahrb., vol. 54, Beibl. no. 120, pp. 21, 22, 1917).

Sometimes grains supposed not to be preserved in peat are fossilized even though, for one reason or another, they are apt to be overlooked. Among these we may cite as an example the grains of *Taxus*. As may be seen from PLATE XXV, FIGS. 437–440, *Taxus* grains have a faint, though characteristic structure, which makes them easy to identify. In order to obtain fossil yew pollen for morphological studies, a number of surface samples were collected from peaty soil from near a large stand of yews (Hejnum Kallgate Burg, Gotland). After chlorination and acetolysis, they yielded a great number of *Taxus* pollen.

Surface samples from a barren area in northern Gotland with a dense growth of stunted junipers contained indisputable juniper pollen in profusion as well as a great number of grains which could not be readily identified on account of their deformed state. Further investigations are needed concerning the power of resistance of *Populus* pollen to decay. The opinion that rosaceous pollen are not preserved in bogs is misleading. The pollen grains of *Rubus chamaemorus* are among the most characteristic and as a rule the most beautifully preserved grains of raised bog peat.

Differential Resistance of Pollen: Individual Resistance to Decay: — It is well known that pollen grains are preserved in perfect shape in certain peats, whilst in other peats a considerable amount is damaged or deformed. The deformative processes may be apparent either as a corrosion or a disorganisation, leading eventually to complete disfiguration or fragmentation of the grains (*cf.* KIRCHHEIMER 1931). Corrosion causes great changes in the contours of the pollen grains by gradually destroying parts of the exine. As a result of disorganisation, the ordinary exine texture disappears, giving place to a secondary, more or less granular texture. At the same time, the exine becomes increasingly darker. Corrosion and disorganisation usually occur simultaneously.

Pollen grains with supposedly weak specific resistance to decay would logically be expected to be absent or under-represented in soils with poor conditions of preservation. But this is by no means always the case. It is equally erroneous to believe that all grains in peats and sediments, etc., with supposedly good conditions of preservation would be particularly well preserved. Presumably the specific resistance of a certain pollen type may not be characteristic of all grains belonging to that pollen type, for the power of resistance of the individual grains may be very varied. Some grains may arrive in good condition at the locality where preservation takes place. These will be well preserved. Others may arrive in a poorer condition and be preserved in a less complete state. Furthermore the preserving layer may not offer uniform conditions of preservation throughout but may present a mosaic of different micromilieus, only the average of which is taken into account when the conditions of preservation in the layer as a whole are considered.

The variation in the individual powers of resistance has also been referred to by FIRBAS and several facts may be advanced in favour of the correctness of this assumed variation. Among these only one will be mentioned here: the observation that alder pollen grains in the older layers of Shetland peats are usually well preserved, while single alder grains in the upper strata, formed after the complete deforestation of the islands, often appear to be somewhat damaged. These grains must have been carried by the wind from Scotland, Norway, or some other country. The greatest part of the well preserved alder pollen grains, on the other hand, was produced in the Shetland Islands prior to their deforestation. Pollen grains of *Empetrum*, *Ericaceae*, etc. (*i.e.* pollen from plants which have grown on the islands all during the peat formation) are usually well preserved throughout.

A general picture of the ideas just advanced and their consequences may be obtained if we imagine the grains of a certain species of pollen classified according to their state of preservation. We may, *e.g.*, imagine that the pollen types usually dealt with in pollen analysis (such as *Alnus*, *Betula*, *Carpinus*, etc.) exhibit ten " classes " of resistance, of which the tenth represents the most resistant and thus most perfectly preserved grains. In less resistant types, the number of classes may be fewer, as, for example, three only, corresponding to the three first classes in *Alnus* pollen, etc. In particularly resistant pollen types, on the other hand, there may be eleven or twelve classes. Some pollen types may be lacking in peat from the start. Those belonging to classes one and two would be destroyed or rendered unidentifiable even by a weak chemical treatment undertaken in preparing the peat for analysis (*cf. Populus*). An ordinary chemical treatment would presumably destroy also most of the grains, never going beyond the third resistance class (*cf. Juniperus*). If, then, the grains of the great majority of the pollen types were divided into ten classes of resistance, radical changes in the relative amounts of the grains could be brought about neither by gradually increasing the severity in the chemical treatment nor by gradually decreasing possibilities for perfect preservation. The few pollen species which would surpass the tenth class would be over-represented only in very extreme cases as a consequence of particularly bad conditions for preservation or of particularly severe chemical treatment. This seems to be in accordance with present observations. Some of the suggestions made may be tested easily by experiments with homogenized standard peats (p. 201).

Downwash of Pollen through Peat: — Pollen spectra from living *Sphagna* in a bog may be suspected of conveying a misleading conception of the actual pollen rains since large grains, such as those of pine, spruce, etc., would not be carried so far down by rain water as comparatively small grains, such as those of deciduous trees. This would manifestly be a matter of no little inconvenience if it actually happened. Fortunately, however, this source of error seems, as a rule, to be entirely imaginary. Erdtman (1921) proved the pollen spectra of living *Sphagna* to be practically identical with those obtained from nearby cushions of *Grimmia* and other mosses growing on rocks and on other places, such as old tree stumps, where a downwash of pollen is impossible. Experimental evidence has been provided by Malmström (1923), who distributed pollen grains of *Lilium bulbiferum* over a small area (0.5 sq. dm) of a bog, added some water and, a week later, collected samples and subjected them to pollen analysis. A downwash, it was stated, occurred within the unconsolidated litter covering the peat, but never within the peat itself. This was also evinced by preliminary experiments with pollen impregnated with colloidal gold. Such pollen grains may be located easily without the aid of a microscope.

In connection with a sharp criticism of some pollen-analytical investigations of raw-humus, sand, etc., DEWERS (1932) and MOTHES, ARNOLDT, and REDMANN (1937) have dealt briefly with the downwash of pollen. The last mentioned authors give an account of an experiment performed to show whether downwash of pollen occurs in sand or not. A glass cylinder, 40 cm. high, was filled with ignited sand. The uppermost part of the sand was mixed with pollen grains of *Quercus* and *Pinus* and water was allowed to trickle through. After twenty hours, a considerable downwash as well as evident segregation had occurred: the pine pollen grains had been carried at least twelve, those of oak at least twenty cm. The result of this particular experiment, however, is not the last word in the matter. Percolating water is probably liable to carry pollen grains deep into certain sands, but these grains are probably soon destroyed. On the other hand, it may be possible that the sand layers besides these transitory grains may contain a number of permanent grains confined in colloidal matter which adheres to the mineral particles.

Pollen grains of different age may sometimes be found side by side not only as a consequence of downwash or resedimentation (*cf.* p. 202), but also as a consequence of the activities of animals, such as earthworms and the fauna of the lake bottoms (KEILHACK 1899, WESENBERG-LUND 1909, LEWIS and COCKE 1929). In beds pierced by cracks pollen grains may be carried by streaming water from higher to lower levels, even, as mentioned by RUDOLPH (1935), from alluvial beds into Tertiary strata.

Pollen Frequency: — In pollen analysis the term "pollen frequency" (PF) originally meant the number of pollen grains in a preparation of a fixed size (VON POST 1918). More recently, however, it has been defined as the number of pollen grains per unit area (one sq. cm) of a preparation (ERDTMAN 1924). The definition is vague, because the PF varies according to the amount of liquid added to the pollen-bearing material, the thickness of the preparation, etc. These sources of error cannot be circumvented, but they can be reduced by uniform and careful technique. Furthermore the PF is not necessarily proportional to the density of the forests that produced the pollen. A peat formed slowly would have a greater PF than one which had been formed quickly, provided that they both had the same capacity for catching and preserving pollen grains. SANDEGREN (1913) and AARIO (1932) showed that the PF of *Sphagnum* peats was usually correlated with humification, the PF of highly humified peats being higher than that of less humified peats. VON POST (1929) has published a diagram showing the PF-fluctuations in different kinds of peats and sediments in southern Sweden. Irregular as these fluctuations are, the abundance of the forests seems to be reflected by the PF-figures only in exceptional cases. Such an exception is the obvious decrease in the tree pollen frequency during the period of the "Boreal hazel forests".

Investigations by ERDTMAN (1924) concerning the average PF-values from barren, sparsely, or from well wooded parts of northwestern Europe seem to demonstrate the general diminution of present and past tree cover progressively across the area of investigation towards the Atlantic. While the PF of the far-off Faröe Islands is less than 1, it rises to 8 in the Shetland Islands, 12 in the Orkneys, 25 in the Isle of Lewis (Hebrides), 270 near Achnasheen in Ross-shire, Scotland, and 335 in northwestern Germany.

A low tree PF may be encountered not only in peat from barren districts, such as the isles north of Scotland, but also in peat from well-wooded countries, if the bog surfaces at the time of pollen shedding were not in a proper condition to catch and preserve the pollen. The values found by ERDTMAN (1935) in peats from Alberta are of some interest in this respect. The highest PF found was 90 (from bogs within the coniferous forests near the foothills of the Rocky Mountains); the lowest 14 (from a muskeg in the parkland, not far from the prairie). The highest PF is only a third as large as that of some bogs near Achnasheen in the poorly wooded areas of northern Scotland, and the lowest is about the same as that from the Orkney Islands which are now practically treeless. On the whole, the pollen frequency of the bogs in Alberta is low.

For a clew to the proper explanation of this fact we must remember that "dead" bogs, *i.e.* those where peat no longer is being formed, cannot catch and preserve pollen grains. This is also the case with living bogs during their inactive periods, when the cold of the winter, exceptional drought, or, in some types of peat deposits, excessive rain puts a temporary stop to the formation of peat. As to the muskegs in Alberta, long, cold winters and rainless springs would tend to retard the annual peat-forming activities to a considerable extent. The spring of 1931, for example, was unusually dry and, in the later part of April and the first weeks of May, when many trees and shrubs, such as hazel, alder, birch, and poplar were in bloom, the ground was still covered with the dead plant remains of the past season. Probably little formation of peat was in progress at that time of the year. Therefore, it might be fair to assume that the comparatively low PF of the muskegs in continental Alberta is due to their low power of catching and preserving pollen grains, probably less than that of the bogs of oceanic western Europe, where the dead season of the bogs is shorter or, in some cases, probably absent altogether.

The low PF may also be connected with the fact that, in spite of the summers being comparatively short, peat in many places is evidently forming very rapidly; furthermore, such trees as aspen and poplar, the pollen of which is not preserved in the bogs at all or else is poorly preserved, play a much more important rôle in Alberta than in most parts of Europe.

Finally, the PF-values may be indicative of changing edaphic conditions through the development of a bog. GODWIN (1934), for example, associates a very low PF with the development of a peat bed into a phase of fen oak-wood when the relatively dry conditions of the forest floor might constitute adequate reason for intense destruction of pollen and consequent low frequencies.

Absolute Pollen Frequency of Peat Samples: — The absolute frequency, or APF, is the total number of pollen grains in one gram (dry-weight) of peat (real APF) or the same number obtainable from one gram of peat by means of a certain method (apparent APF; *cf.* ERDTMAN and ERDTMAN 1933). As indicated by VON POST (1918), APF-figures are hardly ever directly comparable as the samples may represent widely different time-values on account of the varying rate at which peat is formed. Therefore, the APF might reasonably be considered an unnecessary item altogether, without which we could proceed. However, there are at least two reasons for considering it here.

In the first place, it is valuable for a worker in pollen-statistics to obtain a general idea concerning the APF of different kinds of peats, sediments, etc. In this way vague expressions alluding to the supposed APF may be replaced with real facts. In the second place it is important, and in fact indispensable, to secure absolute frequencies when carrying on certain experimental work such as studies of the action of chemicals on the pollen contents of different kinds of pollen-bearing material. Indications of the technical prosecution of such studies are found in the paper by G. and H. ERDTMAN quoted above. They are chiefly as follows:

In order to obtain — for experimental purpose — a homogeneous stock of peat, a slightly humified *Sphagnum* peat was mixed with ten per cent cold sodium hydroxide solution and stirred until a semi-liquid mass was obtained. After a few hours, it was pressed through a metal screen (meshes 4 mm) in order to remove coarse débris. Following acidification with dilute hydrochloric acid (1:1), the peat was filtered by means of suction through a Buechner funnel and washed with water until the filtrate gave only a weak test for chloride ions with silver nitrate. The peat was then spread on glass plates and dried at slightly above room temperature. The dried peat was carefully ground in a mortar and sifted (meshes 0.4 mm). The peat powder thus obtained was then used as a standard peat in the experiments. In this special case, treatment with sodium hydroxide solution was very useful. However, it is not always necessary, and stocks of standard peat may, as a rule, be prepared at once by drying and powdering the material.

One tenth of a gram of standard peat proved to be a sufficient quantity for the experiments. After chemical treatment, the residue was washed, centrifuged, suspended in 5 cc lactophenol or dilute glycerine (1:1), and 0.1 cc of the suspended material transferred to a counting chamber.

The APF must be calculated on dry-weight basis as peat is hygroscopic and, when air-dried, contains much water, given off at high temperatures only. For drying, the peat is heated in an oven at about 110° C until a constant weight is reached. One gram of air-dried, "standardized" *Sphagnum* peat was found to be equivalent to 0.94 g of water-free peat. The apparent APF — after oxidative destruction of the lignin and humic acid components and hydrolysis of the polysaccharides — was about 340,000.

ARMSTRONG, CALVERT, and INGOLD (1930) have also dealt with the calculation of the absolute PF, although in a less elaborate way. In their investigations of the amount of pollen grains in the peat and the percentages of the various tree pollens present, they used a standard weight of peat (3 grams). This quantity was heated for six hours with 10 cc of 10 per cent NaOH at 110° C. Each sample was diluted with distilled water to 20 cc, thoroughly agitated, and allowed to stand for five minutes. The liquid was then decanted and stored for counting. Immediately before counting, this liquid was shaken to obtain a uniform suspension of the pollen, and a standard drop was removed in a capillary tube and placed on a slide. The volume of this drop was determined by a large number of measurements, and it was found to be exactly 0.033 cc. All the known tree pollen grains in this drop were counted, and since all the samples received uniform treatment, the values obtained represented the density of each type of pollen in the sample. The APF of peat samples from the Mourne Mountains (about 500 m above sea level) near Belfast was found to vary from 7.200 to 240.000.

In a peat sample from Greenland, ERDTMAN (1936) found about 28,000 tree pollen grains per gram (calculated on the dry-weight basis), 355 of which were conifer pollen. Evidence in regard to the APF has also been obtained from Norway by ORDING (1934). Special attention was devoted to slightly humified *Sphagnum* peat formed during and after the later part of the Bronze age. Samples were taken at every fifth centimeter and absolute PF-figures were obtained in a simple way adapted to routine work. The technical devices are described in the original paper, which contains a summary in English: " On new methods and facilities concerning pollenanalytical investigations ". The APF-figures obtained seem to be very characteristic, providing indications of minor climatic fluctuations.

Secondary Pollen: — Megascopic fossils, such as *Brasenia* seeds and *Carpinus* nuts, may be washed out from old beds of interglacial age and redeposited in younger (postglacial) layers (*cf.* HARTZ *in* NORDMANN 1910). This, of course, may also happen to microscopic fossils. Thus KRÄUSEL (1920) in diluvial Silesian peats encountered pollen types which were characteristic of the neighbouring Miocene brown-coals. Speaking of spruce pollen in subarctic and arctic deposits in Estonia, THOMSON (1929, footnote p. 83) suggests that these grains may come, partly at least, from interglacial beds.

The fundamental difference between the pollen flora of autochthonous and that of allochthonous sediments is stressed in a later paper by THOMSON (1935, p. 89). The former sediments never contain secondary pollen (*i.e.* pollen grains which previously have been incorporated in another deposit) while the latter, according to circumstances, may or may not contain secondary pollen.

Instructive examples of secondary pollen have been considered by ERNST (1934, p. 287: pollen of *Juglans, Tsuga*, etc., in marine deposits of northwestern Germany) and IVERSEN (1936) who discussed the entire secondary pollen question with particular reference to the pollen flora of Danish arctic clays. JONAS (1935) has found pollen grains of *Alnus* and *Tilia* in the Ortstein layer of certain podsolized soils in northwestern Germany; he regards them as originating from interglacial or interstadial deposits. In Estonian morainic material, THOMSON (1937) has found all the tree and most of the non-tree pollen types of the postglacial deposits. He feels that there would not be much doubt as to their coming from hitherto unknown interglacial beds on the bottom of the Gulf of Finland. Further examples may be found in OVERBECK and SCHNEIDER (1938) and other publications.

According to IVERSEN (*l.c.*), it is often possible to calculate, approximately at least, the admixture of primary and secondary pollen grains and spores in allochthonous sediments. The whole matter may be reduced to its true proportions and will not constitute so serious a problem as might first be expected. However, it demands careful attention, particularly in connection with pollen analyses of subarctic-arctic, interglacial, and still older sediments of lacustrine, brackish, or marine origin.

References: —

AARIO, L., 1932: Pflanzentopographische und paläographische Mooruntersuchungen in N-Satakunta (Comm. Inst. Forest. Fenn. 17:1).

ARMSTRONG, J. I., CALVERT, J., INGOLD, C. T., 1930: The ecology of the mountains of Mourne with special reference to Slieve Donard (Proc. R. Irish Acad., vol. XXXIX B, no. 20).

BRINCKMANN, P., 1934: Zur Geschichte der Moore, Marschen und Wälder Nordwestdeutschlands, III. Das Gebiet der Jade (Bot. Jahrb., vol. LXVI).

DEWERS, F., 1936: Probleme der Flugsandbildung in Nordwestdeutschland (Abh. Nat. Ver. Brem., vol. XXIX).

ERDTMAN, G., 1921: Pollenanalytische Untersuchungen von Torfmooren und marinen Sedimenten in Südwest-Schweden (Ark. f. Bot., vol. 17:10).

—— —— 1924: Studies in the micropalaeontology of postglacial deposits in northern Scotland and the Scotch Isles (Journ. Linn. Soc. (Bot.), vol. XLVI).

—— —— 1935: Pollen statistics (pp. 110–125 *in* WODEHOUSE, R. P.: Pollen Grains. New York, McGraw-Hill).

—— —— 1936: New methods in pollen analysis (Svensk Bot. Tidskr., vol. 29).

—— —— and ERDTMAN, H., 1933: The improvement of pollen-analysis technique (*Ibid.*, vol. 27).

ERNST, O., 1934: Zur Geschichte der Moore, Marschen und Wälder Nordwestdeutschlands, IV. Untersuchungen in Nordfriesland (Schr. Naturw. Ver. Schlesw.-Holst., vol. XX; thesis, Kiel).

GODWIN, H., 1934: Pollen analysis. An outline of the problems and potentialities of the method, Part I (New Phytologist, vol. XXXIII).

IVERSEN, J., 1936: Sekundäres Pollen als Fehlerquelle. Eine Korrektions-methode zur Pollenanalyse minerogener Sedimente (Danm. Geol. Unders., IV. Raekke, vol. 2, no. 15).

JONAS, F., 1935: Klimaschwankungen des Würmglazials und Bodenbildungen des nordwestdeutschen Diluviums (Beitr. zur Emslandskunde, Heft 4).

KEILHACK, K., 1899: Die bodenbildende Tätigkeit der Insekten (Ztschr. D. Geol. Ges., vol. 51).

KIRCHHEIMER, F., 1931: Zur pollenanalytischen Braunkohlenforschung (Braunkohle, Heft 7).

KRÄUSEL, R., 1920: Ein Beitrag zur Kenntnis der Diluvialflora von Ingramsdorf in Schlesien (Neue Jahrb. f. Mineralogie etc.).

LEWIS, I. F. and COCKE, E. C., 1929: Pollen analysis of Dismal swamp peat (Journ. Elisha Mitchell Scient. Soc., vol. 45).

MALMSTRÖM, C., 1923: Degerö Stormyr (Meddel. Statens Skogsförsöksanst., H. 20).

MOTHES, K., ARNOLDT, C. und REDMANN, H., 1937: Zur Bestandesgeschichte ostpreussischer Wälder (Schr. Phys.-ökon. Ges. Königsberg, vol. LXIX).

NORDMANN, V., 1910: Post-Glacial climatic changes in Denmark. Veränderungen des Klimas etc. (Stockholm).

ORDING, A., 1934: Om nye metoder og hjelpemidler ved pollenanalytiske undersøkelser (Meddel. fra det norske Skogsforsøksvesen).

OVERBECK, F. und SCHNEIDER, S., 1938: Mooruntersuchungen bei Lüneburg und bei Bremen und die Reliktnatur von *Betula nana* L. in Nordwestdeutschland (Ztschr. Bot., vol. 33).

von Post, L., 1918: Om skogsträdspollen i sydsvenska torvmosselagerföljder (Forhandl. 16. skand. naturforskermøte, Kristiania 1916).

—— —— 1930: Die Zeichenschrift der Pollenstatistik (Geol. Fören. Förhandl., vol. 51, 1929).

Rudolph, K., 1935: Mikrofloristische Untersuchung tertiärer Ablagerungen im nördlichen Böhmen (Beih. Bot. Centralbl., vol. LIV, Abt. B).

Sandegren, R., 1913: Några drag ur Hornborgasjöns postglaciala utvecklingshistoria (Geol. Fören. Förhandl., vol. 35).

Thomson, P., 1929: Die regionale Entwicklungsgeschichte der Wälder Estlands (Acta et Comm. Univ. Tartuensis, A XVII:2).

—— —— 1937: Der Einfluss des Bruch- und Auenwaldgürtels auf das Pollendiagramm (Schr. Phys.-ökon. Ges. Königsberg, vol. LXIX).

Wesenberg-Lund, C., 1909: Om limnologiens betydning for kvartaergeologien, saerlig med hensyn til postglaciale tidsbestemmelser og temperaturangivelser (Geol. Fören. Förhandl., vol. 31).

Chapter XVI

GEOGRAPHICAL SURVEY — POLLEN-STATISTICAL INVESTIGATIONS IN DIFFERENT COUNTRIES

The great progress of pollen analysis of Quaternary deposits in different countries is shown by the large number of papers which had appeared by the end of 1939 — probably about 2000 which are concerned in some way with pollen analysis. The purpose of the following lines is to offer a cursory geographical conspectus of some of these papers, particularly the earlier pollen-statistical notes together with some of the more important of the later papers. The countries are enumerated alphabetically. The list refers to pre-war conditions and is an excerpt from the following bibliographies: —

ERDTMAN, G., 1927: Literature on Pollen-statistics published before 1927 (Geol. Fören. Förhandl. Stockholm, vol. 49:196–211).

—— —— 1930: Literature on Pollen-statistics published during the years 1927–1929 (*Ibid.*, vol. 52:191–213).

—— —— 1932: Literature on Pollen-statistics and related topics published 1930 and 1931 (*Ibid.*, vol. 54:395–418).

—— —— 1933: Literature on Pollen-statistics and related topics published 1932 and 1933 (*Ibid.*, vol. 56:463–481).

—— —— 1935: Literature on Pollen-statistics and related topics published 1934 (*Ibid.*, vol. 57:261–274).

—— —— 1937: Literature on Pollen-statistics and related topics published 1935 and 1936 (*Ibid.*, vol. 59:157–181).

—— —— 1940: Literature on Pollen-statistics and related topics published 1937–1939 (*Ibid.*, vol. 62:61–97).

GAMS, H., 1927: Die Ergebnisse der pollenanalytischen Forschungen in Bezug auf die Geschichte der Vegetation und des Klimas von Europa (Ztschr. f. Gletscherkunde, vol. XV: 161–190).

—— —— 1929: Nachträge zum Verzeichnis der pollenanalytischen Literatur (*Ibid.*, vol. XVII: 244–248).

—— —— 1929: Zweiter Nachtrag (*Ibid.*, vol. XVII: 389–391).

—— —— 1931: Dritter Nachtrag (*Ibid.*, vol. XIX: 327–334).

—— —— 1933: Vierter Nachtrag (*Ibid.*, vol. XXI: 188–196).

—— —— 1935: Fünfter Nachtrag (*Ibid.*, vol. XXII: 267–274).

Besides the usual data, the bibliographies published by ERDTMAN contain an enumeration of different types of pollen grains and spores which are mentioned in the papers quoted.

Africa: —

DUBOIS, G. et C. (1939).

Asia (Siberia excepted): —

(China) — CHEN (1934), TING (1939);

(Japan) — JIMBO (1932), YAMASAKI (1933?; a publication comprising 128 pages; illustrations of pollen grains; seven diagrams showing the frequencies of the pollen grains of *Abies sachalinensis* and *Picea jezoensis;* Japanese text; no summary in European language);

(Dutch East Indies) — POLAK (1933);

(India, Kashmir) — WODEHOUSE (1935), LUNDQVIST (1936), DEEVEY (1937).

Australia and New Zealand: —

ERDTMAN (1924*e*, 1925*a*; scattered analyses only), CRANWELL and VON POST (1936); Hawaiian peat has been studied by O. SELLING, Stockholm. Very recently Dr. CRANWELL contributed an excellent study of New Zealand conifer pollen (1940, vide p. 144) and a — methodologically very interesting — key for the identification of pollen grains (1942).

Austria: —

Pollen-statistical contributions to the history of Austrian forests have been furnished particularly by FIRBAS (1923 and later), GAMS (1927 and later), and pupils of GAMS, as FEURSTEIN (1933) and VON SARNTHEIN (1936).

Belgium: —

As yet, only a few diagrams have been published; *cf.* ERDTMAN (1927*b*), FLORSCHÜTZ (1937), FLORSCHÜTZ and VAN OYE (1939).

Bulgaria: —

From Mount Vitosa, fourteen diagrams have been published by STOJANOFF and GEORGIEFF (1934).

Canada: —

Papers by AUER (1927, 1930, etc.), BOWMAN (1931), ERDTMAN (1931), etc.

Czechoslovakia: —

Extensive work along pollen-statistical lines has been done by RUDOLPH and FIRBAS (1924). The first results appeared in 1923 (RUDOLPH 1923). *Cf.* also FIRBAS (1927*a*) and SALASCHEK (1935).

Denmark and the Faröes (*see also* Greenland): —

Many papers have been written from 1918 onwards by JESSEN, a pupil of VON POST. The most important of these papers were published in 1920 (pollen-statistics of postglacial deposits) and 1928 (JESSEN and MILTHERS; pollen-statistics of interglacial deposits). Among other workers in this field, IVERSEN (1934*a* and later) should be mentioned.

Estonia: —

Almost all work on pollen-statistics of Estonian peats is that of THOMSON (from 1925 onwards; *cf.* particularly THOMSON 1929).

Finland: —

The results of scattered analyses have been published by LINDBERG in the period 1898–1916, *i.e.* prior to the establishment of pollen analysis as an independent science. The list of pollen-analysts is headed by AUER (1921, 1923 etc.), followed by HELLAAKOSKI (1928 etc.), AARIO (1932 etc.), HYYPPÄ (1932 etc.), BRANDER (1933 etc.), LUKKALA (1933 etc.), BACKMAN (1934 etc.), SAURAMO (1934 etc.), KILPI (1937), AUROLA (1938), and others.

France: —

Scattered analyses have been done by ERDTMAN (1924*d* and later) and G. DUBOIS (1925 and, later, partially in collaboration with C. DUBOIS; *cf. e.g.* G. and C. DUBOIS 1937). First diagrams from Central France: DENIS, ERDTMAN, FIRBAS (1927), FIRBAS (1931); from Corsica: FIRBAS (1927*b*); from the Pyrenees: KELLER (1929); from the Vosges: HATT (1937), OBERDORFER (1937).

Germany: —

The first diagrams were published in 1924 by ERDTMAN (1924*a*; diagrams from northern Germany), BERTSCH and STARK (diagrams from southern Germany). In a number of subsequent papers BERTSCH and STARK have made important contributions (*cf. e.g.* BERTSCH 1931, STARK 1927) to the knowledge of forest history in postglacial times and encouraged others to take up work along the same lines. The stimulating effect of the work " Postglaziale Klimaänderungen und Erdkrustenbewegungen in Mitteleuropa " by GAMS and NORDHAGEN (1923) should also be mentioned although it deals only sparsely with fossil pollen. A total of several hundred papers, or considerably more than in any other country, has been published. Of these only a very few can be referred to here, *viz.* in the order the different workers entered the field of pollen analysis: VON BÜLOW (1926), PAUL and RUOFF (1927, 1932), FIRBAS (1928 and many subsequent publications, *e.g.* 1934, 1935), GISTL (1926; interglacial deposits), HESMER (1928), HUECK (1928), OVERBECK (1928), KOCH (1929), SCHMITZ (1929), GAMS and RUOFF (1929), FRENZEL (1930), SCHROEDER (1930), OBERDORFER (1931), SCHAAF (1931), BAAS (1932), GAMS (1932), JONAS (1932), WILDWANG (1933), GROSS (1933), SCHUBERT (1933), BENRATH (1934), BRINCKMANN (1934), ERNST (1934), F. BERTSCH (1935), GROSCHOPF (1935; *cf.* particularly 1936), JAESCHKE (1935), SCHÜTRUMPF (1935), L. STARK (1936), WERTH and KLEMM (1936), OVERBECK and SCHNEIDER (1938), ZEIDLER (1939), LOSERT (1940), SCHMEIDL (1940), etc.

Great Britain, Ireland, and the North Sea: —

The first diagrams from these areas were made by ERDTMAN (Scotland 1924*b*, Ireland 1924*c*, Dogger Bank 1925*b*, England 1926; summary in papers published 1928*a* and 1929). Among other contributors are: RAISTRICK (from 1932 on), Dr. and Mrs. GODWIN (from 1933 on; *cf.* particularly H. GODWIN 1940), JESSEN (from 1934 on: Ireland); MITCHELL (1940).

Greenland: —

IVERSEN (1934*b*).

Hungary: —

ZOLYOMI (1931), KINZLER (1936).

Iceland: —

THORARINSSON (1941).

Italy: —

Diagrams from northern Italy have been published by Fischer and Lorenz (1931), Keller (1931), and Dalla Fior (since 1932); from the Apennines, etc., by Chiarugi (1935), Marchetti (1936), and others.

Latvia: —

Papers by P. Galenieks and M. Galenieks (Linin) from 1925 on; *cf.* particularly M. Galenieks (1935). A recent contribution has been made by Gilbert (1939).

Lithuania: —

Gams (1929; interglacial material); Thomson (1931).

Netherlands: —

First diagrams by Erdtman (1928*b;* about the same time, investigations were also undertaken by van Baren). Subsequently work along these lines has been carried out by B. Polak (1929), and, particularly, by Florschütz (from 1930) and his pupils. A thesis by Vermeer-Louman (1934) also deals with " moor-log ", or submerged peats, from the North Sea.

Norway: —

Earliest investigations by Holmsen (1920); first diagrams in a note by Erdtman (1927*a*). *Cf.* also Ording (1934) and particularly the papers by Faegri (1936, 1941).

Poland: —

First analyses by Lilpop and Passendorfer (1925) and Szafer (1925); first diagrams by Szafran (1926). Other workers along pollen-statistical lines include Tolpa (since 1927), Dyakowska and Trela (since 1928), Thomaschewski and Tymrakiewicz (since 1929), Dabkowska (since 1932), etc.

Rumania: —

First diagrams published by Peterschilka (1928) and Solacolu (1928). Chief contributions made by Pop (since 1929; *cf.* particularly Pop 1932).

Russia, incl. Siberia: —

First diagrams published by Erdtman (Erdtman and Hultén 1924; Kamtchatka) and Gerasimov (1924; European Russia). For several years, Dokturowsky (from 1925 on) was the most productive pollen analyst. Other workers are: Neustadt (since 1927), Anufriew (*cf.* particularly 1931), Markow and Zerow (*cf.* particularly Zerow 1938), and Sukatchew (since 1932).

South America: —

von Post (1930), Auer (1933), Salmi (1941).

Sweden: —

First diagrams by VON POST (1916, 1918). The most important among VON POST'S subsequent papers were printed in 1924, 1926, 1928 (*a* and *b*), and 1930 (*a* and *b*); besides these, there are many papers and descriptions of maps with Swedish text only; many analyses have been done by Mrs. S. VON POST. Other workers are: TEILING (a few notes in Geol. Fören. Förhandl. and Svensk Botanisk Tidskrift 1909 and later), SANDEGREN (since 1916; first diagram 1920; *cf.* also 1924), HALDEN (since 1917), SUNDELIN (since 1917; first diagrams 1919), ERDTMAN (since 1920; *cf.* particularly 1921), LUNDQVIST (since 1920; *cf.* particularly 1925, 1927, 1928), GRANLUND (since 1924; *cf.* particularly 1932), MALMSTRÖM (since 1923), BOOBERG (since 1924; *cf.* particularly 1930), ASSARSSON, ISBERG, and THOMASSON (since 1927), LARSSON (since 1932), NILSSON (1935), FROMM (1938), and others.

Switzerland: —

First diagrams by SPINNER (1925) and TROLL (HÄRRI 1925). Further contributions by KELLER (since 1926; *cf.* particularly 1928), FURRER (1927), LÜDI (since 1929), RYTZ (since 1931), and others.

United States of America: —

First diagram by DRAPER (1928); further contributions by SEARS (since 1930; *cf.* particularly 1935), VOSS (since 1931), HOUDEK (since 1932), L. R. WILSON (*cf.* particularly 1938), H. P. HANSEN (since 1937), etc. — An enumeration of very recent literature will be found in the *Pollen Analysis Circular*, No. 1 (May 1943), issued by Dr. SEARS of Oberlin College. Dr. VERDOORN sent me a copy of this by air mail. Practically none of the papers listed there are at present available in Sweden. Dr. VERDOORN has therefore listed them, mostly at the end of this book, together with a number of other recent references.

Yugoslavia: —

FIRBAS (1923) and ČERNJAVSKI (*cf.* particularly 1937).

References: —

AARIO, L., 1932: Pflanzentopographische und paläographische Mooruntersuchungen in N-Satakunta (Comm. Inst. Forest. Fenn., 17:1).

ANUFRIEW, G. I., 1931: Der Aufbau der Torfmoore des Leningrader Bezirks (Arb. d. wiss. Torfinst., vol. 9).

AUER, V., 1921: Zur Kenntnis der Stratigraphie der mittelösterbottnischen Moore (Acta Forest. Fenn. 18).

—— —— 1923: Moorforschungen in den Vaaragebieten von Kuusamo und Kuolajärvi (Comm. Inst. quaest. forest. Fenn., 6).

—— —— 1927: Stratigraphical and morphological investigations of peat bogs of southeastern Canada (*Ibid.*, 12).

—— —— 1930: Peat Bogs in southeastern Canada (Geol. Surv. Canada, Mem. 162).

—— —— 1932: Verschiebungen der Wald- und Steppengebiete Feuerlands in postglazialer Zeit (Acta Geogr., 5:2).

AUROLA, E., 1938: Die postglaziale Entwicklung des südwestlichen Finnlands (Bull. Comm. Géol. Finl., no. 121).

BAAS, J., 1932: Eine frühdiluviale Flora im Mainzer Becken (Ztschr. Bot., vol. 25).

BACKMANN, A. L., 1934: Über die Vorgeschichte des Åländischen Waldes (Acta Forest. Fenn., 40).

BENRATH, W., 1934: Untersuchungen zur Pollenstatistik und Mikrostratigraphie von Tonen und Torfen in Randgebieten des Kurischen Haffs unter Berücksichtigung methodischer Fragen (Königsberg).
BERTSCH, F., 1935: Das Pfrunger Ried und seine Bedeutung für die Florengeschichte Südwestdeutschlands (Beih. Bot. Centralbl., vol. LIV).
BERTSCH, K., 1924: Paläobotanische Untersuchungen im Reichermoos (Jahresh. Ver. Naturk. Württemb., 80).
—— —— 1931: Paläobotanische Monographie des Federseerieds (Bibl. Bot. 103).
BOOBERG, G., 1924: Julamossen-Ymsen-Fredsbergsmossen (Svenska Mosskulturfören. Tidskr.).
—— —— 1930: Das Gisselåsmoor. Eine pflanzensoziologische und entwicklungsgeschichtliche Monographie über ein nordschwedisches Silurmoor (Norrländskt Handbibl., XII).
BOWMAN, P., 1933: Study of a peat bog near the Matamek River, Quebec, Canada, by the method of pollen analysis (Ecology, vol. XII).
BRANDER, G., 1933: Über den Dopplerit von Haapamäki in Finnland (C. R. Soc. géol. Finl., no. 6).
BRINCKMANN, P., 1934: Zur Geschichte der Moore, Marschen und Wälder Nordwestdeutschlands, III. Das Gebiet der Jade (Bot. Jahrb., vol. LXVI).
BROCHE, W., 1929: Pollenanalytische Untersuchungen an Mooren des südlichen Schwarzwaldes und der Baar (Ber. Naturf. Ges. Freiburg, vol. XXIX).
ČERNJAVSKI, P., 1937: Pollenanalytische Untersuchungen der Sedimente des Vlasinamoores in Serbien (Beih. Bot. Centralbl., vol. LVI).
CHEN, Y.-C. A., 1934: A preliminary study of the evidence for the taxonomic value of pollen grains in determining the local plants (Lingnan Sci. J. 13:89).
CHIARUGI, A., 1935: Resultati dell'analisi pollinica della torbiera del Lago del Greppo nell'Appennino Etrusco (N. Giorn. Bot. Ital., vol. XLII).
CRANWELL, L. M., 1942: New Zealand Pollen Studies, I. Key to the pollen grains of families and genera in the native flora (Auckland Inst. and Mus. Rec. 2:280–308).
—— —— and VON POST, L., 1936: Post-pleistocene pollen diagrams from the southern hemisphere, I. New Zealand (Geogr. Annaler).
DĄBKOWSKA, F., 1932: Les tourbières de la vallée de la Łania (Acta Soc. Bot. Polon., vol. IX).
DALLA FIOR, G., 1932: Analisi polliniche di torbe e depositi lacustri della Venezia Tridentina (Mem. Mus. Stor. Nat. Venez. Trident., vol. I).
DEEVEY, E. S., 1937: Pollen from interglacial beds in the Pang-gong valley and its climatic interpretation (Amer. Journ. Sci., vol. XXXIII).
DENIS, M., ERDTMAN, G. et FIRBAS, F., 1927: Premières analyses effectuées dans les tourbières auvergnates (Arch. Bot., T. 1, Caen).
DOKTUROWSKY, W. S., 1925: Über die Stratigraphie der russischen Torfmoore (nebst Angaben zur interglazialen Flora) (Geol. Fören. Förhandl., vol. 47).
DRAPER, P., 1928: A demonstration of the technique of pollen analysis (Proc. Okla. Acad. Sci., VIII).
DUBOIS, G., 1925: Examen pollinique d'une tourbe de Lille (Ann. Soc. Géol. Nord, t. XLIX).
—— —— et DUBOIS, C., 1937: Étude paléobotanique de tourbières de la région parisienne (Bull. Soc. géol. France, vol. VII).
—— —— —— —— 1939: Caractères micropaléobotaniques d'une tourbe du Togo (C. R. Acad. Sci., vol. 208).
DYAKOWSKA, J., 1928: Histoire de la tourbière "na Czerwonem" (Spraw. Kom. Fizjogr. Polsk. Akad. Um., t. LXIII).
ERDTMAN, G., 1920: Einige geobotanische Resultate einer pollenanalytischen Untersuchung von südwestschwedischen Torfmooren (Svensk Bot. Tidskr., vol. 14).
—— —— 1921: Pollenanalytische Untersuchungen von Torfmooren und marinen Sedimenten in Südwest-Schweden (Ark. f. Bot., vol. 17).
—— —— 1924*a*: Pollenstatistische Untersuchung einiger Moore in Oldenburg und Hannover (Geol. Fören. Förhandl., vol. 46).
—— —— 1924*b*: Studies in the Micropalaeontology of Postglacial Deposits in Northern Scotland and the Scotch Isles (Linn. Soc. Journ. (Bot.), vol. XLVI).
—— —— 1924*c*: Mitteilungen über einige irische Moore (Svensk Bot. Tidskr., vol. 18).
—— —— 1924*d*: Analyses from Brittany (Finistère) (Geol. Fören. Förhandl., vol. 46).
—— —— 1924*e*: Peat from the Chatham Islands and the Otago District, New Zealand (*Ibid.*).
—— —— 1925*a*: Peat from the Snares, Antipod Island etc. (Svensk Bot. Tidskr., vol. 19, p. 528).
—— —— 1925*b*: Microanalyses of "moorlog" from the Dogger Bank (Essex Naturalist, vol. XXI; *cf.* also Svensk Bot. Tidskr., vol. 19, pp. 115, 116).
—— —— 1926: On the immigration of some British trees (Journ. Bot.).
—— —— 1927*a*: Tapergränsen på Jaederen och dess relation till skogarnas historia i sydvästra Norge (Svensk Bot. Tidskr., vol. 21).
—— —— 1927*b*: Vestiges de l'histoire quaternaire récente des forêts belges (Acad. Roy. Belg., t. XIII).

—— —— 1928*a*: Studies in the Postarctic History of the forests of Northwestern Europe, I. Investigations in the British Isles (Geol. Fören. Förhandl., vol. 50).

—— —— 1928*b*: Studien über die postarktische Geschichte der nordwesteuropäischen Wälder, II. Untersuchungen in Nordwestdeutschland und Holland (*Ibid.*, vol. 50).

—— —— 1929: Some aspects of the postglacial history of British forests (Journ. Ecol., vol. XVII).

—— —— 1931: Worpswede-Wabamun. Ein pollenstatistisches Menetekel (Abh. Nat. Ver. Brem., vol. XXVIII).

—— —— et Hultén, E., 1924: Observations sur quelques tourbières kamtchatiques (Geol. Fören. Förhandl., vol. 46).

Ernst, O., 1934: Zur Geschichte der Moore, Marschen und Wälder Nordwest-Deutschlands, IV. Untersuchungen in Nordfriesland (Schr. Naturwiss. Ver. Schlesw. Holst., vol. XX).

Faegri, K., 1936: Quartärgeologische Untersuchungen im westlichen Norwegen, I. Über zwei präboreale Klimaschwankungen im südwestlichsten Teil (Bergens Mus. Årbok 1935, no. 8).

—— —— 1941: Quartärgeologische Untersuchungen im westlichen Norwegen, II. Zur spätquartären Geschichte Jaerens (Bergens Mus. Årbok 1939–40, Naturvidensk. rekke, no. 7).

Feurstein, P., 1933: Geschichte des Viller Moores und des Seerosenweihers an der Lanser Köpfen bei Innsbruck (Beih. Bot. Centralbl., vol. LI).

Firbas, F., 1923: Pollenanalytische Untersuchungen einiger Moore der Ostalpen (Lotos, vol. 71).

—— —— 1927*a*: Die Geschichte der nordböhmischen Wälder und Moore seit der letzten Eiszeit (Untersuchungen im Polzengebiet) (Beih. Bot. Centralbl., vol. XLIII).

—— —— 1927*b*: Beiträge zur Geschichte der Moorbildungen und Gebirgswälder Korsikas (*Ibid.*, vol. XLIV).

—— —— 1928: Über die Flora und das interglaziale Alter des Helgoländer Süsswassertöcks (Senckenbergiana, vol. 10).

—— —— 1931: Über die Waldgeschichte der Süd-Sevennen und über die Bedeutung der Einwanderungszeit für die nacheiszeitliche Waldentwicklung der Auvergne (Planta, vol. 13).

—— —— 1934: Zur spät- und nacheiszeitlichen Vegetationsgeschichte der Rheinpfalz (Beih. Bot. Centralbl., vol. LII).

—— —— 1935: Die Vegetationsentwicklung des mitteleuropäischen Spätglazials (Bibl. Bot., H. 112).

Fischer, O. und Lorenz, A., 1931: Pollenanalytische Untersuchungen an Mooren der Südostalpen (Zeitschr. f. Bot., vol. 24).

Florschütz, F., 1937: De pollenanalyse als hulpmiddel bij de studie van de geschiedenis der bosschen (Nederl. Boschbouw-Tijdschr.).

—— —— et van Oye, 1939: Recherches analytiques de pollen dans la région des Hautes-Fagnes Belges (Biol. Jaarb. Dodonaea, 6. jaarg.).

Frenzel, H., 1930: Entwicklungsgeschichte der sächsichen Moore und Wälder seit der letzten Eiszeit (Abh. Sächs. Geol. Landesamt, H. 9).

Fromm, E., 1938: Geochronologisch datierte Pollendiagramme und Diatomeenanalysen aus Ångermanland (Geol. Fören. Förhandl., vol. 60).

Furrer, E., 1927: Pollenanalytische Studien in der Schweiz (Vierteljahrsschr. Naturf. Ges. Zürich, vol. 72).

Galenieks, M., 1935: The development of bogs and forests in the post-glacial period in Latvia (Acta Univ. Latv.).

Galenieks, P., 1925: Interglacial peat-bed at Dēsele, Kurzeme (Latvia) (*Ibid.*, vol. XII).

Gams, H., 1927: Die Geschichte der Lunzer Seen, Moore und Wälder (Int. Rev. Hydrobiol. u. Hydrogr., vol. XVIII).

—— —— 1929: Ein interglazialer Tannenfund aus Litauen (Sukatschew-Festschr.; Russ.).

—— —— 1932: Zur Geschichte der Moore der Kurischen Nehrung und des Samlands (Schr. Phys.-ökon. Ges. Königsb., vol. LXVII).

—— —— und Nordhagen, R., 1923: Postglaziale Klimaänderungen und Erdkrustenbewegungen in Mitteleuropa (Landeskundl. Forsch. Geogr. Ges. München, H. 25).

—— —— und Ruoff, S., 1929: Geschichte, Aufbau und Pflanzendecke des Zehlaubruches (Schr. phys.-ökon. Ges. Königsb., vol. LXVI).

Gerasimov, D. A., 1924: Einige wissenschaftliche und praktische Ergebnisse geobotanischer Mooruntersuchungen (Torfwirtschaft, Moscow; Russ.).

Gilbert, M., 1939: Pollenanalytische Untersuchungen im Gebiet der reichen Moränenböden in NO-Kurzeme (Korr.-bl. Naturf.-Ver. Riga, Vol. LXIII).

Gistl, R., 1928: Die letzte Interglazialzeit der Lüneburger Heide pollenanalytisch betrachtet (Bot. Arch., vol. 21).

Godwin, H., 1940: Studies of the Post-Glacial History of British Vegetation, III. Fenland pollen diagrams; IV. Post-Glacial changes of relative land- and sea-level in the English Fenland (Phil. Trans. Roy. Soc., ser. B., no. 570, vol. 230).

—— —— and M. E., 1933: Pollenanalyses of Fenland peats at St. Germans, near King's Lynn (Geol. Mag., vol. LXX).

—— —— 1940: Pollen analysis and forest history of England and Wales (New Phytologist 39:370–400; 13 Textfig.)
GRANLUND, E., 1922: Torvmarker. Beskrivning till kartbladet Mjölby (Sveriges Geol. Unders., ser. Aa, no. 150).
—— —— 1932: Die Geologie der schwedischen Hochmoore, ihre Bildungsbedingungen, Entwickelungsgeschichte und Verbreitung, sowie der Zusammenhang von Hochmoorbildung und Versumpfung (*Ibid.*, ser. C, no. 373).
GROSCHOPF, P., 1936: Die postglaziale Entwicklung des Grossen Plöner Sees in Ostholstein auf Grund pollenanalytischer Sedimentuntersuchungen (Arch. Hydrobiol., vol. XXX).
GROSS, H., 1933: Zur Frage des Weberschen Grenzhorizontes in den östlichen Gebieten der ombrogenen Moorregion (Beih. Bot. Centralbl., vol. LI).
HALDEN, B., 1917: Om torvmossar och marina sediment inom norra Hälsinglands litorinaområde (Sveriges Geol. Unders., ser. C, no. 280).
HÄRRI, H., 1925: Prähistorisches und Naturwissenschaftliches vom Hallwilersee (Mitt. Aarg. naturf. Ges., H. XVII).
HATT, J.-P., 1937: Contribution à l'analyse pollinique des tourbières du Nord-Est de la France (Bull. Serv. Carte Géol. Als. Lorr., vol. 4).
HELLAAKOSKI, A., 1928: Puulan järvi ryhmän kehityshistoria (Fennia, vol. 51).
HESMER, H., 1928: Die Waldgeschichte der Nacheiszeit des nordwestdeutschen Berglandes auf Grund von pollenanalytischen Mooruntersuchungen (Zeitschr. Forst- u. Jagdwesen).
HOLMSEN, G., 1920: Naar indvandret granskogen til Kristianiatrakten? (Tidsskr. f. Skogbruk).
HOUDEK, P., 1932: Pollen statistics for two Indiana bogs (Proc. Ind. Acad. Sci., vol. 42, 1933).
HUECK, K., 1928: Zur Kenntnis der Hochmoore des Thüringer Waldes (Beitr. z. Naturdenkmalpfl., vol. XII).
HYYPPÄ, E., 1932: Die postglazialen Niveauverschiebungen auf der karelischen Landenge (Ann. Acad. Sc. Fenn., ser. A, tom. XXXVII).
ISBERG, O., 1927: Beitrag zur Kenntnis der postarktischen Landbrücke (Geogr. Ann.).
IVERSEN, J., 1934*a*: Fund af Vildhest (*Equus caballus*) fra Overgangen mellem Sen- og Postglacialtid i Danmark (Meddel. Dansk Geol. Foren., vol. 8).
—— —— 1934*b*: Moorgeologische Untersuchungen auf Grönland (*Ibid.*).
JAESCHKE, J., 1935: Zur Waldgeschichte des Odenwaldes und des Taunus (Forstwiss. Centralbl., vol. 57).
JESSEN, K., 1918: Bidrag til Vegetationens Historie i Randers Fjord-dal (*in* "Randers Fjords Naturhistorie", by A. C. JOHANSEN).
—— —— 1920: Bog investigations in North East Sjaelland (Danm. Geol. Unders., II. Raekke, no. 34).
—— —— 1934: Pollenstatistical data (*in* MAHR, A., A wooded cauldron from Altartate, Co. Monaghan. Proc. Roy. Irish Acad., vol. XLII).
—— —— and MILTHERS, V., 1928: Stratigraphical and paleontological studies of interglacial fresh-water deposits in Jutland and northwest Germany (Danm. Geol. Unders., II. Raekke, no. 48).
—— —— and RASMUSSEN, R., 1922: Section of a bog in the Faröe Islands (*Ibid.*, IV. Raekke, vol. 1).
JIMBO, T., 1932: Pollen-Analytical Studies of Peat formed on Volcanic Ash (Sci. Rep. Tôhoku Imp. Univ., 4. ser., vol. VII).
JONAS, F., 1932: Die ältesten Urkunden unserer Heimat (Mein Emsland).
KELLER, P., 1926: Pollenanalytische Untersuchungen an einigen thurgauischen Mooren (Mitt. Thurg. Naturforsch. Ges.).
—— —— 1928: Pollenanalytische Untersuchungen an Schweizer-Mooren und ihre florengeschichtliche Deutung (Veröff. Geobot. Inst. Rübel, H. 5).
—— —— 1929: Analyse pollinique de la Tourbière de Pinet (Arch. Bot., t. III).
KILPI, S., 1937: Das Sotkamo-Gebiet in spätglazialer Zeit (Bull. Comm. Géol. Finl., no. 117).
KOCH, H., 1929: Paläobotanische Untersuchungen einiger Moore des Münsterlandes (Beih. Bot. Centralbl., vol. XLVI).
LARSSON, C., 1932: Fossilt pollen av *Abies alba* och *Pinus cembra* (?) i Skåne (Geol. Fören. Förhandl., vol. 54).
LILPOP, J. and PASSENDORFER, E., 1925: The interglacial formations near Sulejów on the Pilica (Bull. Serv. Géol. Pologne, vol. III).
LINDBERG, H., 1896: Botanisk undersökning af Isosuo mosse i Sakkola socken (Finska Mosskulturföreningens Årsbok) (and other papers in later volumes of the same journal).
LOSERT, H., 1940: Beiträge zur spät- und nacheiszeitlichen Vegetationsgeschichte Innerböhmens, I–III (Beih. Bot. Centralbl., vol. LX, Abt. B).
LÜDI, W., 1929: Das Siehenmoos bei Eggiwil im Emmental und seine Geschichte (Mitt. Naturforsch. Ges. Bern).
LUKKALA, O. J., 1933: Vollzieht sich gegenwärtig Versumpfung von Waldboden? (Comm. Inst. Forest. Fenn., 19).
LUNDQVIST, G., 1920: Pollenanalytiska åldersbestämningar av flygsandsfält i Västergötland (Svensk Bot. Tidskr., vol. 14).

—— —— 1925: Utvecklingshistoriska insjöstudier i Sydsverige (Sveriges Geol. Unders. Ser. Aa, no. 330).
—— —— 1927: Bodenablagerungen und Entwicklungstypen der Seen (Die Binnengewässer, vol. II).
—— —— 1928: Studier i Ölands myrmarker (Sveriges Geol. Unders., Ser. C, no. 353).
MALMSTRÖM, C., 1923: Degerö Stormyr. En botanisk, hydrologisk och utvecklingshistorisk undersökning över ett nordsvenskt myrkomplex (Meddel. Statens Skogsförsöksanstalt, h. 20).
MARCHETTI, M., 1936: Analisi pollinica della torbiera di Campotosto (Appenino Abruzzese) (N. Giorn. Bot. Ital., vol. XLIII).
MARKOW, K., 1931: Die Ausbildung des Reliefs im nordwestlichen Teil des Leningrader Gebiets (Trav. Geol. Prosp. Serv., no. 177).
MITCHELL, G. F., 1940: Studies in Irish Quaternary deposits: some lacustrine deposits near Dunshaughlin, county Meath (Proc. R. Irish Acad., vol. XLVI, sect. B, no. 2).
NEUSTADT, M. I., 1927: Die Entwicklungsgeschichte des Sees Somino. Versuch der Synchronisation der Seeablagerungen (Arch. f. Hydrobiol. vol. XVIII).
NILSSON, T., 1935: Die pollenanalytische Zonengliederung der spät- und postglazialen Bildungen Schonens (Geol. Fören. Förhandl., vol. 57).
OBERDORFER, E., 1931: Die postglaziale Klima- und Vegetationsgeschichte des Schluchsees (Schwarzwald) (Ber. Naturf. Ges. Freib., vol. XXXI).
—— —— 1937: Zur spät- und nacheiszeitlichen Vegetationsgeschichte des Oberelsasses und der Vogesen (Zeitschr. f. Bot., vol. 30).
ORDING, A., 1934: On new methods and facilities concerning pollenanalytical investigations (Medd. norske Skogsforsøksvesen).
OVERBECK, F., 1928: Studien zur postglazialen Waldgeschichte der Röhn (Zeitschr. f. Bot., vol. 20).
—— —— und SCHNEIDER, S., 1938: Mooruntersuchungen bei Lüneburg und bei Bremen und die Reliktnatur von *Betula nana* L. in Nordwestdeutschland (*Ibid.*, vol. 33).
PAUL, H. und RUOFF, S., 1927: Pollenstatistische und stratigraphische Mooruntersuchungen im südlichen Bayern, I. Teil (Bayer. Bot. Ges., vol. XIX).
PETERSCHILKA, F., 1928: Pollenanalyse einiger Hochmoore Neurumäniens (Ber. Deutsch. Bot. Ges., vol. XLVI).
POLAK, B., 1929: Een onderzoek naar de botanische samenstelling van het hollandsche veen (Amsterdam).
—— —— 1935: Über Torf und Moor in Niederländ. I. Indien (Verh. Akad. Wet. Amsterdam, d. XXX).
POP, E., 1929: Pollenanalyse einiger Moore der Ostkarpathen (Dorna-Lucina) (Bul. Grad. Bot. Cluj, vol. IX).
—— —— 1932: Beiträge zur quaternären Pflanzengeschichte Siebenbürgens (*Ibid.*, vol. XII).
VON POST, L., 1916: Einige südschwedischen Quellmoore (Bull. Geol. Inst. Upsala, vol. XV).
—— —— 1918: Skogsträdpollen i sydsvenska torvmosselagerföljder (Forhandl. 16. skand. naturforskermøte 1916, Kristiania).
—— —— 1924: Some features of the regional history of the forests of southern Sweden in post-arctic time (Geol. Fören. Förhandl., vol. 46).
—— —— 1926: Einige Aufgaben der regionalen Moorforschung (Sveriges Geol. Unders., Ser. C, no. 337).
—— —— 1928*a*: Pollenanalyse (Reallexikon der Vorgeschichte, Berlin).
—— —— 1928*b*: The geological age of the Svea River. A pollen-analytical study Ancylus Time Geography (*Ibid.*, Ser. C, no. 347).
—— —— 1930*a*: Die Zeichenschrift der Pollenstatistik (Geol. Fören. Förhandl., vol. 51, 1929).
—— —— 1930*b*: Die postarktische Geschichte der europäischen Wälder nach den vorliegenden Pollendiagrammen (C. R. Congr. Int. rech. forest., Stockholm 1929).
—— ——, VON WALTERSTORFF, und LINDQVIST, S., 1925: Die Bronzezeitliche Mantel vom Gerumsberget in Västergötland (Monografiserie utg. av K. Vitt. Hist. Antikv. Akad. Stockholm, no. 15).
RAISTRICK, A., 1932: The pollen analysis of peat (Naturalist).
RUDOLPH, K., 1929: Die Ergebnisse der bisherigen Mooruntersuchungen in Böhmen (Beih Bot. Centralbl., vol. XLV).
—— —— und FIRBAS, F., 1923: Pollenanalytische Untersuchungen böhmischer Moore (Ber. Deutsch. Bot. Ges., vol. XI).
—— —— —— —— 1924: Die Hochmoore des Erzgebirges. Ein Beitrag zur postglazialen Waldgeschichte Böhmens (Beih. Bot. Centralbl., vol. XLI).
RYTZ, W., 1931: *in* BECK, RYTZ, STEHLIN und TSCHUMI: Der neolithische Pfahlbau Thun (Mitt. Naturf. Ges. Bern 1930).
SALASCHEK, H., 1935: Paläofloristische Untersuchungen mährisch-schlesischer Moore (Beih. Bot. Centralbl., vol. LIV).
SALMI, M., 1941: Die postglazialen Eruptionsschichten Patagoniens und Feuerlands (Ann. Acad. Scient. Fenn., Ser. A, III, 2).
SANDEGREN, R., 1916: Hornborgasjön. En monografisk framställning av dess postglaciala utvecklingshistoria (Sveriges Geol. Unders., Ser. Ca, no. 14).

—— —— 1920: *Najas flexilis* i Fennoskandia under postglacialtiden (Svensk Bot. Tidskr., vol. 14).
—— —— 1924: Ragundatraktens postglaciala utvecklingshistoria enligt den subfossila florans vittnesbörd (Sveriges Geol. Unders., Ser. Ca, no. 12).
von Sarnthein, R., 1936: Moor- und Seeablagerungen aus den Tiroler Alpen in ihrer waldgeschichtlichen Bedeutung, I. Teil: Brennergegend und Eisacktal (Beih. Bot. Centralbl., vol. IV).
Sauramo, M., 1934: Zur spätquartären Geschichte der Ostsee (C. R. Soc. géol. Finl., no. 8).
Schaaf, G., 1931: Blütenstaubzählungen an Hohenloher Mooren (Veröff. Staatl. Stelle f. Naturschutz Württemb. Landesamt f. Denkmalpfl., H. 8).
Schmeidl, H., 1940: Beitrag zur Frage des Grenzhorizontes im Sebastiansberger Hochmoor (Beih. Bot. Centralbl., vol. LX, Abt. B).
Schmitz, H., 1929: Beiträge zur Waldgeschichte des Vogelsbergs (Planta, vol. 7).
Schubert, E., 1933: Zur Geschichte der Moore, Marschen und Wälder Nordwestdeutschlands, II. Das Gebiet an der Oste und Niederelbe (Mitt. Prov.-st. f. Naturdenkmalpfl., Hannover, H. 4).
Sears, P., 1930: A record of post-glacial climate in northern Ohio (Ohio Journ. Sci., vol. XXX).
—— —— 1930: Common Fossil Pollen of the Erie Basin (Bot. Gaz. 89:95).
—— —— 1935: Types of North American pollen profiles (Ecology, vol. XVI).
Solacolu, T., 1928: Aplicarea analizei polenului la turbăriile din România (Cult. Natională Bucuresti).
Spinner, H., 1925: Analyse pollinique de la tourbe de deux marais de la Vallée de la Brévine (Bull. soc. neuch. Sci. nat., vol. L).
Stark, P., 1924: Pollenanalytische Untersuchungen an zwei Schwarzwaldhochmooren (Zeitschr. f. Bot., vol. XVI).
—— —— 1927: Die Moore des Badischen Bodenseegebiets, II (Ber. Naturf. Ges. Freib., vol. XXVIII).
Stojanoff, N. und Georgieff, T., 1934: Pollenanalytische Untersuchungen auf dem Vitosa-Gebirge (Spis. Bulg. Akad. Nauk., vol. XLVII).
Sukatschew, W., 1932: Die phytopaläontologische Irtysch-Expedition (Exp. Acad. USSR 1931).
Sundelin, U., 1917: Fornsjöstudier inom Stångåns och Svartåns vattenområden (Sveriges Geol. Unders., Ser. Ca, no. 16).
—— —— 1919: Über die spätquartäre Geschichte der Küstengegenden Östergötlands und Smålands (Bull. Geol. Inst. Upsala, vol. XVI).
Szafer, W., 1925: Über den Charakter der Flora und des Klimas der letzten Interglazialzeit bei Grodno in Polen (Bull. Acad. Polon. Sc. Lettr.).
Szafran, B., 1926: Der Bau und das Alter des Moores von Pakoslaw bei Ilza in Mittelpolen (*Ibid.*).
Thomaschewski, M., 1929: Pollenanalytische Untersuchung des Torfmoores Kalmusen in Pomerellen (*Ibid.*).
Thomasson, H., 1927: Baltiska tidsbestämningar och baltisk tidsindelning vid Kalmarsund (Geol. Fören. Förhandl., vol. 49).
Thomson, P., 1925: Die Pollenflora der Torflager in Estland (Bot. Archiv).
—— —— 1929: Die regionale Entwickelungsgeschichte der Wälder Estlands (Acta Comm. Univ. Tart., A XVII:2).
—— —— 1931: Beitrag zur Stratigraphie der Moore und zur Waldgeschichte S. W. Litauens (Geol. Fören. Förhandl., vol. 53).
Thorarinsson, L., 1941: Mot eld och is (Ymer, vol. 61, Stockholm).
Tołpa, S., 1927: Pollenanalytische Studien über Janower-Torfmoor (Kosmos, vol. 52).
Trela, J., 1928: Pollenanalytische Untersuchung des Torfmoores bei Wolbrom in Mittelpolen (Acta Soc. Bot. Polon., vol. 3).
Tymrakiewicz, W., 1929: Pollenanalytische Studien über Biłohorszcza-Torfmoor (Kosmos, vol. 53).
Vermeer–Louman, G., 1934: Pollen-analytisch onderzoek van den westnederlandschen bodem (Amsterdam).
Wildwang, D., 1933: Versuch einer stratigraphischen Eingliederung der ostfriesischen Marschmoore ins Alluvialprofil und die sich dabei ergebenden Folgerungen in Bezug auf Bodenschwankungen (Jahrb. Preuss. Geol. Landesanst., vol. 54).
Wilson, L. R., 1938: The postglacial history of vegetation in northwestern Wisconsin (Rhodora, vol. 40).
Wodehouse, R., 1935: The pleistocene pollen of Kashmir (Mem. Connecticut Acad., vol. IX).
Yamasaki (1933?): *cf.* Asia, p. 205.
Zeidler, H., 1939: Untersuchungen an Mooren im Gebiet des mittleren Mainlaufs (Zeitschr. f. Bot., vol. 34).
Zerow, D. K., 1931: Fossile Torflager im Dnieprufergebiet, I. Interglaziale Torflager (Die Quartärperiode, Lief. 3).
—— —— 1938: Die Moore der Ukrainischen SSR. Vegetation und Stratigraphie (Akad. nauk USSR, Kiew).

Chapter XVII

TERTIARY DEPOSITS

The first notes on pollen grains in Tertiary deposits, their morphology and identification were published more than a century ago by EHRENBERG (1838) and GÖPPERT (1841). However, the first attempt to study Tertiary pollen from a pollen-analytical point of view was made by LAGERHEIM. He examined samples of " brown coal-mud " from Jutland, collected by the Danish geologist N. HARTZ. After treating the mud with a boiling NaOH-solution LAGERHEIM was able to isolate — besides " many queer pollen grains which I never have seen in Quaternary deposits ", as he says in a letter to HARTZ — pollen grains of *Abies* (or *Picea*), *Betula*, *Corylus*, *Pinus*, *Tilia*, *Ulmus*, *Caryophyllaceae*, *Ericaceae*, *Gramineae*, and *Umbelliferae*, as well as mycelia and spores of *Pyrenomycetes* and remains of an alga, *Botryococcus* (HARTZ 1909, pp. 54, 55). The investigation was only qualitative and, cursory as it was, its importance lies in the fact that it was shown how easy a task it is to isolate and determine, at least to some extent, the microfossils of certain brown coals.

It is surprising in view of these discoveries that the possibility of their wide application was not realized, and that the progress was at first so slow. It was not until 1928 that two palaeobotanists and specialists in brown coal study [of whom at least one had previously contributed to the knowledge of Tertiary pollen grains (KRÄUSEL 1920)] stressed the desirability and importance of the pollen-analytical investigations of Tertiary deposits (JURASKY 1928, KRÄUSEL 1928). At the same time, the first contribution to this phase of pollen analysis appeared (HECK 1927). HECK's contribution contains a number of quantitative pollen analyses and a pollen diagram of brown coal and diatomite from Beuern in Vogelsberg, Germany. This start was, however, not entirely successful since the pollen determinations were too narrow (" *Phragmites* ", " *Sium* ", etc.) and the conclusions as to climatic and biotic conditions, etc., too wide (*cf. e.g.* FIRBAS 1929). Later, KIRCHHEIMER (1929 and later), WODEHOUSE (1933), ROB. POTONIÉ (1934) with pupils, and RUDOLPH (1935) made important contributions to the knowledge of the fossil pollen and spore content of tertiary deposits.

In order to secure evidence as to the trustworthiness of Tertiary pollen spectra, etc., experimental investigations have been carried out by KIRCHHEIMER. Judging from the evidence produced by him, it is beyond a doubt that the chemicals employed in isolating the pollen grains may cause a considerable alteration of the pollen flora. Consequently, great changes may be induced in a pollen flora which may already have been subjected to much bacterial activity, influence from heat, and pressure etc. Pressure may mechanically damage the exines on account of the gradual disappearance of their remarkable elasticity during the transformation of the brown-coal mother-substance into brown coal proper. Many grains, particularly those with thin walls, may be corroded or disorganized and, consequently, more easily destroyed by the chemicals than undamaged grains. The effect of this is that the final pollen spectrum may be very different from that of the original pollen flora. This may occur even in Quaternary deposits, but the risk of encountering similar trouble when dealing with Tertiary deposits is, obviously, still greater.

Identification of Pollen Grains and Spores: — " Like other palaeobotanical identifications, those of pollen grains and spores are usually made with some degree of uncertainty. The value of such identifications is exactly proportionate to the degree of probability of their correctness. In the last analysis, the identification of pollen must always be based upon comparisons with living species, or with previously recorded fossil forms, as is the identification of leaves, stems, seeds, and various other parts of plants. The botanist of fossil pollen does not have available reference collections comparable in extent to the large herbaria which are at the disposal of other paleobotanists. Nor has he the knowledge of the structures of pollen grains and their phylogenetic significance which is in any way comparable to that which is the common heritage of other paleobotanists with their material. It is to be hoped, therefore, that the value of such pollen-grain identifications as these will be made sufficiently apparent by workers in this field to warrant the building up of collections of permanently mounted pollen slides comparable in extent to the great herbaria of the world, and that capable workers will be attracted to study and to portray the pollen forms, as has been done with other plant structures. Until this gap is filled, the fossil-pollen botanist must build up and interpret his own reference collections to meet his needs as the work of identification progresses, and the degree of reliability of his identifications will depend largely upon the extent of his collections and his understanding of them " (WODEHOUSE 1933, pp. 480, 481).

There is not much literature on this subject. WODEHOUSE (1932) has published an account of the pollen grains of the living representatives of the Eocene Green River Flora of Western United States, which may profitably be consulted. KIRCHHEIMER has stressed the importance of a thorough knowledge of modern pollen and the necessity of working from top to bottom when dealing with Tertiary deposits: *i.e.* to start with the youngest Pliocene beds, and then, using the knowledge thus obtained, to deal with the older beds in proper sequence. The older beds contain many species which do not exist at the present day, the identification of which, by means of pollen analysis, always will be a matter of great difficulty, unless the pollen grains are directly identified by the discovery of them within the mother plant. R. POTONIÉ and his pupils have systematically described a great number of Tertiary pollen and spore species. These " species " may be valuable from a geological point of view. Their botanical relationships, however, are often a matter of conjecture. In describing pollen grains and spores, POTONIÉ (1934) used a number of symbols (*cf.* also ARMBRUSTER-OENIKE, Die Pollenformen als Mittel zur Honigherkunftsbestimmung

Handb. Bienenkunde, vol. X, Neumünster 1929) to convey an idea of the shape, the number of furrows or pores, the type of ornamentation, etc. A standardization of this kind would be useful in the routine work of a specialized institution, but would be less so in common practice.

The generic name *Sporonites* (R. Potonié 1931) should be used when dealing with fossil fungus spores (*Eumycetes; cf.* Ibrahim 1933 and Potonié and Gelletich 1933) while fossil bryophyte and pteridophyte spores, whose determinations cannot be made exactly, may be called *Sporites*. This name should also be used whenever any doubt may arise as to whether a spore be a spore *sensu stricto* or a pollen grain. Such situations may occur now and then, probably more frequently than a botanist with but a cursory knowledge of pollen and spore morphology would be inclined to believe. It is thus not to be wondered at that fossil pollen grains, later identified as belonging to the *Sciadopitys*-type, were once considered to be fern spores. It is likewise not doubtful that pollen grains of *Bennettitales*, *Cycadofilicales* etc. on account of their shape have been described as *Sporites*.

The group *Sporites* may be subdivided into three smaller groups, *Aletes*, *Monoletes*, and *Triletes*, embracing respectively alete, monolete, and trilete spores (*cf.* pp. 44 *seq.*). These groups may be further subdivided according to the character of the surface of the spores. There is a number of different systems in use and we must limit ourselves quoting only one of them as an example, *viz.* the system proposed and adopted by Ibrahim (1933).

SPORITES R. Potonié 1893

A. *Aletes*

I. *Punctata-sporites*
II. *Apiculata-sporites*
III. *Zonola-sporites*
IV. *Reticulata-sporites*

B. *Monoletes*

I. *Laevigato-sporites*
II. *Punctato-sporites*
III. *Zonato-sporites*

C. *Triletes* Reinsch 1881

I. *Laevigati-sporites* (*Laevigati* Bennie and Kidston 1886)
II. *Punctati-sporites*
III. *Granulati-sporites*
IV. *Tuberculati-sporites*
V. *Apiculati-sporites* (*Apiculati* Bennie and Kidston, *l.c.*)
VI. *Verrucosi-sporites*
VII. *Setosi-sporites*
VIII. *Zonales-sporites* (*Zonales* Bennie and Kidston, *l.c.*)
IX. *Alati-sporites*
X. *Valvati-sporites*
XI. *Reticulati-sporites*.

The chief character of each group is made evident by the group names. A closer description is to be found in Ibrahim, *l.c.*

Nomenclature of Pollen Grains: — In the works of Rob. Potonié and others, pollen species are designated as *Pollenites* or *Pollinites*, and this name is followed by the specific name proposed by the author and preceded by the generic name when this is known or suspected. The use of the word *Pollinites* has simply the value of pointing out the fact that the object so designated is a fossil pollen grain. As *Pollinites* designates all pollen, it becomes so universal that it means almost nothing.

To avoid this, WODEHOUSE (1933) has proposed the following system of nomenclature. The word *Pollinites* is retained for its slight value in indicating the general nature of the object under consideration, but it is contracted to "*-pites*" and is used as a suffix which is applied to the specific designation if the genus of the pollen-species is known with the usual degree of accuracy; otherwise, to its generic designation. Thus a grain which is certainly that of *Pinus*, and which resembles most closely that of the living *Pinus strobus*, is called *Pinus strobipites*. But if the genus is not accurately or certainly known, the termination "*-pites*" is applied to the generic designation of the fossil grain instead of to its specific designation. For example, a grain that matches the living species of *Smilax* but also equally well those of some other genera, is called *Smilacipites mollioides*, its specific designation referring to its further resemblance to the grains of the living *Smilax mollis*. If at some later day, any of those genera which bear the termination "*-pites*" should become more closely defined or prove to have been accurately determined, the termination may then be transferred to the specific names.

The advantages of this system are that complete freedom is allowed in the use of descriptive adjectives as specific names without the introduction of trinomials; some idea is conveyed of the closeness or reliability of the determination; further, the fact that the determination is based on a fossil pollen grain is always shown. Long and unwieldy names may also appear in WODEHOUSE's nomenclature, exemplified by its progenitor himself: *e.g.* by octosyllables as *tetraforaminipites*, and crypto-trinomials as *Caprifoliipites viridi-fluminis*. However, this is unessential, and the system ought, on the whole, to be preferred to a nomenclature of the type *Nyssa-pollenites pseudocruciatus*, *Abies-pollenites absolutus*, etc. (*cf.* POTONIÉ 1934, THIERGART 1937, etc.).

When dealing particularly with early Tertiary or still older deposits which contain a good many species, genera, or even families now extinct, it will probably be difficult fully to apply the nomenclature proposed by WODEHOUSE. In such cases, it seems to be necessary to return to the artificial group name *Pollinites*. But since the group *Sporites*, as mentioned above, has been divided into several artificial sub-groups, the same could be done with the *Pollinites*-group. We may speak of *Tetradopites* (grains united in tetrads), *Dyadopites* (grains in dyads), *Acolpopites* (grains without furrows and pores), *Monocolpopites* (grains with one furrow), *Di-*, *Tri-*, *Tetra-*, *Hexa-*, *Octo-* and *Polycolpopites* etc. as well as of *Monoporipites* (grains with one pore), *Di-*, *Tri-*, *Tetra-*, *Hexa-*, *Octo-* and *Polyporipites* etc.

The naming of fossil pollen grains and spores must be done carefully to avoid the impression of overstressing the denomination. The following species are imaginary spore (pollen) species where the names have been chosen with a consideration of the increase of characteristic details in their construction:

1. *Sporites* (*cf. Monoletes*) *ovalis:* probably a monolete spore; systematic position unknown.
2. *Monocolpopites reticulatus:* a monocolpate grain, either of a type unknown in recent plants or of an indistinct habit, making it impossible to detect any apparent similarities with the pollen grains of a special family or group of plants.
3. *Magnoliaceopites longus:* family not accurately known; grains similar to those of the *Magnoliaceae* family.
4. *Liriodendropites formosus:* genus not accurately known; grains similar to those of *Liriodendron.*
5. *Liriodendron tulipiferopites:* genus accurately known on basis of pollen grain characters; pollen grains similar to or identical with those of *L. tulipifera;* megascopical remains of *L. tulipifera* not encountered in the bed investigated.
6. *Liriodendron cf. tulipifera:* as in no. 5, with the exception that the bed investigated has yielded megascopical remains of *L. tulipifera.*
7. *Liriodendron tulipifera:* this name may be retained — if the beds are Tertiary or older — in exceptional cases when it can be proved directly — by isolation of pollen grains from fossil flowers — that the pollen grains do belong to *Liriodendron tulipifera.*

Tertiary Spores and Pollens: — The following list includes those plants which are supposed to have been found in Tertiary beds as a result of pollen or spore analysis. The list is somewhat arbitrary and is by no means complete. Its sole purpose is to convey some idea of the botanical elements on which the interest of research workers has been focused in connection with pollen and spore analysis of Tertiary deposits. In making this enumeration no personal opinion is expressed with regard to the determinations. While some of these can not be doubted, others ought to be recognized at once as sheer absurdities. With regard to the views expressed above, the author's attitude in the matter of identifications would be more conservative than that of most of the authors cited.

MONOCOTYLEDONEAE

Araceae: —

Peltandripites Davisii (Wodehouse 1933, p. 498)

Cyperaceae: —

Cyperaceae pollen (Rudolph 1935)

Gramineae: —

Gramineae pollen (Lagerheim, *in* Hartz 1909)
cf. Gramineae (Kirchheimer 1930)
Gramineae-pollenites sp. (Thiergart 1937, p. 309)

Liliaceae: —

Smilacipites echinatus (Wodehouse 1933, p. 500)
S. herbaceoides (*ibid.*)
cf. Smilax (Simpson 1936, p. 96: "Attention might also be drawn to the somewhat similar pollen of *Sassafras* sp. While recognizing the uncertainty that these resemblances introduce, the probable relationship of the fossil grains to *Smilax* has been empha-

sized, since actual comparison with the various forms most strongly suggest this conclusion").

Smilax-pollenites setarius (THIERGART 1937, p. 309)

Palmae: —

Arecipites punctatus (WODEHOUSE 1933, p. 497)
A. rugosus (*ibid.*)
Sabal-pollenites convexus (THIERGART 1937, p. 308)
According to KIRCHHEIMER (1938), these species are dubious.

Potamogetonaceae: —

Potamogeton hollickipites (WODEHOUSE 1933, p. 496)

Sparganiaceae: —

Sparganiaceae-pollenites polygonalis (THIERGART 1937, p. 307)

Typhaceae: —

cf. Typha (POP 1936)

DICOTYLEDONEAE

Aceraceae: —

Aceraceae-type (KOSTYNIUK 1938; = *Pollenites laesus* R. Pot.?)

Anacardiaceae: —

Rhoipites bradleyi (WODEHOUSE 1933, p. 513)
Rhus-type (RUDOLPH 1935; BACMEISTER 1936)
Rhus?-pollenites pseudocingulum (THIERGART 1937, p. 320)

Aquifoliaceae: —

Ilex-type (RUDOLPH 1935)
Ilex-pollenites iliacus (THIERGART 1937, p. 321)
Ilex-pollenites margaritatus (*ibid.*, p. 321)
Ilex-pollenites propinquus (*ibid.*, p. 322)

Betulaceae: —

cf. Alnus (KIRCHHEIMER 1933)
Alnus sp. (LAGERHEIM *in* HARTZ 1909, KIRCHHEIMER 1932, RUDOLPH 1935, POP 1936, SIMPSON 1936)
Alnus-pollenites verus (POTONIÉ 1931, p. 229)
Alnus speciipites (WODEHOUSE 1933, p. 508)
cf. Betula, *Betula*-type, *Betula* sp. (LAGERHEIM *in* HARTZ 1909, KIRCHHEIMER 1932, RUDOLPH 1935, BACMEISTER 1936; = *Pollenites bituitus* R. Pot.?)
Betula claripites (WODEHOUSE 1933, p. 509)
Betula-pollenites microexcelsus (THIERGART 1937, p. 315)
Carpinus (KIRCHHEIMER 1934, RUDOLPH 1935)
Carpinus ancipites (WODEHOUSE 1933, p. 510)
Carpinus?-pollenites granifer megagranifer (THIERGART 1937, p. 315)
Corylus-type, *Corylus* sp. (LAGERHEIM *in* HARTZ 1909, RUDOLPH 1935, BACMEISTER 1936, POP 1936, SIMPSON 1936)
Corylus-pollenites coryphaeus (R. POTONIÉ 1934)
Pollenites granifer (R. POTONIÉ 1934)
Pollenites laevis (*ibid.*)

Caprifoliaceae: —

Caprifoliipites viridi-fluminis (WODEHOUSE 1933, p. 518)

Caryophyllaceae: —

Caryophyllaceae (LAGERHEIM *in* HARTZ 1909)

Centrospermae: —

Centrospermae-type (RUDOLPH 1935, BACMEISTER 1936)

Compositae: —

Compositae-type (RUDOLPH 1935, KOSTYNIUK 1938)
Pollenites echinatus (WOLFF 1934, p. 77)
Pollenites setiger (*ibid.*, p. 76)

Cornaceae: —

cf. Mastixioideae (KIRCHHEIMER 1938)
Pollenites ortholaesus (POTONIÉ 1934)

Dipsacaceae: —

Succisa-type (RUDOLPH 1935)

Ericaceae: —

Calluna-type (RUDOLPH 1935)
cf. Erica (KIRCHHEIMER 1933)
cf. Ericaceae, Ericaceae spp. (LAGERHEIM *in* HARTZ 1909, KIRCHHEIMER 1930, 1933)
Ericaceae-pollenites ericius (THIERGART 1937, p. 324)
Ericaceae-pollenites roboreus (*ibid.*, p. 325)
Ericipites brevisulcatus (WODEHOUSE 1933, p. 518)
Ericipites longisulcatus (*ibid.*, p. 517)

Fagaceae: —

Castanea-pollenites exactus (THIERGART 1937, p. 313; = *Castanea*-type, BACMEISTER 1936?)
Castanea-pollenites facetus (*ibid.*)
Castanopsis-pollenites cingulum (*ibid.*, p. 314)
Fagus (RUDOLPH 1935)
Pollenites laesus (POTONIÉ 1931, p. 3; BACMEISTER refers this species to *Salicaceae*, while KIRCHHEIMER refers it to *cf. Cornaceae-Mastixioideae*)
Pollenites pseudolaesus (POTONIÉ 1931)
cf. Quercus (RUDOLPH 1935, POP 1936)
Quercus?-pollenites Henrici (POTONIÉ 1931, p. 229).

Haloragidaceae: —

Myriophyllum-type (BACMEISTER 1936, SIMPSON 1936)
Myriophyllum ambiguipites (WODEHOUSE 1933, p. 516)

Hamamelidaceae: —

Bucklandia sp. (SIMPSON 1936)
Corylopsis sp. (*ibid.*)
Dicoryphe spp. (*ibid.*)
Distylium sp. (*ibid.*)
Fortunearia sp. (*ibid.*)
cf. Hamamelidaceae-type (BACMEISTER 1936)
Liquidambar-pollenites stigmosus (THIERGART 1937, p. 319)
Loropetalum sp. (SIMPSON 1936)

Juglandaceae: —

Hicoria juxtaporipites (WODEHOUSE 1933, p. 504)
Hicoria viridi-fluminipites (*ibid.*, p. 503)
Juglandaceae-type (BACMEISTER 1936)
Juglans aff. *regia?* (POP 1936)
Juglans nigripites (WODEHOUSE 1933, p. 504)
Pterocarya, cf. Pterocarya (KIRCHHEIMER 1933, 1936)
Pterocarya-pollenites stellatus (THIERGART 1937, p. 311)

Leguminosae: —

Pollenites quisqualis (POTONIÉ 1933)

Magnoliaceae: —

Liriodendron psilopites (WODEHOUSE 1933, p. 501)
Magnolia sp. (KIRCHHEIMER, SIMPSON 1936)

Myricaceae: —

Myrica, Myrica-type (KIRCHHEIMER 1935, p. 411, and 1938, p. 14). „Die mitteleozäne Braunkohle der Grube Cecilie im Geiseltal lieferte einen pollenführenden Blütenstand. Der ihm entnommene Pollen entspricht morphologisch dem Blütenstaub von *Myrica* und liegt in sehr verschiedener Erhaltung vor. ... In meinem einwandfrei auf eine Stammpflanze zurückgehende Material glaube ich besonders folgende der Potoniéschen „Arten" zu erkennen: *Coryli?-pollenites coryphaeus* R. Pot., *Pollenites granifer mega-*

granifer R. Pot., *P. g. granifer* R. Pot., *P. g. orbicularis* R. Pot., *P. g. bituitus* R. Pot., *Pollenites bituitus* R. Pot." (KIRCHHEIMER 1935, p. 411).
Myricipites dubius (WODEHOUSE 1933, p. 506)

Nymphaeaceae: —

Nymphaea sp. (KIRCHHEIMER 1930)

Nyssaceae: —

Nyssa sp., *Nyssa*-type (KIRCHHEIMER 1933, 1934, RUDOLPH 1935)
Nyssa-pollenites pseudocruciatus (THIERGART 1937, p. 322)
Pollenites vestibulum (POTONIÉ 1931)

Oenotheraceae (Onagraceae): —

Onagraceae-type (RUDOLPH 1935)

Rhamnaceae: —

Rhamnus sp. (KIRCHHEIMER 1930)

Rosaceae: —

cf. Prunus? (KIRCHHEIMER 1935)

Salicaceae: —

Pollenites gertrudae (POTONIÉ 1931, p. 229)
Salicaceae-type A (BACMEISTER 1936); identical with *Pollenites laesus* Pot.?
Salicaceae-type B (BACMEISTER 1936)
Salix-type (RUDOLPH 1935)
Salix discoloripites (WODEHOUSE 1933, p. 506); according to WODEHOUSE, probably identical with *Pollenites fraudulentus* Pot.

Sapindaceae: —

Talisiipites Fischeri (WODEHOUSE 1933, p. 513)

Simarubaceae: —

Ailanthipites berryi (WODEHOUSE 1933, p. 512)

Symplocaceae: —

Symplocaceae? (KIRCHHEIMER 1938, p. 17; proper identification so far impossible "on account of the occurrence of similar pollen types in a number of other families").

Tiliaceae: —

Tilia-type, *cf. Tilia* sp., *Tilia* sp. (LAGERHEIM *in* HARTZ 1909, KIRCHHEIMER 1932, 1933, 1936, 1938)
Tilia crassipites (WODEHOUSE 1933, p. 515)
T. type *platyphyllos* (POP 1936)
T. tetraforaminipites (WODEHOUSE 1933, p. 516)
T. vescipites (*ibid.*)
T.-pollenites instructus (THIERGART 1937, p. 323)

Ulmaceae: —

Momipites coryloides (WODEHOUSE 1933, p. 511; = *Celtis Debequensis?*)
Planera sp. (SIMPSON 1936)
cf. Pteroceltis sp. (*ibid.*)
Ulmus sp., *Ulmus*-type (LAGERHEIM *in* HARTZ 1909, RUDOLPH 1935)
Ulmus-pollenites undulosus (WOLFF 1934)

Umbelliferae: —

Sium (HECK 1927)
Umbelliferae (LAGERHEIM *in* HARTZ 1909)

Vitaceae: —

Vitipites dubius (WODEHOUSE 1933, p. 514)

GYMNOSPERMAE

Abies sp., *cf. Abies* (KIRCHHEIMER 1933, RUDOLPH 1935, BACMEISTER 1936, SIMPSON 1937)
Abies type *alba* (POP 1936)
Abies concolipites (WODEHOUSE 1933, p. 490)

Abies-pollenites absolutus (THIERGART 1937, p. 306)
Abiepites antiquus (WODEHOUSE 1933, p. 491)
Cedripites eocenicus (*ibid.*, p. 490)
Cedrus sp. (SIMPSON 1937, p. 94)
Cunninghamia concedipites (WODEHOUSE 1933, p. 495)
Cupressinae-type (RUDOLPH 1935)
Cycadopites (WODEHOUSE 1933, p. 483)
Dioonipites (*ibid.*, p. 484)
Ephedra eocenipites (*ibid.*, p. 495)
Ginkgo (SIMPSON 1936)
Glyptostrobus vacuipites (WODEHOUSE 1933, p. 494)
Gnetaceae-pollenites ellipticus (THIERGART 1937, p. 307)
cf. Keteleeria? (RUDOLPH 1935, POP 1936, KIRCHHEIMER 1938, KOSTYNIUK 1938)
Larix?-pollenites magnus (THIERGART 1937, p. 304)
Phyllocladus? (*ibid.*, p. 300)
Picea (RUDOLPH 1935, BACMEISTER 1936)
Picea or *Abies* sp. (LAGERHEIM *in* HARTZ 1909)
Picea grandivescipites (WODEHOUSE 1933, p. 488)
Picea-pollenites alatus (POTONIÉ 1931)
Pinus sp. (LAGERHEIM *in* HARTZ 1909, KIRCHHEIMER 1935, SIMPSON 1937)
Pinus haploxylon-type (RUDOLPH 1935, BACMEISTER 1936)
Pinus cf. sect. *Pinaster* (KIRCHHEIMER 1934)
Pinus-pollenites labdacus (THIERGART 1937, p. 305; = *P. silvestris*-type)
Pinus-pollenites microalatus (*ibid.*, = *P. haploxylon*-type)
Pinus scopulipites (WODEHOUSE 1933, p. 487)
Pinus silvestris-type (RUDOLPH 1935, BACMEISTER 1936)
Pinus strobipites and *P. tuberculipites* (WODEHOUSE 1933, pp. 487, 488)
Pinus sp. (DARRAH 1939, p. 224)
cf. Podocarpus and *Podocarpus* sp. (SIMPSON 1937, THIERGART 1937, KOSTYNIUK 1938)
Pollenites magnus dubius (POTONIÉ 1934; different *Cupressaceae* and *Taxaceae*)
cf. Sciadopitys? (RUDOLPH 1935, p. 261)
Sciadopitys-pollenites serratus (THIERGART 1937, p. 302)
Sequoia-pollenites poliformosus (*ibid.*, p. 301)
Sporites macroserratus (WOLFF 1934); = *Tsuga*, according to KIRCHHEIMER 1934
Taxodiineae, Cupressinae, Taxeae (SIMPSON 1936)
Taxodium-type (BACMEISTER 1936)
Taxodium hiatipites (WODEHOUSE 1933, p. 493; identical with *Pollenites hiatus* R. Pot. according to THIERGART 1937)
Tsuga (KIRCHHEIMER 1933, RUDOLPH 1935, BACMEISTER 1936)
Tsuga type *canadensis* (RUDOLPH 1935, POP 1936)
Tsuga type *diversifolia* (RUDOLPH 1935, POP 1936)
Tsuga moenana (KIRCHHEIMER 1933; = *Tsuga-pollenites igniculus* THIERGART 1937)
Tsuga viridi-fluminipites (WODEHOUSE 1933, p. 491)

PTERIDOPHYTA and BRYOPHYTA

Cyatheaceae: —

Cyatheaceae-sporites cf. adriennis (THIERGART 1937, p. 295)
Sporites neddeni (POTONIÉ 1931)

Lycopodiaceae: —

Lycopodium (KIRCHHEIMER 1936)
Lycopodium annotinum-type (RUDOLPH 1935; = *Lycopodium-sporites* aff. *agathoecus* THIERGART 1937, p. 293?)
Lycopodium clavatum-type (*ibid.*; POP 1936)
Lycopodium inundatum (RUDOLPH 1935)
Lycopodium?-sporites primarius (THIERGART 1937, p. 293)

Osmundaceae: —

cf. Osmunda (RUDOLPH 1935)

Polypodiaceae: —

aff. *Aspidium filix-mas* (RUDOLPH 1935)
cf. Aspidium thelypteris (*ibid.*)

cf. Athyrium (KIRCHHEIMER 1930)
Polypodium cf. vulgare (RUDOLPH 1935)
Polypodium-sporites alienus (THIERGART 1937, p. 296)
P.-s. favus (*ibid.*, p. 295; = *P. vulgare?*)
P.-s. hardtii and *P.-s. hardtii minor* (*ibid.*, p. 297)
cf. Pteridium aquilinum (RUDOLPH 1935)
Sporites secundus (POTONIÉ 1934)

Psilotaceae: —

Sporites ligneolus (POTONIÉ 1934)

Sphagnaceae: —

Sphagnum (KIRCHHEIMER 1934, RUDOLPH 1935)
Sphagnum-sporites stereoides (THIERGART 1937, p. 292)

A still greater number of types have been described as *Pollenites* spp. and *Sporites* spp. They are not quoted here inasmuch as nothing can be said with any certainty as to their botanical position.

Of many plants, such as *Amentotaxus*, *Callitris*, *Libocedrus*, *Pseudotsuga*, *Aldrovandia*, *Brasenia*, *Ceratophyllum*, *Cinnamomum*, *Dulichium*, *Eucommia*, *Gleditschia*, *Populus*, *Spirematospermum*, *Stratiotes*, *Trapa*, etc., only megascopical remains have been found (*cf.* KIRCHHEIMER 1937). Further micropalaeontological investigations of Tertiary material may reveal the presence also of pollen grains of at least some of these plants, while, as in the study of Quaternary deposits, we probably must rely entirely on megascopical remains for the history of the remaining species.*

The results of pollen and spore analysis of Tertiary and older deposits are to a great extent vague and may be much modified by future work. Indeed, quantitative analyses have only just begun. In due course of time, this line of research no doubt will develop into an important part of palaeobotany, and micropalaeontologists will meet with problems of paramount interest such as the origin and early development of Angiosperms. Intensive action on the micropalaeontological front can, however, only be undertaken on the basis of a solid foundation of morphological knowledge and technical skill.

References: —

BACMEISTER, A., 1936: Pollenformen aus den obermiozänen Süsswasserkalken der "Öhninger Fundstätten" am Bodensee (Ber. Geobot. Inst. Rübel 1935).

BENNIE, I. and KIDSTON, R., 1886: On the occurrence of spores in the Carboniferous formation of Scotland (Proc. Roy. Soc. Edinb., vol. IX).

BODE, H., 1931: Die Pollenanalyse in der Braunkohle (Internat. Bergwirtsch. u. Bergtechnik, vol. 24).

DARRAH, W. C., 1939: Textbook of Paleobotany (New York).

—— —— 1941: Observations on the Vegetable Constituents of Coals (Econ. Geol. vol. 36).

—— —— 1941: Changing Views of Petrifaction (Pan-Amer. Geol. vol. 76).

EHRENBERG, C. G., 1838: Beobachtungen über neue Lager fossiler Infusorien und das Vorkommen von Fichtenblütenstaub neben deutlichem Fichtenholz, Haifischzähnen, Echinoiden und Infusorien in volhynischen Feuersteinen der Kreide (Monatber. Berliner Acad.).

FIRBAS, F., 1929: Einige Bemerkungen zur heutigen Anwendung der Pollenanalyse (Centralbl. f. Min. etc., Abt. B, no. 9).

GÖPPERT, H. R., 1841: Über das Vorkommen von Pollen im fossilen Zustande (Neues Jahrb. f. Min. etc.).

HARTZ, N., 1909: Bidrag til Danmarks tertiaere og diluviale flora (Danm. Geol. Unders., II. Raekke, no. 20).

* The application of pollen analytical methods to lignites and bituminous coals of Mesozoic and Paleozoic age has been undertaken by various workers (RAISTRICK 1934, SCHOPF 1938). The results promise much in the way of improvement over the older purely descriptive pollen and spore studies. The general distribution and varied occurrences of these structures has been demonstrated by many investigators (DARRAH, KIRCHHEIMER).

HECK, H. L., 1927: Die tertiäre Kieselgur und Braunkohle von Beuern im Vogelsberg und ihre Flora, I. Teil: Geologische Untersuchungen der Kieselgur und Braunkohle von Beuern und ihre Mikroflora (Notizblatt Ver. Erdkunde u. d. Hess. Landesanst. f. 1927; printed 1928?).

IBRAHIM, A. C., 1930: Sporenformen des Aegirhorizonts des Ruhr-Reviers (Inaug.-Diss., Würzburg).

JURASKY, K., 1928: Aufgaben und Ausblicke für die paläobotanische Erforschung der niederrheinischen Braunkohle (Braunkohle).

KIRCHHEIMER, F., 1929: Braunkohlenumformung und Pollenverteilung (Ber. Oberhess. Ges. Natur- u. Heilkunde, Giessen, vol. 13).

—— —— 1930: Braunkohlenforschung und Pollenanalytik (Braunkohle).

—— —— 1931*a*: Zur pollenanalytischen Braunkohlenforschung, I (*Ibid.*).

—— —— 1931*b*: Zur pollenanalytischen Braunkohlenforschung, II (*Ibid.*).

—— —— 1931*c*: Ein Beitrag zur Kenntnis von Pollenformen der Eozänbraunkohle des Geiseltales ("Die Wirbeltierfundstelle im Geiseltal", Halle).

—— —— 1933: Die Erhaltung der Sporen und Pollenkörner in den Kohlen sowie ihre Veränderungen durch die Aufbereitung (Bot. Arch., vol. 35).

—— —— 1934: Fossile Sporen und Pollenkörner als Thermometer der Inkohlung (Brennstoff-Chemie, vol. 15, 1934).

—— —— 1935: Die Korrosion des Pollens (Beih. Bot. Centralbl., Abt. A, vol. LIII).

—— —— 1937: Grundzüge einer Pflanzenkunde der deutschen Braunkohlen (Halle, W. Knapp).

—— —— 1938: Bemerkungen über die botanische Zugehörigkeit von Pollenformen aus den Braunkohlenschichten (Planta, vol. 28).

KOSTYNIUK, M., 1938: Über die tertiären Pollen und Koniferenhölzer von einigen Gegenden Polens (Kosmos, Journ. Soc. Pol. Nat. Kopernik, vol. LXIII).

KRÄUSEL, R., 1920: Nachträge zur Tertiärflora Schlesiens, I (Jahrb. preuss. geol. Landesanst. f. 1918).

—— —— 1928: Paläobotanische Braunkohlenstudien (Abhandl. Naturf. Ges. Görlitz, vol. 30).

MINER, E. L., 1935: Paleobotanical Examinations of Cretaceous and Tertiary Coals. II. Cretaceous and Tertiary Coals from Montana (Amer. Midl. Nat. vol. 16).

POP, E., 1936: Die pliozäne Flora von Borsec (Univ. Reg. Ferdinand. fac. stiinte, no. 1. Cluj.)

POTONIÉ, H., 1893: Die Flora des Rothliegenden von Thüringen (Berlin).

POTONIÉ, R., 1931: Zur Mikroskopie der Braunkohlen. Tertiäre Blütenstaubformen, I. Mitteil. (Braunkohle).

—— —— 1934: Zur Mikrobotanik der Kohlen und ihrer Verwandten, II. Zur Mikrobotanik des eozänen Humodils des Geiseltals (Arb. Inst. Paläobot. Petrogr. Brennsteine, vol. IV).

POTONIÉ, R. und GELLETICH, J., 1933: Über Pteridophyten-Sporen einer eocänen Braunkohle aus Dorog in Ungarn (Sitz.-Ber. naturf. Ges.).

—— —— und VENITZ, H., 1934: Zur Mikrobotanik des miocänen Humodils der niederrheinischen Bucht (Arb. Inst. Paläobot. Petrogr. Brennsteine, vol. 5).

RAISTRICK, A., 1934: The correlation of coal-seams by microspore-content, Part I. The seams of Northumberland (Trans. Inst. Min. Engin., vol. LXXXVIII).

REINSCH, P. F., 1881: Neue Untersuchungen über die Mikrostruktur der Steinkohle (Leipzig).

REISSINGER, A., 1939: Die "Pollenanalyse", ausgedehnt auf alle Sedimentgesteine der geologischen Vergangenheit (Palaeontogr. 84B:1–20).

RUDOLPH, K., 1935: Mikrofloristische Untersuchung tertiärer Ablagerungen im nördlichen Böhmen (Beih. Bot. Centralbl., vol. LIV, Abt. B).

—— —— 1930: Grundzüge der nacheiszeitlichen Waldgeschichte Mitteleuropas (Beih. z. Bot. Centralbl. 47, Abt. 2).

SCHOPF, J. M., 1938: Spores from the Herrin (No. 6) Coal Bed in Illinois (State Geol. Surv. Ill. Report of Investigations no. 50).

SIMPSON, J. B., 1936: Fossil pollen in Scottish tertiary coals (Proc. Roy. Soc. Edinb., vol. LVI, pt. II).

THIERGART, F., 1937: Die Pollenflora der Niederlausitzer Braunkohle, besonders im Profil der Grube Marga bei Senftenberg (Jahrb. Preuss. Geol. Landesanst., vol. 58).

WODEHOUSE, R., 1932: Tertiary pollen, I. Pollen from the living representatives of the Green River flora (Bull. Torr. Bot. Club., vol. 59).

—— —— 1933: Tertiary pollen, II. The oil shales of the eocene Green River formation (*Ibid.*, vol. 60).

WOLFF, H., 1934: Mikrofossilien des pliocänen Humodils der Grube Freigericht bei Dettingen a.M. und Vergleich mit älteren Schichten des Tertiärs sowie posttertiären Ablagerungen (Arb. Inst. Paläobot. Petrogr. Brennsteine, vol. 5).

ZALESSKY, M. D., 1939: Sur la question de la classification des spores fossiles (Probl. of Paleont. 5:325–327).

ZETZSCHE, F. und KÄLIN, O., 1932: Eine Methode zur Isolierung des Polymerbitumens (Sporenmembranen, Kutikulen usw.) aus Kohlen (7. Mitt. zu Unters. üb. d. Membran der Sporen u. Pollen) (Braunkohle).

Chapter XVIII

POLLEN ANALYSIS OF HONEY AND DRUGS

Pollen analysis is frequently mentioned in connection with honey investigations (GRIEBEL 1930–31, MAURIZIO 1939, etc.). Specialists in pollen analysis in its original, geological and botanical, sense may object to this and wonder why something so entirely different from ordinary pollen analysis should be mentioned, if only briefly, in a book of this kind. Logically there should not be any objection since in dubious cases a study of the pollen flora of a honey may help to determine its country of origin: Swiss, Californian, South African, Australian, etc. It may also even help to reveal dishonest practices: a supposed German honey with an abundance of *Bombacaceae* grains and other exotic pollens may readily be shown to be Mexican, etc. Sometimes the pollen flora may also give hints as to the time of year the honey was produced.

There is an extensive literature on honey pollen. The greater part comes from Germany, Switzerland, and Czechoslovakia. "Beiträge zur Herkunftsbestimmung bei Honig" (3 volumes) by E. ZANDER is a standard work along this line of pollen research. It contains extensive references and more than a thousand pollen microphotographs. This makes it useful also to ordinary pollen analysts, looking for pollen in peat, and to the more exclusive guild of palaeomelitologists, looking for pollen in ancient honeys from Egyptian graves (*cf.* ZANDER 1941) or in dry metheglin remains which may occur in old drinking horns.*

In preparing honey for microscopical examination, several methods are used. As far as the study of pollen grains and spores is concerned beautiful permanent slides may be made after acetolysis. The procedure involves: —

1. dissolution of the sugar;
2. collection of the insoluble material on a filter;
3. acetolysis of the filter;
4. removal of wax and similar substances,

and may be carried out as follows:

Fifty grams of honey are dissolved in 100 cc of hot water. The solution is filtered through a Buechner funnel, and the filter is washed

* In this connection attention may well be drawn to the numerous papers recently published by the Div. of Bee Culture of the U.S. Dept of Agriculture (chief, JAMES I. HAMBLETON, Beltsville, Md.) several of which deal with problems of pollen analysis, pollen chemistry, etc. *Cf. e.g.* E. OERTEL 1939: Honey and Pollen Plants of the U.S. (U.S.D.A. Circular, No. 554); F. E. TODD and O. BRETHERICK 1942: The Composition of Pollens (J. Econ. Entomology 35:312); C. W. SCHAEFER and C. L. FARRAR 1941: The Use of Pollen Traps and Pollen Supplements in developing honeybee colonies (Contrib. E. 531, Bureau Entomol. and Plant Quar.); F. E. TODD 1941: The role of Pollen in the Economy of the Hive (*idem* E. 536); F. E. TODD and G. H. VANSELL 1942: Pollen Grains in Nectar and Honey (J. Econ. Entomology 35:728); F. E. TODD and R. K. BISHOP 1941: Trapping Honeybee-Gathered Pollen and Factors Affecting Yields (J. Econ. Entomol. 33:866); U.S.D.A. Div. of Bee Cult. 1942: The Dependence of Agriculture on the Beekeeping Industry (Contrib. E. 584, Bureau Entomol. and Plant Quar.).

first with water and then with glacial acetic acid. The filter is then put in a centrifuge tube to which is added a mixture of 9 cc of acetic anhydride and 1 cc of concentrated sulphuric acid. The tube with the reaction mixture is then placed in a water bath which is brought to a boil. When boiling has begun, the tube is transferred to the centrifuge. After centrifuging the acetolyzing chemicals are decanted, and the sediment is thoroughly washed with ethyl acetate, then with acetone, and finally with distilled water. The sediment is then suspended in glycerine, allowed to settle, and embedded in glycerine jelly (ERDTMAN 1935).

Pollen grains sometimes are very important in qualitative analyses of drug powders (KNELL 1914), in showing up adulterations and substitutions. Liquid drugs may be prepared for microscopical investigation by filtering followed by acetolysis of the filter (*cf.* above), while dry drugs are treated in the same way as herbarium material (p. 29).

References: —

ERDTMAN, G., 1935: Investigation of honey pollen (Svensk Bot. Tidskr., vol. 29).

GRIEBEL, C., 1930/31: Zur mikroskopischen Pollenanalyse des Honigs, I–IV (Zeitschr. Unters. Lebensmittel, vols. 59, 61).

KNELL, 1914: Die Pollenkörner als Diagnostikum in Drogenpulvern (Thesis, Würzburg).

MAURIZIO, A., 1939: Untersuchungen zur quantitativen Pollenanalyse des Honigs (Mitt. Geb. Lebensmittelunters., Eidg. Gesundheitsamt Bern, vol. XXX).

—— 1941: Pollenanalytische Beobachtungen 1–9 (Ber. Schweizer Bot. Ges. 51:77 *et ante*).

ZANDER, E., 1935/37/41: Pollengestaltung und Herkunftsbestimmung bei Blütenhonig, I–III (Berlin und Leipzig).

Some recent publications: —

(These have not been seen by the author as a result of war conditions)

BENNINGHOFF, W. S., 1942: The pollen analysis of the Lower Peat (Papers Peabody Mus. Harvard Univ. 2:96–104).

BUELL, M. F., 1939: Peat formation in the Carolina Bays (Bull. Torrey Bot. Club 66:483–487).

COOPER, W. S., 1942: Vegetation of the Prince William Sound region, Alaska; with a brief excursion into post-Pleistocene climatic history (Ecol. Monogr. 12:1–22).

FIRBAS, F., LOSERT, H., und BROIHAN, FR., 1940: Untersuchungen zur jüngeren Vegetationsgeschichte im Oberharz (Planta 30:422–456).

FLORSCHÜTZ, F., 1939: Spätglaziale Torf- und Flugsandbildungen in den Niederländen als Folge eines dauernden Frostbodens (Abhandlungen herausgegeben vom naturwissenschaftlichen Verein zu Bremen, XXXI. Band, Heft 2).

FULLER, G. D., 1939: Interglacial and post-glacial vegetation of Illinois (Trans. Ill. State Acad. Sci. 32:5–15).

GAMS, H., 1935: Beiträge zur Mikrostratigraphie und Paläontologie des Pliozäns und Pleistozäns von Mittel- und Osteuropa und West-sibirien. (Eclogae geologicae Helvetiae, vol. 28, No. 1).

—— 1937: Die Seen Europas im Eiszeitalter (Internationale Revue der gesamten Hydrobiologie und Hydrographie, Band 35, Heft 4/6).

HAMP, F. A., 1940: A fossil pollen study of two northern Indiana bogs (Butler Univ. Bot. Studies 4:217–225).

HANSEN, H. P., 1939: Pollen Analysis of a bog in northern Idaho (American Journ. of Bot. 26:225–228).

—— 1939: Paleoecology of a central Washington bog (Ecology 20:563–568).

—— 1939: Postglacial vegetation of the Driftless Area of Wisconsin (Amer. Midland Nat. 21:742–762).

—— 1940: Paleoecology of two peat bogs in southwestern British Columbia (Amer. Jour. Bot. 27:144–149).

—— 1940: Paleoecology of a montane peat deposit at Bonaparte Lake, Washington (Northwest Sci. 14:60–68).

—— 1941: Paleoecology of a peat deposit in west centra Oregon (Amer. Jour. Bot. 28:206–212).

——— 1941: A Pollen Study of Post-Pleistocene Lake Sediments in the Upper Sonoran Life Zone of Washington (American Journal of Science 239:503–522).
——— 1941: Further pollen studies of post Pleistocene bogs in the Puget Lowland of Washington (Bull. Torrey Bot. Club 68:133–148).
——— 1941: Paleoecology of a Bog in the Spruce-Hemlock Climax of the Olympic Peninsula (The American Midland Naturalist 25:290–297).
——— 1941: Paleoecology of a montane peat deposit near Lake Wenatchee, Washington (Northwest Sci. 15:53–65).
——— 1942: The influence of volcanic eruptions upon post-Pleistocene forest succession in eastern Oregon (Amer. Jour. Bot. 29:214–219).
——— 1942: A pollen study of lake sediments in the lower Willamette Valley of western Oregon (Bull. Torrey Bot. Club 69:262–280).
——— 1942: Post-Mount Mazama forest succession on the east slope of the central Cascades of Oregon (Amer. Midland Nat. 27:523–534).
——— and MACKIN, J. HOOVER, 1940: A further study of interglacial peat from Washington (Bull. Torrey Bot. Club 67:131–142).
KOLUMBE, E., und BEYLE, M., 1940: Pollenanalytische Untersuchungen des Satrupholmer Moores (Veröff. Inst. Volks- u. Landwirtsch. Forsch. Kiel 1940: 11–14).
LANE, G. H., 1941: Pollen analysis of interglacial peats of Iowa (Ann. Rept. Iowa Geol. Surv. 37:237–262).
MITCHELL, G. F., 1942: A composite pollen diagram from Co. Meath, Ireland (New Phytol. 41:257–261).
MOSS, B. W., 1940: A comparative pollen analysis of two bogs within boundaries of the late Wisconsin glaciation in Indiana (Butler Univ. Bot. Studies 4:207–216).
PEARSON, P. B., 1942: Pantothenic acid content of pollen (Soc. Expt. Biol. and Med. Proc. 51:291–292).
POTZGER, J. E., 1942: Pollen spectra from four bogs on the Gillen Nature Reserve along the Michigan-Wisconsin state line (Amer. Midland Nat. 28:501, 511).
——— and RICHARDS, R. R., 1942: Forest succession in the Trout Lake, Vilas County, Wisconsin area; a pollen study (Butler Univ. Bot. Studies 5:179–189).
——— and WILSON, IRA T., 1941: Post-pleistocene forest migration as indicated by sediments from three deep inland lakes (Amer. Midland Nat. 25:270–289).
SEARS, P. B., 1941: A submerged migration route (Science 94:301).
——— 1941: Postglacial vegetation in the Erie-Ohio area (Ohio Jour. Sci. 41:225–234).
——— 1942: Postglacial migration of five forest genera (Amer. Jour. Bot. 29:684–691).
——— 1942: Forest sequences in the north central states (Bot. Gaz. 103:751–761).
SWICKARD, D. A., 1941: Comparison of pollen spectra from bogs of early and late Wisconsin glaciation in Indiana (Butler Univ. Bot. Studies 5:67–84).
TREWARTHA, G. T., 1940: The vegetal cover of the Driftless Guestaform hill land: Presettlement record and postglacial evolution (Trans. Wisconsin Acad. Sci. Arts and Lett. 32:361–382).
VOSS, J., 1939: Forests of Yarmouth and Sangamon interglacial periods in Illinois (Ecology 20:517–528).
WILSON, L. R. and COE, E. A., 1940: Descriptions of some unassigned plant microfossils from the Des Moines Series of Iowa (Amer. Midland Nat. 23:182–186).
——— and KOSANKE, R. M., 1940: The microfossils in Pre-Kansan peat deposit near Belle Plaine, Iowa (Torreya 40:1–5).
——— and WEBSTER, R. M., 1942: Fossil evidence of wider post-pleistocene range for butternut and hickory in Wisconsin (Rhodora 44:409–414).

AUTHOR INDEX

E. J. Lennart von Post

— *cf. p. 1* —

NILS GUSTAF LAGERHEIM

—*cf. p. 3*—

INDEX *of* PLANT NAMES

N. O. Holst

— *cf. p. 4* —

Lightning Source UK Ltd.
Milton Keynes UK
UKOW050104190212

187542UK00001B/19/P